The Inhibitors of Hematopoiesis

Les Inhibiteurs de l'Hématopoïèse

Colloques INSERM
ISSN 0768-3154

Other *Colloques* published as co-editions by John Libbey Eurotext and INSERM

133 Cardiovascular and Respiratory Physiology in the Fetus and Neonate. *Physiologie Cardiovasculaire et Respiratoire du Fœtus et du Nouveau-né.*
Scientific Commitee : P. Karlberg,
A. Minkowski, W. Oh and L. Stern;
Managing Editor : M. Monset-Couchard.
ISBN : John Libbey Eurotext 0 86196 125 0
   INSERM 2 85598 340 1

134 Porphyrins and Porphyrias. *Porphyrines et Porphyries.*
Edited by Y. Nordmann.
ISBN : John Libbey Eurotext 0 86196 087 4
   INSERM 2 85598 281 2

137 Neo-Adjuvant Chemotherapy. *Chimiothérapie Néo-Adjuvante.*
Edited by C. Jacquillat, M. Weil and D. Khayat.
ISBN : John Libbey Eurotext 0 86196 125 0
   2 85598 340 1

139 Hormones and Cell Regulation (10th European Symposium). *Hormones et Régulation Cellulaire (10ᵉ Symposium Européen).*
Edited by J. Nunez, J.E. Dumont and R. J.B. King.
ISBN : John Libbey Eurotext 0 86196 125 0X
   INSERM 2 85598 340 1

147 Modern Trends in Aging Research. *Nouvelles Perspectives de la Recherche sur le Vieillissement.*
Edited by Y. Courtois, B. Faucheux, B. Forette,
D.L. Knook and J.A. Tréton.
ISBN : John Libbey Eurotext 0 86196 125 0X
   INSERM 2 85598 340 1

149 Binding Proteins of Steroid Hormones. *Protéines de liaison des Hormones Stéroïdes.*
Edited by M.G. Forest and M. Pugeat.
ISBN : John Libbey Eurotext 0 86196 125 0
   INSERM 2 85598 340 1X

151 Control and Management of Parturition. *La Maîtrise de la Parturition.*
Edited by C. Sureau, P. Blot, D. Cabrol,
F. Cavaillé and G. Germain.
ISBN : John Libbey Eurotext 0 86196 125 0
   INSERM 2 85598 340 1

153 Hormones and Cell Regulation (11th European Symposium). *Hormones et Régulation Cellulaire (11ᵉ Symposium Européen).*
Edited by J. Nunez and J.E. Dumont
ISBN : John Libbey Eurotext 0 86196 104 8
   INSERM 2 85598 324 X

The Inhibitors of Hematopoiesis

Les Inhibiteurs de l'Hématopoïèse

Proceedings of the First International Symposium on Inhibitory Factors in the Regulation of Hematopoiesis, Paris (France), 26-28 April 1987

Organized by the Department of Hematology of the Faculté de Médecine Saint-Antoine (Université Pierre et Marie Curie, Paris, France) with the help of J.-Y. Mary (INSERM U263 Paris VII, France)

Under the auspices of Monsieur le Ministre Chargé de la Recherche et de l'Enseignement Supérieur

Sponsored by the Association pour la Recherche sur le Cancer and the Institut National de la Santé et de la Recherche Médicale

Edited by

Albert Najman
Martine Guigon
Norbert-Claude Gorin
Jean-Yves Mary

British Library Cataloguing in Publication Data
International Symposium on Inhibiting Factors in the
Regulation of Hematopoiesis (1st : 1987 : Paris)
The inhibitors of hematopoiesis : First International
Symposium on Inhibiting Factors in the Regulation of
Hematopoiesis (Paris, France, 26-28 April, 1987). —
(Colloques INSERM, ISSN 0768-3154; 162).
1. Hematopoiesis 2. Cell differentiation
I. Title II. Najman, A. III. Series
599.01'16 QP 92

ISBN 0 86196 125 0
ISSN 0768-3154

First published in 1987 by

John Libbey & Company Ltd
80/84 Bondway, London SW8 1SF, England. (01) 582 5266
John Libbey Eurotext Ltd
6 rue Blanche, 92120 Montrouge, France. (1) 47 35 85 52
ISBN 0 86196 125 0

Institut National de la Santé et de la Recherche Médicale
101 rue de Tolbiac, 75654 Paris Cedex 13, France.
(1) 45 84 14 41
ISBN 2 85598 340 1

ISSN 0768-3154
© 1987 Colloques INSERM/John Libbey Eurotext Ltd, All rights reserved
Unauthorised duplication contravenes applicable laws

Preface

The studies carried out since more than 20 years in animals and humans have shown that the regulation of hematopoiesis is under the control of several stimulatory and inhibitory factors. Progress in the knowledge of stimulatory factors has already allowed the purification of some of them which can now be obtained by molecular cloning.

Work on inhibitory factors has developed more slowly, partly because of the lack of reliable bioassays. Small peptides have now been identified which inhibit the proliferation of some hematopoietic stem cells. Known molecules such as metalloproteins, interferons, some lymphokines and monokines may play a role in the negative regulation of hematopoiesis. Other factors, not yet purified, have been found in normal and/or pathological situations. Some of these inhibitors may have possible clinical applications in protecting bone marrow during chemotherapy or during the treatment of certain leukemias and related diseases.

The main purpose of this Symposium was to arrive at an international review of the State of the Art in this rapidly developing and fascinating field, which may interest all those involved in experimental and clinical Hematology and Oncology, and to present orientations by some of the best specialists. This Symposium has been the first opportunity for the participants working in this area all over the world to discuss the most recent facts about molecules involved in the negative control of normal and pathological hematopoiesis. It was followed by a workshop which aimed at an international register of the inhibitors of hematopoiesis.

This book, which is published a few months after the Symposium, provides the reader with an up-to-date and exhaustive overview of this promising area. Invited speakers have offered "State of the Art" reviews on the main topics which have led to the papers included in this book. Each session was followed by a general discussion which was recorded and typed immediately. Posters were also presented and are published as brief reports. However the discussions of the poster sessions have not been recorded and thus will not appear in this book.

A. Najman
M. Guigon
N.-C. Gorin
J.-Y. Mary

Préface

Les études réalisées depuis plus de 20 ans chez l'animal et chez l'homme ont montré que la régulation de l'hématopoïèse est sous le contrôle de nombreux facteurs stimulants et inhibiteurs. Les progrès dans la connaissance des facteurs stimulants ont déjà permis de purifier plusieurs d'entre eux qui peuvent être maintenant obtenus par clonage moléculaire.

La recherche de facteurs inhibiteurs a progressé plus lentement en raison notamment de l'absence d'une méthodologie adéquate. On a maintenant identifié des peptides de petite taille qui inhibent la prolifération des cellules souches hématopoïétiques. Des molécules connues telles que des métalloprotéines, les interférons, certaines lymphokines et monokines peuvent jouer un rôle dans la régulation négative de l'hématopoïèse. D'autres facteurs non encore purifiés ont été mis en évidence dans des tissus hématopoïétiques normaux et/ou pathologiques. Certains de ces inhibiteurs pourraient avoir un intérêt clinique dans la protection de la moelle osseuse au cours de la chimiothérapie ou lors du traitement des leucémies.

L'objectif de ce Symposium a été de faire le point sur les données actuelles dans ce domaine passionnant et en évolution constante qui présente un grand intérêt aussi bien pour les chercheurs que pour les cliniciens en hématologie et en cancérologie. Pour la première fois, des spécialistes de renommée mondiale ont eu l'occasion de présenter et de discuter les données les plus récentes sur des molécules impliquées dans l'inhibition de l'hématopoïèse normale et pathologique. Le Symposium a été suivi d'une table ronde dans le but d'établir un registre international des inhibiteurs de l'hématopoïèse.

Ce livre publié dans les mois suivant le Symposium est une mise à jour et une revue exhaustive des travaux effectués dans un domaine très prometteur. Les conférences plénières regroupées par thème ainsi que les discussions qui les ont suivies figurent dans cet ouvrage. Les communications affichées sont également publiées sous forme d'articles brefs; les discussions qui leur ont fait suite, n'ayant pu être enregistrées, ne figurent pas dans ce livre.

A. Najman
M. Guigon
N.-C. Gorin
J.-Y. Mary

Acknowledgments

We wish to thank the **Association pour la Recherche sur le Cancer** (J. Crozemarie, President) and the **Institut National de la Santé et de la Recherche Médicale** (Ph. Lazar, Director) for providing financial support. We also appreciated the contribution of the following pharmaceutical companies : Unicet, Lilly (France), Beecham (France), Merck-Sharp-Dohme (France), Glaxo (France), Roger Bellon, Upjohn, Hoechst, (France) and Travenol (France).

We are grateful to M. Schwartz, Deputy Director of the Institut Pasteur, and to our colleagues, G. Milon and G. Marchal, who made it possible to run the Symposium in the Grand Amphithéâtre of this famous place and to the Département des Relations Extérieures (C. Volkerick) for the efficient help in the organization.
We would like to thank J.P. Lévy (INSERM) and Ph. Thibault (Faculté de Médecine St-Antoine) for introducing the Symposium, G. Milhaud (Faculté de Médecine St-Antoine) for his constant support.
We also thank the members of the Department of Hematology of the CHU St-Antoine and especially Marie-Christine Bataille, who was in charge of the Secretariat of the Symposium. Mélanie Moussé, Christian Nourry and Nicolas Denoix (Institut Gustave Roussy, Villejuif) were also very helpful.

C. Hamburger is thanked for her excellent welcome at La Maison de la Recherche (Fondation pour la Recherche Médicale).

We thank the Institut National de la Santé et de la Recherche Médicale, in particular Y. Coadou (Bureau des Colloques), and John Libbey Eurotext who have permitted the publication of this volume.
We would like to express our utmost gratitude to the speakers who not only presented excellent papers, but who willingly put together a manuscript which made possible the publication of this book.
We would also like to thank the chairpersons who organized interesting discussions (D. Metcalf, E. Frindel, N. Young, B. Varet, A.A. Axelrad, P. Boivin, W.R. Paukovits, J. Breton-Gorius, L.A. Rozenszajn and G. Milon), who were in charge of the poster sessions (N. Dainiak, C. Dresch, E. Wright and M. Lenfant) and who prepared the workshop (H. Broxmeyer, W. R. Paukovits and M. Guigon), the participants who corrected the texts of their intervention and Claire Cocheteux for her excellent typing of the discussions from the tapes and their corrections.

Remerciements

Nous remercions l'**Association pour la Recherche sur le Cancer** (Président, J. Crozemarie) et l'**Institut National de la Santé et de la Recherche Médicale** (Directeur, Ph. Lazar) pour leur aide financière. Nous avons aussi apprécié la contribution des laboratoires pharmaceutiques suivants : Unicet, Lilly (France), Beecham (France), Merck-Sharp-Dohme (France), Glaxo (France), Roger Bellon, Upjohn, Hoechst (France) et Travenol (France).

Nous sommes reconnaissants à M. Schwartz, Sous-Directeur de l'Institut Pasteur, et à nos collègues G. Milon et G. Marchal grâce à qui ce Symposium a pu avoir lieu dans le Grand Amphithéâtre de cette célèbre maison ainsi qu'au Département des Relations Extérieures (C. Volkerick) pour son aide efficace dans l'organisation.
Nous tenons à remercier J.P. Lévy (INSERM) et Ph. Thibault (Faculté de Médecine Saint-Antoine) qui ont accepté d'introduire le Symposium ainsi que G. Milhaud (Faculté de Médecine Saint-Antoine) pour son soutien constant.
Merci à tous les membres du Département d'Hématologie du CHU Saint-Antoine et particulièrement à Marie-Christine Bataille, qui a assuré avec compétence, efficacité et gentillesse le secrétariat du Symposium. Merci aussi à Mélanie Moussé, Christian Nourry et Nicolas Denoix (Institut Gustave Roussy, Villejuif) pour leur aide.

Que C. Hamburger trouve ici l'expression de nos remerciements pour l'excellent accueil qu'elle nous a réservé à la Maison de la Recherche (Fondation pour la Recherche Médicale).

Nous remercions vivement l'Institut National de la Santé et de la Recherche Médicale et, en particulier, Y. Coadou (Bureau des Colloques), ainsi que les Editions John Libbey Eurotext qui ont permis la publication de ce livre. Nous tenons à exprimer notre vive reconnaissance aux orateurs qui ont non seulement fait d'excellentes présentations, mais aussi préparé dans les délais un manuscrit pour cet ouvrage.
Nous remercions les modérateurs des différentes sessions (D. Metcalf, E. Frindel, N. Young, B. Varet, A.A. Axelrad, P. Boivin, W.R. Paukovits, J. Breton-Gorius, L.A. Rozenszajn, et G. Milon), des séances de communications affichées (N. Dainiak, C. Dresch, E. Wright et M. Lenfant) et de la table ronde (H. Broxmeyer, W.R. Paukovits et M. Guigon).
Merci aux participants qui ont bien voulu corriger les textes de leur intervention et à Claire Cocheteux qui a réalisé la frappe des discussions enregistrées et de leurs corrections.

Légende de la photo

1. W.R. Paukovits; 2. N.C. Gorin; 3. G.D. Roodman; 4. M. Guigon; 5. N. Young; 6. D. Metcalf; 7. H.N. Steinberg; 8. A. Najman; 9. N. Dainiak; 10. H. Broxmeyer; 11. M. Aglietta; 12. A.A. Axelrad; 13. L.A. Solberg Jr; 14. O.D. Laerum; 15. B.I. Lord.
* C. Hamburger : La Maison de la Recherche (Fondation pour la Recherche Médicale, Paris).

List and address of contributors
Liste et adresse des auteurs

Aglietta M., Universita di Torino, Dipartimento di Scienze Biomediche E Oncologia Umana, Sezione Clinica, Via Genova 3, 10126 Torino, Italie.

Axelrad A.A., Department of Anatomy, Faculty of Medicine, University of Toronto, Medical Sciences Building, Toronto, Ontario M5S 1A8, Canada.

Broxmeyer H., Department of Medicine, Indiana University School of Medicine, Clinical Bldg, Room 379, 541 Clinical Drive, Indianapolis, IN 46223, États-Unis.

Carlo Stella C., Clinica Medica II, Università di Pavia, Policlinico S. Matteo, 27100 Pavia, Italie.

Dainiak N., Departments of Medicine and Biomedical Research, St. Elizabeth's Hospital of Boston, Tufts University School of Medicine, Boston, MA 02135, États-Unis.

Dautry F., Laboratoire d'Oncologie Moléculaire, Pavillon de Recherche, Institut Gustave Roussy, 39-53 rue Camille Desmoulins, 94805 Villejuif Cedex, France.

Dezza L., Clinica Medica II, Università di Pavia, Policlinico S. Matteo, 27100 Pavia, Italie.

Eriksen J., Nycomed FoU-4, Nycoveien 2, PO Box 4220 Torshov N-0401, Oslo 4, Norvège.

Fasciotto B., Unité de Virologie, INSERM/CNRS, 1 Place du Pr. J. Renaut, 69371 Lyon Cedex 08, France.

Fetsch J., Institut für Pharmazie der Freien Universität Berlin, D-1000 Berlin 33, RFA.

Frickhofen N., Department of Hematology/Oncology, University of Ulm, Steinhovelstrasse 9, D-7900 Ulm, RFA.

Gaggioli L., Istituto di Istologia, University of Bologna, Via Belmeloro 8, 40126 Bologna, Italie.

Geissler D., Universitätsklinik für Innere Medizin-Innsbruck, Anichstr. 35, A-6020 Innsbruck, Autriche.

Greher J., Zentrum Innere Medizin, Abt. Hematologie, Universität Frankfurt, Theodor-Stern-Kai 7, D-6000 Frankfurt 70, RFA.

Guigon M., Laboratoire d'Hématologie, Faculté de Médecine Saint-Antoine, 27 rue de Chaligny, 75012 Paris, France.

Hofmann M.C., Med. Klinik, Lab. für Onkologie, Universitätsspital Zürich, Zürich, Suisse.

Irvine A., Northern Ireland Leukaemia Research Fund, Royal Victoria Hospital, Belfast BT 12, Irlande du Nord.

Laerum O.D., Department of Pathology, The Gade Institute, Haukeland Hospital, 5016 Bergen, Norvège.

Lauria F., Istituto di Istologia, University of Bologna, Via Belmeloro 8, 40126 Bologna Italie.

Lombard M.-N., Institut Curie, Bat 110, INSERM U22, Faculté des Sciences, 91045 Orsay, France.

Lord B.I., Paterson Laboratories, Christie Hospital and Holt Radium Institute, Wilmslow Road, Manchester, M20 9BX, Royaume-Uni.

Lunardi-Iskandar Y., INSERM U 248, Hôpital Paul Brousse, BP 200, 94804 Villejuif Cedex, France.

Mathiot C., Laboratoire d'Hématologie, Institut Curie, 26 rue d'Ulm, 75005 Paris, France.

Metcalf D., The Walter and Eliza Hall Institute of Medical Research, Royal Melbourne Hospital, POB 3050, Victoria, Australie.

Morris T.C.M., Department of Hematology, Belfast City Hospital, Belfast BT9 7AB, Irlande du Nord.

Najman A., Laboratoire d'Hématologie, Faculté de Médecine Saint-Antoine, 27 rue de Chaligny, 75012 Paris, France.

Necas E., Department of Pathophysiology, Faculty of General Medicine, Charles University, 12853 Prague, Tchécoslovaquie.

Olofsson T., Department of Medicine, Division of Hematology, University of Lund, Lund Hospital, Lund, Suède.

Paukovits W.R., Institute for Tumor Biology and Cancer Research, University of Vienna, Borschekegasse 8a, 1090 Vienna, Autriche.

Pavlović-Kentera V., Institute for Medical Research, PO Box 721, Belgrade, Yougoslavie.

Rich I.N., Department of Tranfusion Medicine, University of Ulm, DRK Blutspendezentrale, Oberer Elselsberg, Postfach 1564, D-7900 Ulm/Donau, RFA

Riches A., Department of Anatomy and Experimental Pathology, University of St Andrews, Bute Medical Buildings, St Salvator's College, St Andrews, KY16 9TS, Ecosse.

Roodman G.D., Research 151, Audie Murphy Va Hospital, 7400 Merton Minter Bvd, San Antonio, TX 78284, États-Unis.

Slater K., Medical Research Centre, City Hospital, Nottingham, NG5 1PB, Royaume-Uni.

Solberg L.A. Jr., Hematology and Internal Medicine, Mayo Clinic, Rochester, MN 55905, États-Unis.

Steinberg H.N., Department of Medicine, Beth Israel Hospital, 330 Brookline Ave, Boston, MA 02215, États-Unis.

Tubiana M., Institut Gustave Roussy, 39-53 rue Camille Desmoulins, 94805 Villejuif Cedex, France.

Vainchenker W., INSERM U91, Hôpital Henri Mondor, 51 rue du Maréchal de Lattre de Tassigny, 94000 Créteil, France.

Vinci G., INSERM U. 91, Hôpital Henri Mondor, 51 rue du Maréchal de Lattre de Tassigny, 94000 Créteil, France.

Völkers B., Zentrum Innere Medizin, Abt. Hematologie, Universität Frankfurt, Theodor-Stern-Kai 7, D-6000 Frankfurt 70, RFA.

Wolpe S.S., The Rockfeller University, New York NY, États-Unis.

Wu K., Institute of Hematology, Chinese Academy of Medical Sciences, 288 Nanjing Road, Tianjin, Chine.

Young N., Cell Biology Section, Clinical Hematology Branch, National Heart, Lung and Blood Institute, ACRF 7C103 Bethesda, MD 20892, États-Unis.

Contents
Sommaire

V   Preface
VI *Préface*
XI List of contributors
    Liste des auteurs

INTRODUCTION

3   **D. Metcalf**
    The molecular nature of stimulators and inhibitors of granulocyte and macrophage production
    La nature moléculaire des facteurs stimulants et inhibiteurs de la production des granulocytes et des macrophages

NEGATIVE REGULATION OF GRANULOPOIESIS/*INHIBITEURS DE LA GRANULOPOÏÈSE*

Invited papers/Conférences plénières

11 **M. Aglietta, W. Piacibello and F. Gavosto**
    Growth factors and growth inhibitors: their interaction in the regulation of human myelopoiesis
    Facteurs de croissance et facteurs inhibiteurs : leur interaction dans la régulation de la myélopoïèse humaine

21 **O.D. Laerum and W.R. Paukovits**
    Biological effects of myelopoiesis inhibitors
    Effets biologiques des inhibiteurs de la myélopoïèse

31 **W.R. Paukovits, O.D. Laerum, J.B. Paukovits, M. Guigon and J.S. Schanche**
    Regulatory peptides inhibiting granulopoiesis
    Peptides régulateurs inhibant la granulopoïèse

Discussion

45 Chairpersons/*Modérateurs*: **D. Metcalf** (Australie), **E. Frindel** (France)

Brief reports/Communications brèves

51 **J.A. Eriksen, J.S. Schanche, K. Hestdal, S.E. Jakobsen, T. Tveteraas, J.H. Johansen, W.R. Paukovits and O.D. Laerum**
    Hemoregulatory peptide synthesis, purification of tritium labelled peptide and uptake of peptide in hematopoietic tissues *in vitro*
    Synthèse d'un peptide hémorégulateur, purification du peptide marqué au tritium et incorporation du peptide dans les tissus hématopoïétiques in vitro

55 **J. Fetsch and H.R. Maurer**
A specific, low molecular mass granulopoiesis inhibitor, isolated from calf spleen
Un inhibiteur de la granulopoïèse spécifique et de faible poids moléculaire isolé de la rate de veau

59 **L. Dezza, M. Cazzola, G. Bergamaschi, C. Carlo Stella, P. Pedrazzoli, M. Aglietta and E. Ascari**
Inhibitory activity of recombinant human H-subunit ferritin on the *in vitro* growth of human granulocyte-macrophage progenitor cells
Activité inhibitrice de la sous-unité H de la ferritine recombinante humaine sur la croissance in vitro *des progéniteurs granulomacrophagiques humains*

63 **I.N. Rich and G. Sawatzki**
The role of lactoferrin in regulating colony stimulating factor production
Le rôle de la lactoferrine dans la régulation de la production de facteur stimulant la croissance des colonies

67 **K. Slater and J. Fletcher**
Lactoferrin inhibits Interleukin-2 production in mixed lymphocyte culture
La lactoferrine inhibe la production d'Interleukine-2 dans des cultures mixtes de lymphocytes

69 **T.C.M. Morris, A.E. Irvine and A. French**
Lymphocyte mediated suppression of granulopoiesis
Suppression de la granulopoïèse par l'intermédiaire des lymphocytes

73 **A.E. Irvine, V. Craig, A. Thomson and T.C.M. Morris**
Autoimmune neutropenia and haemolytic anaemia
Neutropénie auto-immune et anémie hémolytique

NEGATIVE REGULATION OF ERYTHROPOIESIS AND MEGAKARYOPOIE-SIS/*INHIBITEURS DE L'ÉRYTHROPOÏÈSE ET DE LA MÉGACARYOPOÏÈSE*

Invited papers/ Conférences plénières

79 **A.A. Axelrad, H. Croizat, D. del Rizzo, D. Eskinazi, G. Pezzutti, S. Stewart and H. Van der Gaag**
Properties of a protein NRP that negatively regulates DNA synthesis of the early erythropoietic progenitor cells BFU-E
Propriétés d'une protéine qui inhibe la synthèse d'ADN des progéniteurs érythroïdes précoces (BFU-E)

93 **N. Dainiak, S. Kreczko and P.R. Strauss**
Suppression of human erythropoiesis by inactivation of topoisomerases
Suppression de l'érythropoïèse chez l'homme par inactivation des topoisomérases

101 **W. Vainchenker, M.T. Mitjavila, G. Vinci, N. Kieffer, J.L. Villeval, A. Henri and J. Breton-Gorius**
In vitro inhibition of human megakaryocyte colony formation by platelet products : its relationship to TGF-beta
Inhibition in vitro *de la formation des colonies de mégacaryocytes humains par un produit plaquettaire : sa relation probable avec le TGF-bêta*

111 **L.A. Solberg Jr., R.F. Tucker, B.W. Grant, K.G. Mann and H.L. Moses**
Transforming growth factor-beta inhibits colony formation from human megakaryocytic, erythroid and multipotent stem cells
Le « transforming growth factor-beta » inhibe la formation des colonies de cellules souches mégacaryocytaires, érythroïdes et pluripotentes humaines

Discussion

125 Chairpersons/*Modérateurs* : **N. Young** (États-Unis), **B. Varet** (France)

Brief reports/Communications brèves

133 **V. Pavlović-Kentera, L. Djukanović, L. Biljanović-Paunović, N. Stojanović and P. Milenković**
Inhibitors of erythropoiesis in patients with chronic renal failure
Inhibiteurs de l'érythropoïèse chez des malades atteints d'insuffisance rénale chronique

SUPPRESSION OF HEMATOPOIESIS IN ACUTE LEUKEMIAS/*SUPPRESSION DE L'HÉMATOPOÏÈSE DANS LES LEUCÉMIES AIGUËS*

Invited papers/Conférences plénières

139 **H.E. Broxmeyer, D.E. Williams, L. Lu, S. Vadhan, S. Cooper, D.C. Bicknell, P. Ralph, J. Gutterman and G. Tricot**
Biomolecules associated with suppression of myelopoiesis in normal conditions and during myeloid leukemia and other related disorders
Biomolécules associées à la suppression de la myélopoïèse à l'état normal et dans les leucémies myéloïdes et autres maladies voisines

151 **A. Najman, L. Kobari, N. Dainiak, C. Baillou, J.P. Laporte, L. Douay and N.C. Gorin**
Erythropoiesis in human acute non lymphoblastic leukemias : place of stimulating and inhibitory factors
L'érythropoïèse dans les leucémies aiguës non lymphoblastiques : place des facteurs stimulants et inhibiteurs

163 **H.N. Steinberg**
Suppression of normal hemopoiesis in leukemia : *in vivo* and *in vitro* studies
Suppression de l'hématopoïèse normale dans la leucémie : études in vivo *et* in vitro

177 **T.B.J. Olofsson**
Leukemia associated inhibitor (LAI) : biological characterization and purification of the active subunit
L'inhibiteur associé à la leucémie (LAI) : caractérisation biologique et purification de la sous-unité active

Discussion

191 Chairpersons/*Modérateurs* : **A.A. Axelrad** (Canada), **P. Boivin** (France)

Brief reports/Communications brèves

197 **S.D. Wolpe, S. Sassa and A. Cerami**
Macrophages secrete an inhibitory factor for mouse erythroleukemic cell differentiation : characterization and partial purification

Les macrophages sécrètent un facteur inhibant la différenciation des cellules érythroleucémiques de la souris : caractérisation et purification partielle

201 **B. Fasciotto, D. Kanazir, J.P. Durkin, J.F. Whitfield and V. Krsmanovic**
Blockage of erythroleukemia cell induced differentiation by autocrine growth factor(s)
Blocage de la différenciation induite des cellules érythroleucémiques par un ou des facteur(s) de croissance autocrine(s)

205 **F. Lauria, G.P. Bagnara, D. Raspadori, M. Buzzi, A. Guarini, P.L. Zinzani, L. Catani, L. Gaggioli, M. Marini and M.A. Brunelli**
Hairy cell leukemia (HCL) : 1) Inhibitory effect of serum from HCL-patients on the normal bone marrow colony growth possibly removed by a prolonged alpha interferon treatment
Leucémie à tricholeucocytes : 1) L'effet inhibiteur du sérum de patients atteints de leucémie à tricholeucocytes sur la formation de colonies par la moelle humaine normale pourrait être supprimé par un traitement prolongé par l'interféron alpha

209 **L. Gaggioli, L. Bonsi, L. Valvassori, L. Catani, A. Guarini, P.L. Zinzani, M. Buzzi, D. Raspadori, M. Marini, F. Lauria, G.P. Bagnara and C. Rizzoli**
Hairy cell leukemia (HCL) : 2) Hairy cells produce factor(s) affecting the *in vitro* growth of normal hemopoietic stem cells
Leucémie à tricholeucocytes : 2) Les tricholeucocytes produisent un ou des facteur(s) affectant la croissance in vitro *des cellules souches hématopoïétiques normales*

213 **N. Frickhofen, A. Raghavachar, I.N. Rich, W. Heit and H. Heimpel**
Inhibition of hematopoietic progenitor cells by a variant of the L1210 leukemia cell line
Inhibition des progéniteurs hématopoïétiques par une variante de la lignée leucémique L1210

217 **K. Wu, L. Liu, J. Chu, J. Wan, Y. Song and W. Chen**
Negative autocrine activity of fractions of leukemic ascites and conditioned medium
Activité autocrine négative de fractions d'ascite leucémique et du milieu conditionné

221 **M.C. Hofmann, S. Arrenbrecht and C. Sauter**
Stimulatory and inhibitory factors of myelopoiesis in normal and AML human serum
Facteurs stimulants et inhibiteurs de la myélopoïèse dans le sérum humain normal et leucémique

NEGATIVE REGULATION OF PLURIPOTENT STEM CELL (CFU-S) PROLIFERATION/*INHIBITEURS DE LA PROLIFÉRATION DES CELLULES SOUCHES PLURIPOTENTES (CFU-S)*

Invited papers/Conférences plénières

227 **B.I. Lord, L. Fu-Lu, Z. Pojda and E. Spooncer**
Inhibitor of haemopoietic CFU-S proliferation : assays, production sources and regulatory mechanisms
Un inhibiteur de la prolifération des cellules souches pluripotentes hématopoïétiques : essais, sources et mécanismes régulateurs

241 **M. Guigon**
Biological properties of low molecular weight pluripotent stem cell (CFU-S) inhibitors
Propriétés biologiques d'inhibiteurs de l'entrée des CFU-S en synthèse d'ADN de faible poids moléculaire

Discussion

255 Chairpersons/*Modérateurs*: **W.R. Paukovits** (Autriche), **J. Breton-Gorius** (France)

Brief reports/Communications brèves

263 **E. Nečas and V. Znojil**
Importance of the measurement error in demonstrating inhibition of stem cell proliferation
Importance des erreurs de mesure dans la démonstration de l'inhibition de la prolifération des cellules souches

267 **A. Riches and M. Cork**
The effect of CFU-S proliferation inhibitor on stimulator production *in vitro* by human foetal liver cells
Effet d'un inhibiteur de la prolifération des CFU-S sur la production d'un stimulant in vitro *par les cellules du foie fœtal humain*

271 **M-N. Lombard, J. Wdzieczak-Bakala, D. Sotty, M. Lenfant, C. Nadal and M. Guigon**
Effect on hepatocyte G1-S transition of an inhibitor of CFU-S entry into DNA synthesis
Effet sur la transition G1-S des hépatocytes d'un inhibiteur de l'entrée des CFU-S en synthèse d'ADN

LYMPHOKINES IN THE NEGATIVE REGULATION OF HEMATOPOIESIS/*RÔLE DES LYMPHOKINES DANS LA RÉGULATION NÉGATIVE DE L'HÉMATOPOÏESE*

Invited papers/Conférences plénières

279 **N.S. Young**
Interferons and other lymphokines in bone marrow suppression *in vitro* and *in vivo*: implications for the pathogenesis of aplastic anemia
Rôle des interférons et autres lymphokines dans la suppression médullaire in vitro *et* in vivo *: implications dans la pathogénie de l'aplasie médullaire*

289 **G.D. Roodman**
Suppression of hematopoietic cells by TNF-Alpha
Inhibition des cellules hématopoïétiques par le TNF-Alpha

301 **F. Dautry and D. Weil**
Expression of the TNF genes in response to hematopoietic growth factors
Expression des gènes du TNF en réponse aux facteurs de croissance hématopoïétiques

Discussion

311 Chairpersons/Modérateurs: **L.A. Rozenszajn** (Israël), **G. Milon** (France)

Brief reports/Communications brèves

317 **J. Greher, A. Ganser, B. Völkers, C. Carlo Stella and D. Hoelzer**
Inhibition of the growth of normal megakaryocytic progenitor cells by recombinant human alpha and gamma-Interferons
Inhibition de la croissance des progéniteurs mégacaryocytaires normaux par les Interférons humains alpha et gamma recombinants

321 **C. Carlo Stella, M. Cazzola, A. Ganser, B. Völkers, L. Dezza, F. Meloni, P. Pedrazzoli and E. Ascari**
Recombinant alpha and gamma-Interferons synergistically inhibit the in vitro growth of hemopoietic progenitors (CFU-GEMM, CFU-MK, BFU-E, CFU-GM) from patients with myelofibrosis with myeloid metaplasia
Les Interférons alpha et gamma recombinants inhibent de façon synergique la croissance in vitro des progéniteurs hématopoïétiques (CFU-GEMM, CFU-MK, BFU-E, CFU-GM) des patients atteints de myélofibrose avec métaplasie myéloïde

325 **D. Geissler, G. Gastl, W. Aulitzky, H. Tilg, G. Konwalinka and C. Huber**
In vitro sensitivity of hematopoietic precursor cells to recombinant Interferon alpha (rIFN-alpha), recombinant Interferon gamma (rIFN-gamma) and recombinant tumor necrosis factor alpha (rTNF-alpha) in normal controls and in patients with CML: relationship to the in vivo response
Sensibilité in vitro des précurseurs hématopoïétiques aux interférons alpha et gamma recombinants et au « tumor necrosis factor » alpha recombinant chez des sujets normaux ou atteints de LMC: relation avec la réponse in vivo

331 **B. Völkers, A. Ganser, C. Carlo Stella and D. Hoelzer**
Decrease of in vitro colony formation of the hematopoietic progenitor cells (CFU-GEMM, CFU-MK, BFU-E and CFU-GM) in the acquired immunodeficiency syndrom (AIDS)
Diminution de la formation des colonies de progéniteurs hématopoïétiques in vitro (CFU-GEMM, CFU-MK, BFU-E et CFU-GM) dans le syndrome d'immunodéficience acquise (SIDA)

335 **Y. Lunardi-Iskandar, V. Georgoulias, D. Vittecocq, M.T. Nugeyre, A.M. Bertoli, A. Ammar, C. Clémenceau, F. Barre-Sinoussi, J.C. Chermann, L. Schwarzenberg and C. Jasmin**
Peripheral blood adherent cells from AIDS patients inhibit normal T colony growth through decreased expression of Interleukin-2 receptors and production of Interleukin-2
Les cellules adhérentes du sang périphérique des patients atteints de SIDA inhibent la croissance de colonies T normales par diminution de l'expression des récepteurs à l'Interleukine 2 et de la production d'Interleukine 2

339 **G. Vinci, J.P. Vernant, M. Zohar, A. Henri, H. Rochant, J. Breton-Gorius and W. Vainchenker**
In vitro inhibition of hematopoiesis by HNK_1, CD8, DR positive T cells from allogeneic bone marrow transplantation and normal subjects
Inhibition de l'hématopoïèse in vitro par des cellules T, HNK_1, CD8 et DR positives, provenant de greffe de moelle allogénique et de sujets normaux

343 **C. Mathiot, S. Amigorena, J. Moncuit, J.L. Teillaud and W.H. Fridman**
Immunoglobulin-binding factors (IBF) are potent inhibitors of the growth of tumor B cells
Les facteurs liant les immunoglobulines (IBF) sont de puissants inhibiteurs de la croissance des cellules tumorales B

PERSPECTIVES

351 **M. Tubiana**
Inhibitors of hemopoiesis : the accomplishments and the prospects
Les inhibiteurs de l'hématopoïèse : résultats et perspectives

355 Author Index
Index des auteurs

Introduction

The molecular nature of stimulators and inhibitors of granulocyte and macrophage production

Donald Metcalf

Walter and Eliza Hall Institute of Medical Research, Melbourne, Australia

ABSTRACT

The glycoprotein CSF's are well characterized as a group of molecules stimulating the proliferation of granulocyte-macrophage populations but current information on possible inhibitory molecules is less satisfactory. The candidate inhibitors are widely diverse in nature and possible action. The present incomplete information is capable of being corrected using existing technology and is an important area for study since such molecules may be of therapeutic value in the management of myeloid leukemia.

KEYWORDS

Colony stimulating factors, inhibitors, granulocytes, macrophages.

THE COLONY STIMULATING FACTORS

Analysis of the proliferative behavior in vitro of progenitors of granulocytes and macrophages has indicated that these cells require positive stimulation for proliferation. The only agents known to stimulate the proliferation of granulocytes and macrophages are the group of glycoproteins known as the colony stimulating factors (see reviews Metcalf, 1986, 1987). In man and the mouse four distinct CSF's have been detected, purified and molecularly cloned. Three of the CSF's (GM-CSF, G-CSF and Multi-CSF (IL-3) are monomers in the size range 18 - 28,000 while the fourth (M-CSF) is a dimer of two identical polypeptide subunits. The carbohydrate of the CSF's is somewhat variable but comprises about 30% of the molecular mass. It is not necessary for biological activity either in vitro or in vivo and is not involved in receptor binding. Each CSF is able to stimulate cellular proliferation at $10^{-10} - 10^{-12}$M concentrations and the three-dimensional configuration of the molecules is important as indicated by the presence of mandatory disulphide bridges and the inactivity of CSF peptide fragments.

Most granulocyte-macrophage progenitors simultaneously express receptors for all four CSF's and all can stimulate both granulocyte and macrophage formation. However, for two there is a strong bias towards forming cells of only a single differentiation lineage - granulocytes with G-CSF and macrophages with M-CSF. Conversely, both GM-CSF and Multi-CSF also have proliferative effects on related hemopoietic cells outside the granulocyte-macrophage lineage. The advantage served by this overlapping control system is still unclear although the different CSF's induce differing response patterns in vivo. For example, G-CSF induces high blood granulocyte levels in the mouse but not high peritoneal cell levels while GM-CSF has the reverse effects (Metcalf et al, 1986, 1987).

The major surprise from the full amino acid sequence data for the four CSF's was that the four are totally unrelated molecules. Thus four molecules, with no common ancestry, induce quite similar responses in a common set of target cells. There is some homology between corresponding murine and human CSF's, permitting cross-species reactivity for G-CSF but not for GM-CSF or Multi-CSF.

While the actual regulator molecules differ, there is evidence of coordination of the CSF's at the genomic, transcriptional and receptor levels: (a) In man, the genes for GM-CSF, Multi-CSF, M-CSF and M-CSF-receptor are tightly grouped on chromosome 5. (b) In the mouse, the genes for GM-CSF and Multi-CSF are again located close together on chromosome 11. (c) Common decanucleotide regions (potential signal sites) precede the genes for GM-CSF, Multi-CSF and IL-2. All are transcribed coordinately, if not entirely synchronously, by T-lymphocytes following mitogenic or receptor stimulation. (d) The four distinct CSF receptors are functionally linked on target populations. For example, occupancy of Multi-CSF receptors leads to down-modulation of receptors for GM-CSF, G-CSF and M-CSF and all four receptors are arranged in a down-modulation hierarchy.

The CSF's are not simply proliferative stimuli for granulocyte-macrophage populations. Each CSF possesses four quite distinct actions: (a) survival enhancement, (b) proliferative stimulation, (c) differentiation commitment and (d) stimulation of mature end cell functional activity. This pattern of combined functional actions seems to hold true for comparable regulators in other hemopoietic lineages.

The availability of recombinant CSF's has allowed the demonstration that the CSF's exert similar actions in vivo to those documented in vitro. This, coupled with the indirect evidence indicating the presence of adequate CSF concentrations in the circulation and tissues and variations in these levels during perturbations of granulocyte-macrophage formation, allow the conclusion that the CSF's function as genuine regulators of granulocyte-macrophage production in the intact animal.

REGULATION OF CSF PRODUCTION

An important question in the context of positive and negative control systems for the granulocyte-macrophage population is the nature of the stimuli eliciting CSF production and the factors controlling CSF levels and fate in the body.

The dominant factor determining levels of CSF production in adult life is exposure to microorganisms. Within minutes of such exposure, cells release CSF and begin active synthesis of increased amounts of CSF. High CSF levels are sustained during acute infections but, with resolution of the infection, CSF levels rapidly return to normal, a process aided by the short serum half-life of the molecules.

The CSF system is therefore highly labile and dominantly influenced by exposure to external stimuli in the form of invading microorganisms and their products; CSF levels passively return to normal following the removal of the stimuli inducing the increased synthesis of CSF.

IS AN INHIBITORY SYSTEM NECESSARY FOR THE CONTROL OF GRANULOCYTE AND MACROPHAGE PRODUCTION?

Given the existence of a system in which granulocyte and macrophage production is able to be increased and decreased solely by changing CSF levels, is it necessary to postulate the existence of a matching inhibitory system for modulating granulocyte-macrophage production?

Theoretical biologists are of the view that stable control of a proliferating population cannot be achieved solely by the use of positive proliferative stimuli because such a system would display a marked tendency for cyclical fluctuations. If it is assumed that inhibitors do exist, what properties might be expected of candidate inhibitors? One possibility is that they might closely resemble the CSF's in basic pattern. If so, multiple inhibitors might be expected that could be glycoprotein in nature and be multi-organ in origin. An obvious alternative is that the inhibitors might be products of mature granulocytes and macrophages and represent direct feedback loops whose strength directly reflects existing numbers of granulocytes and macrophages. Comparable inhibitors might originate from other body tissues able to monitor and respond to circulating levels of granulocytes and macrophages. A wide variety of mechanisms of action would be open to such inhibitors: (a) direct inhibition of granulocyte-macrophage precursors by toxic or cytostatic effects, (b) induction of unresponsiveness to CSF stimulation, (c) blockade of CSF binding to membrane receptors, (d) suppression of CSF production or (e) enhancement of CSF clearance.

INHIBITORS OF GRANULOCYTE-MACROPHAGE PRODUCTION

Prostaglandin E
This molecule with diverse physiological actions has been well documented to selectively inhibit macrophage proliferation in vitro. PGE either fails to inhibit or enhances the proliferation of erythroid or multipotential precursor cells. As a product of macrophages whose synthesis is enhanced by CSF stimulation it is an interesting candidate for a selective inhibitor at least of macrophage production and acts apparently as an S-phase specific inhibitor of granulocyte-macrophage progenitor cells (Pelus et al, 1986).

Lactoferrin
Lactoferrin as extractable from polymorphs has the ability to suppress CSF production by at least some cells in vitro and has been reported to reduce circulating levels of CSF in vivo (Broxmeyer et al, 1978). Again the system has a certain intrinsic elegance in providing a possible inhibitory feedback control of granulocyte-macrophage production. However, in bacterial infections where both excess CSF production and granulocyte production are able to be sustained, it is difficult to envisage the operation of a lactoferrin inhibitory system.

Hemoregulatory Peptide
A pentapeptide pGlu-Glu-Asp-Cys-Lys has been isolated and synthesized (Paukovits and Laerum, 1982; Foa et al, 1986) that may be the previously described granulocyte chalone. It is unclear from published evidence whether a similar peptide is extractable from other cells. It is inhibitory for CSF-stimulated granulocyte-macrophage colony formation, inhibition not being seen with higher concentrations of the peptide which suggests that it is not cytotoxic for granulocyte-macrophage precursor cells (Laerum et al, 1986). This pentapeptide appears to be much less inhibitory for erythroid and multipotential colony formation (Foa et al, 1986).

Serum Lipoproteins
Much work was performed some years ago on light density serum lipoproteins because

of their capacity to inhibit granulocyte-macrophage colony formation in vitro (see review, Metcalf, 1984). Amongst hemopoietic populations able to be cloned in vitro, no specificity of action is evident since the lipoproteins are equally inhibitory for lymphoid, erythroid and other cells. It is of interest that levels of such inhibitors are low or undetectable in the sera of many patients with acute myeloid leukemia and fall in the serum prior to marrow regeneration after irradiation. However, in mice, there are marked strain differences in serum inhibitor levels yet mice of such strains exhibit essentially identical levels of granulocyte and macrophage formation and respond equally well to the injection of CSF. The evident lack of specificity of these agents makes them very doubtful candidates for genuine granulocyte-macrophage inhibitors.

Stem Cell Inhibitors
A low molecular weight (less than 2000) inhibitor has been extracted from fetal marrow or liver that suppresses entry of CFU-S into cycle both in vitro and in vivo and is protective for hemopoietic regeneration following cytosine arabinoside treatment (Frindel and Guigon, 1977; Guigon et al, 1982). An inhibitory factor of larger apparent molecular weight has also been extracted from normal, but not regenerating, marrow possibly of macrophage origin that is also capable of blocking entry of CFU-S into cell cycle (see review, Lord, 1986).

It is difficult to fit such inhibitors into a specific control system for granulocyte-macrophage production as no obvious mechanism presents itself for achieving selective suppression only of granulocyte-macrophage precursors. There are however occasions where all hemopoietic populations may need to be regulated simultaneously e.g. following regeneration, and general inhibitors could operate in such situations.

Tumor Cell Products
In acute myeloid leukemia there is suppression of normal progenitors of multiple hemopoietic lineages so although coculture experiments have shown that leukemic cells can suppress normal granulocyte-macrophage colony formation, the suppression of normal hemopoiesis in acute leukemic patients is either not lineage-specific or might be mediated by several independent mechanisms.

Inhibitors of Leukemic Cells
The best characterized biological inhibitors of myeloid leukemic cells are G-CSF and the differentiation-inducing factor, DIF. The suppressive effects of G-CSF originate from the ability of the CSF's to induce differentiation commitment and in essence G-CSF blocks the capacity of responsive myeloid leukemic stem cells to self-generate and induces differentiation to granulocyte and macrophage formation (Metcalf, 1986). Suppression of stem cell self-generation need not be accompanied by morphological differentiation as exemplified by the action of G-CSF on HL60 cells (Begley et al, 1987). The action of G-CSF on human myeloid leukemic cells is potentially quite complex since G-CSF is a strong proliferative stimulus for most primary human myeloid leukemic populations.

DIF of human cell origin has recently been identified as TNFα (Takeda et al, 1986), but available evidence suggests that the murine equivalent is not murine TNFα.

COMMENTS ON CANDIDATE INHIBITORS

None of the inhibitors so far documented fits very well the requirements needed for specific modulating inhibitors of the CSF's. Since adequate in vitro cloning systems now exist for all hemopoietic populations, it should now be routine to establish the range of inhibitory action of all candidate inhibitors. For any exhibiting a promising selectivity of action, the general mode of action (i.e. whether anti-cell or anti-CSF) can now readily be established because of the availability of recombinant CSF's.

From the biology of other regulatory molecules, it is not necessary to establish an exclusive granulocyte or macrophage origin of such inhibitors and it may be useful therefore in searching for sources of granulocyte-macrophage inhibitors to greatly increase the range of tissues being screened. A cardinal requirement of a genuine inhibitor of granulocyte-macrophage production should be that it is either demonstrably humoral (that is present in significant concentrations in the serum) or demonstrable as a product of hemopoietic tissue, particularly the bone marrow.

Ultimately, as was the case for the CSF's, progress with inhibitors depends either on mass-production of purified candidate molecules or the use of molecular biology to isolate cDNA clones for candidate regulators.

REFERENCES

Begley, C.G., Metcalf, D. and Nicola, N.A. (1987): Purified colony stimulating factors (G-CSF and GM-CSF) induce differentiation in human HL60 cells with suppression of clonogenicity. Int.J.Cancer 39, 99-105.

Broxmeyer, H.E., Smithyman, A., Eger, R.R., Meyers, P.A. and De Sousa, M. (1975): Identification of lactoferrin as the granulocyte-derived inhibitor of colony-stimulating activity production. J.Exp.Med. 148, 1052-1067.

Foa, P., Lu, L., Broxmeyer, H.E., Chillemi, F., Maiolo, A.T., Lombardi, L., Lonati, S., Ciani, A. and Polli, E.E. (1986): A synthetic pentapeptide candidate for the role of granulocytic chalone. In Biological Regulation of Cell Proliferation, ed R. Baserga et al, pp 103-109. New York, Raven Press.

Frindel, E. and Guigon, M. (1977): Inhibition of CFU entry into cycle by a bone marrow extract. Exp.Hematol. 5, 74-76.

Guigon, M., Mary, J.Y., Enouf, J. and Frindel, E. (1982): Protection of mice against lethal doses of 1-beta-D-arabinofuranosylcytosine by pluripotent stem cell inhibitors. Cancer Res. 42, 638-641.

Laerum, O.D., Paukovits, W.R. and Sletvold, O. (1986): Hemoregulatory peptide: Biological aspects. In Biological Regulators of Cell Proliferation, ed. R. Baserga et al, pp 121-129. New York: Raven Press.

Lord, B. (1986): Interactions of regulatory factors in the control of haemopoietic stem cell proliferation. In Biological Regulation of Cell Proliferation, ed R. Baserga et al, pp 167-177. New York:Raven Press.

Metcalf, D. (1984): The hemopoietic colony stimulating factors. Elsevier, Amsterdam.

Metcalf, D. (1986): The molecular biology and functions of the granulocyte-macrophage colony-stimulating factors. Blood 67, 257-267.

Metcalf, D. (1987): The molecular control of normal and leukaemic granulocytes and macrophages. Proc.Roy.Soc. (B) (in press).

Metcalf, D., Begley, C.G., Johnson, G.R., Nicola, N.A., Lopez, A.F. and Williamson, D.J. (1986): Effects of purified bacterially-synthesized murine Multi-CSF (IL-3) on hemopoiesis in normal adult mice. Blood 68, 46-57.

Metcalf, D., Begley, C.G., Williamson, D.J., Nice, E.C., DeLamarter, J., Mermod, J-J., Thatcher, D. and Schmidt, A. (1987): Hemopoietic responses in mice injected with purified recombinant murine GM-CSF. Exp.Hematol. 15, 1-9.

Paukovits, W.R. and Laerum, O.D. (1982): Isolation and synthesis of a hemoregulatory peptide. Z.Naturforsch. C. 37, 1297-1300.

Pelus, L.M., Lu, Li and Gentile, P. (1986): Prostaglandin E. Multiphasic control of hematopoiesis. In Biological Regulation of Cell Proliferation, ed R Baserga et al, pp 75-83. New York, Raven Press.

Takeda, K., Iwamoto, S., Sugimoto, H., Takuma, T., Kawatani,N., Noda, M., Masaki, A., Morise, H., Arimura, H. and Konno, K. (1986): Identity of differentiation inducing factor and tumour necrosis factor. Nature 323, 338-340.

Résumé

Les glycoprotéines ayant une action CSF sont bien caractérisées comme un groupe de molécules stimulant la prolifération des populations granulo-macrophagiques alors que les données actuelles sur d'éventuelles molécules inhibitrices sont moins satisfaisantes. Les substances présentées comme des inhibiteurs sont très diverses par leur nature et leur mécanisme d'action. Des progrès sont à attendre dans ce domaine important car ces molécules pourraient avoir une application thérapeutique dans les leucémies myéloïdes.

Negative regulation of granulopoiesis
Inhibiteurs de la granulopoïèse

Growth factors and growth inhibitors: their interaction in the regulation of human myelopoiesis

Massimo Aglietta, Wanda Piacibello and Felice Gavosto

Clinica Medica A, Dipartimento di Scienze Biomediche ed Oncologia Umana, Università di Torino, Via Genova 3, 10126 Torino, Italy

Key words: hemopoietic progenitors, growth factors, inhibitors, transforming growth factor-beta, interferons, isoferritins, tumor necrosis factor, colony stimulating factors.

INTRODUCTION

Myelopoiesis is a dynamic process. Mature cells circulating in the peripheral blood have a short half life. Moreover, their level fluctuates in response to physiological and pathological needs of the organisms. Since these mature cells are incapable of further division, their production is ensured by more immature cells endowed, in human adults, in the bone marrow. Among these immature cells, the compartment of hemopoietic progenitors is responsible for the maintenance of a normal myelopoiesis since these cells, in addition to the potential for proliferation and differentation are self-maintaining, i.e. they are able to reproduce themselves, thus avoiding the depletion of the compartment. In vivo and in vitro studies, in humans as well as in animals, have demonstrated that the microenvironmental structures of the bone marrow are essential for sustaining a normal myelopoiesis. In addition to direct cell to cell interactions, hardly understood, microenvironmental cells cooperate in the regulation of myelopoiesis by the production of specific growth factors and growth inhibitors. These substances are also the product of other cells, not necessarily endowed within the bone marrow. They are released in the plasma and through it they reach target bone marrow cells.

In vitro culture techniques have allowed to identify myelopoietic progenitors and to study the molecules which regulate their proliferation and differentiation as well as their potential for self maintenance.

Mature granulocytes and monocytes derive from the proliferation and differentiation of a class of progenitors named CFU-GM (colony forming units granulocyte-macrofage). They can be divided, on the basis of cycling status, bouyant density and antigenic properties in day 7 and day 14 CFU-GM, the former representing the progeny of the latter. Their in vitro, and probably in vivo, proliferation requires the presence of specific growth factors, which have been purified to hamogeneity and whose genes have been cloned, named G-CSF, GM-CSF, and IL-3 (23,31).

Many cell types, like activated T lymphocytes, monocytes, endothelial cells, are responsible for the production of these molecules (12). Their stimulatory action can be counteracted by a series of inhibitory substances acting at different levels. Some inhibit granulo-monopoiesis by directly influencing progenitors, thus apparently competing with the specific growth factors. The inhibitory action of other molecules is indirect in that they decrease production and/or release by other cells of specific growth factors.

In the present paper we will summarize our present knowledges on some molecules which may act to inhibit the proliferation of granulo-monopoietic progenitors and present some evidence for a potential role of transforming growth-beta in the negative and positive regulation of myelopoiesis.

PROSTAGLANDIN E

Prostaglandins, a family of mediators active in many cell systems, act through activation or inhibition of the adenyl-cyclase. Prostaglandins of the E series (PGE) seem the most relevant in the regulation of myielopoiesis. At concentrations of 10^{-9} M or higher, PGE inhibit in vitro proliferation of murine and human CFU-GM (1,20,28,) with a preferential inhibition for CFU-GM committed towards macrophagic differentiation (2,28). The mechanism of interaction of PGE with myeloid progenitors is matter of debate. A strict link between expression of HLA class II DR antigens by myeloid progenitors and inhibitory action of PGE has been suggested by some authors (27) but not confirmed by others (3,15). PGE are active also in vivo. In mice treated with a long acting PGE analog a sustained inhibition of granulo-monopoiesis has been observed (16).

Monocytes seem the most important cells responsible for the production of PGE in the bone marrow (20). High levels of CSF stimulate monocytes to produce PGE. This suggests possible mechanisms to limit the excessive monocytopoiesis. The observation that PGE have a stimulatory effect on erythroid progenitors (BFU-E) and are inactive on multipotential (CFU-GEMM) cells (22,32) indicate the specifity of their action.

IRON BINDING PROTEINS

Leukemic cells produce high levels of inhibitors for normal but not for leukemic progenitors. These molecules have been identified as acidic isoferritins (AIF), probably produced, inside the bone marrow, by monocytes and macrophages(6). They act on all myelopoietic progenitors, since they inhibit the growth of CFU-GM, BFU-E and CFU-GEMM (21). AIF with a M.W. of about 550.000 daltons are polymers composed in a different proportion of L (M.W. 19,000) and H (M.W. 21,000) subunits. The inhibitory action seems linked to the predominance of H chains (14). Another inhibitor, named LAI, will be described in detail elsewhere in this symposium. Although its biological properties are similar to those of AIF, its biochemical characteristics as well as the cell responsible for its production are different (26).

Lactoferrin (LF) is released by secondary granules of granulocytes upon stimulus with endotoxin or after phagocytosis. When saturated with iron it inhibits granulopoiesis indirectly, by blocking the release of monokines (10). Monocytes have receptors with high affinity for the molecule (5). Its specific mechanism of action is still matter of debate. Broxmeyer has suggested that LF inhibits the release of CSF by DR positive monocytes (7). Bagby, on the other hand, suggests that LF acts by bloking the release of monokines needed for inducing CSF release from fibroblasts and activated T limphocytes (4). Whatever is the predominant way of action, LF seems a specific inhibitors, since it does not suppresses the production of monokines active on B lymphocytes nor is active on lymphokine production. LF has no direct effect on myelopoietic progenitors. Some evidence of an in vivo action has been put forward (17).

Transferrin, a molecule which shows many similarities whith lactoferrin, is produced by monocytes, granulocytes and activated T lymphocytes. It has been suggested that transferrin, produced by CD8+ lymphocytes inhibits CSF production by CD4+ activated T lymphocytes (8).

INTERFERONS

Interferons (IFN) have a direct inhibitory effect on myelopoietic progenitors (9,25,36,). All IFN have a similar potency on CFU-GEMM, BFU-E, CFU-GM, with the exception of IFN-γ which seems more active towards CFU-GM. There is synergism in the inhibitory activity of IFN-α and IFN-γ (9). The action of IFN-γ on myelopoiesis in not limited to its direct effect on hemopoietic progenitors. In fact at low doses IFN-γ indirectly can influence myelopoiesis in two opposite ways. On one side it may increase the suppressive effect of natural killer cells (18). On the other side it may increase CSF release from monocytes and activated T lymphocytes thus acting as a stimulator (29,30).

TUMOR NECROSIS FACTORS

Tumor necrosis factor alpha (TNF-α), whose gene has been cloned, may be either an inhibitor either a stimulator of myelopoiesis, the former action beeing direct, the latter indirect. Added to semisolid cultures of normal bone marrow cells, in the presence of specific growth factors, it inhibits the proliferation of myeloid progenitors (CFU-GEMM, BFU-E, CFU-GM) (11). This inhibitory action appears synergistic with that produced by IFN- . Opposite to this is its ability to increase CSF production by many cell types (24).

TRANSFORMING GROWTH FACTOR BETA

Transforming growth factor beta (TGF-β) is a 25,000 M.W. acid stable molecule composed of two identical polipeptide chains linked by disulphide bridges. It is synthesized by many cell types and, in high concentration, by degranulating platelets. High affinity receptors are expressed by essentially all types of cells. TGF- can have both growth enhancing and growth inhibitory properties. This effect depends on the particular cell type and on the specific growth factors present(33,34,35,). As far as hemolymphopoiesis is concerned, recently it has been shown that interleukin 2 (IL-2) stimulated T lymphocytes secrete high amounts of TGF-β. The secreted TGF-β has an inhibitory effect on the proliferation of T lymphocytes, probably by decreasing the expression of IL-2 and transferrin receptors on activated cells (19). This suggests an autocrine negative loop in the regulation of lymphocyte proliferation.

On the basis of these premises, we decided to investigate whether TGF-β can influence the growth of hemopoietic progenitors. The results we obtained (Aglietta et al., Submitted for publication) can be summarized as follows.
TGF-β in the presence of the supernatant of the 5637 cell lines as a source of growth factors stimulates the growth of day 7 CFU-GM from ficoll isolated normal bone marrow cells. The maximum stimulation (172% of control) is observed with 2,5 ng/ml TGF-β. The effect on progenitor cells seems direct since it can not be abolished by eliminating T cell and monocytes from the bone marrow suspension. When cloned CSF are used as source of growth factors, we observed that TGF-β does not stimulate day 7 CFU-GM proliferation when either G-CSF or GM-CSF are used. However, if maximally active doses of both molecules are added togheter to the culture dish, the enhancing effect of TGF-β shows up again. Thus, it seems likely that TGF-β can enhance day 7 CFU-GM growth only when maximally active doses of CSF are present. In contrast to this stimulatory effect of TGF-β on day 7 bone marrow CFU-GM, TGF-β either does not affect, either inhibits the growth of other classes of hemopoietic progenitors. In particular day 14 bone marrow CFU-GM growth is virtually unchanged and a strong significant inhibition is determined by TGF-β on the growth of normal bone marrow BFU-E, and on peripheral blood CFU-GM. In conclusion our data would point out that also inside myelopoiesis, as in other cell systems, TGF- action is complex: it may be

either a stimulus or an inhibition depending on the target cell and on specific growth factor which is used in conjunction with TGF-β.

CONCLUSIONS

From in vitro data, it appears more than likely that inhibitory molecules can interact with specific growth factors in the regulation of normal myelopoiesis. In translating these in vitro observation into in vivo models for myeloid proliferation one should, however consider few relevant points. Many in vitro observations are made in culture systems where maximally active concentrations of growth factors are present. This situation is unlikely to happen in vivo. Probably the recent availability of cloned growth factors which can be used at defined concentrations will help to define how variations in growth factors levels modify the action of inhibitory molecules. A second relevant point is that actions apparently opposite can be exploited by the same molecule, depending on the target cells. A clear example is represented by TNF-α and IFN-γ which can enhance CSF production and indirectly granulo-monopoiesis, at relatively low doses and when mature cells are their target, whereas at higher doses they directly inhibit progenitor cells. So far it not clear which effect predominates in vivo. It might be that in different physiological and pathological conditions, as well as in some pharmacological situations (i.e. when the compound is administered as a drug), either the stimulatory either the inhibitory action can predominate. At the present time, we do not have experimental model apt to elucidate these situations.

REFERENCES

1) Aglietta M., Piacibello W., Gavosto F. (1980): Insensitivity of chronic myeloid leukemia cells to inhibition of growth by prostaglandin. E. Cancer Res. 40, 2507-2511.

2) Aglietta M., Piacibello W., Gavosto F. (1983): Responsiveness to prostaglandin E1 of different subtypes of normal and pathological committed granulo-monopoietic precusors. Acta Haematol. 69, 376-381.

3) Aglietta M., Piacibello W., Stacchini A., Dezza L., Sanavio F., Malavasi F., Infelise V., Resegotti L., Gavosto F. (1986): Espression of HLA class II (DR,DQ) determinants by normal and chronic myeloid leukemia granulo-monocyte progenitors. Cancer Res. 46, 1783-1787.

4) Bagby G.C., Mc Call E., Layman D.L. (1983): Regulation of colony stimulating activity production. Interaction of fibroblasts, mononuclear phagocytes and lactoferrin. J. Clin Invest. 71, 340-344.

5) Birgens H.S., Hansen N.B., Karle H., Ostergaard-Kristensens L. (1983): Receptor binding of lactoferrin by human monocytes. Br. J. Haematol 54, 383-391.

6) Broxmeyer H.E., Bognacki J., Dorner M.H., De Sousa M. (1981): The identification of leukemia-associated inhibitor activity (LIA) as acidic isoferritins. A regulatory role for acidic isoferritins in the production of granulocytes and macrophages. J. Exp.Med. 153, 1426-1444.

7) Broxmeyer H.E., De Sousa M., Smithyman A., Ralph P., Hamilton J., Kurland J. E., Bognacki J.(1980): Specificity and modulation of the action of lactoferrin, a negative feed-back regulator of myelopoiesis. Blood 55, 324-333.

8) Broxmeyer H.E., Lu L., Bognacki J. (1983): Transferrin, derived from OKT8-positive subpopulations of T lymphocytes, suppresses the production of granulocyte-macrophage colony stimulating factors from mitogen activated T lymphocytes. Blood 62, 37-50.

9) Broxmeyer H.E., Lu L., Platez E., Feit C., Juliano L., Rubin B. (1983): Comparative analysis of the influence of human gamma, alpha and beta interferons on human multipotential (CFU-GEMM), erythroid (BFU-E) and granulocyte-macrophage (CFU-GM) progenitor cells. J. Immunol. 131, 1300-1305.

10) Broxmeyer H.E., Smithyman A., Eger R.R., Meyers P.A., De Sousa M. (1978): Identification of lactoferrin as the granulocyte-derived inhibitor of colony stimulating activity production. J.Exp.Med. 148, 1052-1067.

11) Broxmeyer H.E., Williams D.E., Lu L., Cooper S., Anderson S.L., Beyer G.S., Hoffman R., Rubin B.Y. (1986): The suppressive influences of human tumor necrosis factors on bone marrow hemopoietic progenitor cells from normal donors and patients with leukemia : synergism of tumor necrosis factor and gamma interferon. J. Immunol. 136, 4487-4495.

12) Burgess A.W., Nicola N. (1983): Growth factors and stem cells. New. York: Academic press.

13) Degliantoni G., Perussia B., Mangoni L., Trinchieri G. (1985): Inhibition of bone marrow colony formation by human natural killer cells and by natural killer cell-derived colony inhibiting activity. J. Exp. Med. 161, 1152-1168.

14) Dezza.L., Cazzola M., Piacibello W., Levi S., Arosio P., Aglietta M.(1986): Effect of acidic and basic isoferritins on in vitro growth of human granulo-monocyte progenitors. Blood 67, 789-795.

15) Falkenburg J.H.F., Jansen J., Van Der Vaart-Duinkerken N., Veenhof F.J., Blotkamp J., Goselink H.M., Parlevliet J., Van Rood J.J. (1984): Polymorphic and monomorphic HLA-DR determinants on human hemopoietic progenitors cells. Blood 63, 1125-1132.

16) Gentile P., Byers D., Pelus L.M. (1983): In vivo modulation of murine myelopoiesis following intravenous administration of prostaglandin E2. Blood 62, 1100-1107.

17) Gentile P., Broxmeyer H.E. (1983): Suppression of mouse myelopoiesis by administration of human lactoferrin in vivo and the comparative action of human trasferrin. Blood 61, 982-993.

18) Hanson M., Beran M., Andersson B., Kiessling R. (1982):Inhibition of in vitro granulopoiesis by autologous and allogeneic human NK cells. J. Immunol. 129, 126-132.

19) Kehrl J.H., Wakefield L.M., Roberts A.B., Jakowlew S., Alvarez-mon M., Derynck R., Sporn M.B., Fauci A.S. (1986): Production of transforming growth factor beta by human T lymphocytes and its potential role in the regulation of T cell growth. J. Exp. Med. 163, 1037-1050.

20) Kurland J.I., Broxmeyer H.E., Pelus L.M., Bockman R.S., Moore M.A.S.(1978): Role for monocyte-macrophage derived colony stimulating factor and prostaglandin E in the positive and negative feed-back control of myeloid stem cell proliferation. Blood 52, 388-407.

21) Lu L., Broxmeyer H.E., Meyers P.A., Moore M.A.S., Thaler H.T. (1983): Association of cell cycle expression of Ia-like antigenic determinants on normal human multipotential (CFU-GEMM) and erythroid (BFU-E) progenitor cells with regulation in vitro by acidic isoferritins. Blood 61, 250-256.

22) Lu L., Pelus L.M., Broxmeyer H.E. (1984): Modulation of the expression of HLA-DR (Ia) antigens and the proliferation of human erythroid (BFU-E) and multipotential (CFU-GEMM) progenitors cells by prostaglandin E. Exp. Haematol. 12, 741-748.

23) Metcalf D. (1986): The molecular biology and functions of the granulocyte-macrophage colony stimulating factors. Blood 67, 257-267.

24) Munker R., Gasson J., Ogawa M., Koeffler H.P. (1986): Recombinant human TNF induces production of granulocyte-monocyte colony stimulating factor. Nature 323, 79-82.

25) Neumann H.A., Fauser A.A. (1982): Effect of interferon on pluripotential hemopoietic progenitors (CFU-GEMM) derived from human bone marrow. Exp. Hematol. 10, 587-590.

26) Olofsson T., Nilsson E., Olsson I. (1984): Characterization of the cells in myeloid leukemia that produce leukemia associated inhibitor (LAI) and demonstration of LAI producing cells in normal bone marrow. Leuk. Res. 8, 387-396.

27) Pelus L.M. (1982): Association between CFU-GM expression of Ia-like (HLA-DR) antigen and control for granulocyte and macrophage production. A new role for prostaglandin E. J. Clin. Invest. 70, 568-578.

28) Pelus L.M., Broxmeyer H.E., Moore M.A.S. (1981): Regulation of human myelopoiesis by prostaglandin E and lactoferrin. Cell Tissue Kinet. 14, 515-526.

29) Piacibello W., Lu L., Williams D., Aglietta M., Rubin B., Cooper S., Watcher M., Gavosto F., Broxmeyer H.E. (1986): Human gamma interferon enhances release from phytohemagglutinin stimulated T4+ lymphocytes of activities that stimulate colony formation by granulocyte-macrophage, erythroid, and multipotential progenitor cells. Blood 68, 1339-1347.

30) Piacibello W., Lu L., Watcher M., Rubin B., Broxmeyer H.E. (1985): Release of granulocyte macrophage colony stimulating factors from major hystocompatibility complex class II-antigen positive monocytes is enhanced by human gamma interferon. Blood 66, 1343-1351.

31) Quesenberry P.J. (1986): Synergistic hematopoietic growth factors. Int. J. Cell. Cloning 4, 3-15.

32) Rossi G.B., Migliaccio A.R., Migliaccio C., Lettieri F., Di Rosa M., Mastroberardino G., Peschle C. (1980): In vitro interactions of PGE and cAMP with murine and human erythroid precursors. Blood 56, 74-79.

33) Roberts A.B., Anzano M.A., Wakefield L.M., Roche N.S., Stern D.F., Sporn M.B. (1985): Tipe- trasforming growth factor: a bifunctional regulator of cellular growth. Proc. Nath. Acal. Scc. (U.S.A.) 82, 119-123.

34) Scott Goustin A., Leof E.B., Shipley G.D., Moses H.L. (1986): Growth factors and cancer. Cancer Res. 46, 1015-1029.

35) Sporn M.B., Roberts A.B., Wakefield L.M., Assoian R.K. (1986): Trasforming growth factor-β : biological function and chemical structure. Science 233, 532-534.

36) Van't Hull E., Schellekens H., Lowenberg B., De Vries M. (1978): Influence of interferon preparations on the proliferative capacity of human and mouse bone marrow cells in vitro. Cancer Res. 38, 911-914.

ACKNOWLEDGEMENTS: This work was supported with grants from CNR, Special Project "Oncology" and Italian Association for Cancer Research.

Résumé

Les auteurs font une mise au point sur le rôle connu des inhibiteurs dans la régulation de la myélopoïèse. Ils discutent le méchanisme d'action de ces molécules. On peut distinguer deux modes d'action :
- une action directe au niveau des progéniteurs, en une apparente compétition avec des facteurs de croissance spécifiques.
- une action indirecte en empêchant la libération des facteurs de croissance par des cellules plus mûres.

Les auteurs présentent également des données nouvelles sur l'action du TGF beta dans la régulation de la myélopoïèse.

Biological effects of myelopoiesis inhibitors

Ole D. Laerum[1] and Walter R. Paukovits[2]

[1]Department of Pathology, The Gade Institute, University of Bergen, Haukeland Hospital, 5016-Bergen, Norway and [2]Institute for Tumor Biology/Cancer Research, University of Vienna, Borschekegasse 8a, A-1090, Vienna, Austria

ABSTRACT

A review of inhibitors with selective effects on myelopoietic cells is given. This both includes substances which act by interfering with the production of colony stimulating factors, such as lactoferrin and prostaglandins as well as factors with a more direct action on the cells, such as interferons, retinoic acid and chemically unidentified protein factors. In addition, mature granulocytes seem to contain and secrete low molecular weight factors which selectively inhibit myelopoiesis, both in vitro and in vivo. One such factor, a pentapeptide named hemoregulatory peptide, has been chemically identified and synthesized and has specific inhibitory effects both on human, rat and mouse myelopoietic cells. The main effect seems to be on the committed stem cells for myelopoiesis (CFU-GM), where in vivo administration leads to prolonged reduction of granulocytes in peripheral blood. The peptide easily forms a dimer through disulfide bridges in cystein, which is a potent stimulator of myelopoiesis. A theory for an equilibrium between the monomer and dimer peptide as regulator of myelopoiesis is proposed.

KEY WORDS

Myelopoiesis, stem cells, hemoregulatory peptide.

INTRODUCTION

In recent years a series of different substances with more or less selective inhibitory effects in the myelopoietic system have been identified. Some of these are general regulating molecules in the body which are secreted by different cell types and act on many different cell functions, but at the same time with significant effects on myelopoietic cells. These include proteins of high molecular weight such as lactoferrin (Broxmeyer et al., 1980), interferons (Verma et al., 1981) and factors of small molecular weight such as prostaglandin E_1 and E_2 (Aglietta et al., 1983; Pelus, 1984), and retinoic acid (Bradley et al., 1983). Lactoferrin and PGE seem to act by interfering with the production of colony stimulating factors in macrophages, by which their stimulatory action on myelopoietic cells decreases. The other factors seem to act more directly on both proliferation and differentiation of the maturing cell compartment.

Although striking effects have been seen on mice in vivo and in very low doses (see e.g. Broxmeyer, 1983), it is at present not clear to what extent they act as modifiers or are the rate limiting factors for maintaining steady state conditions of granulopoiesis in an intact organism.

In parallel with this several workers have reported the existence of factors of small molecular weight that are associated with mature granulocytes or bone marrow cells and which seem to supress myelopoiesis with high specificity. Recently, the active principle in such inhibitory extracts has been identified as a pentapeptide named hemoregulatory peptide. A short review on biological properties of this molecule as well as of more or less purified extracts containing this inhibitor will here be given.

CRUDE AND SEMIPURIFIED GRANULOCYTE EXTRACTS

It is now twenty years ago since Rytömaa and Kiviniemi (1967 and 1968 a and b) described that extracts of mature granulocytes specifically inhibited rat myelopoietic cell proliferation in coverslip cultures. Later they were able to show that the factor, which had a molecular weight below 3000 was able to induce regression of a transplantable rat granulocytic leukemia (Rytömaa and Kiviniemi, 1969). They also showed that such extracts could retard the growth of leukemia cells in humans (Rytömaa et al., 1976).

In parallel with this other workers made similar observations. Paukovits (1971) extracted a similar factor from rat bone marrow cells and showed that it inhibited 3H-thymidine uptake in the same cells. Laerum and Maurer (1973), Benestad et al. (1973) and MacVittie and McCarthy (1974) showed that similar extracts inhibited mouse myelopoietic cell proliferation in diffusion chamber culture in vivo. Later, Benestad and Rytömaa (1979) showed that this factor inhibited rapidly proliferating committed progenitor cells for myelopoiesis in the same type of culture.

Boll et al. (1979) were the first to demonstrate inhibitory effects of rat granulocyte extracts on human bone marrow cells in culture. Muller-Bérat et al. (1973) and later Aardal et al. (1977) found that semi-purified granulocytic extracts of different sources specifically inhibited myelopoietic colony formation in agar (Fig. 1). This action was non-toxic and could be reversed by washing the cells in excess of medium prior to plating.

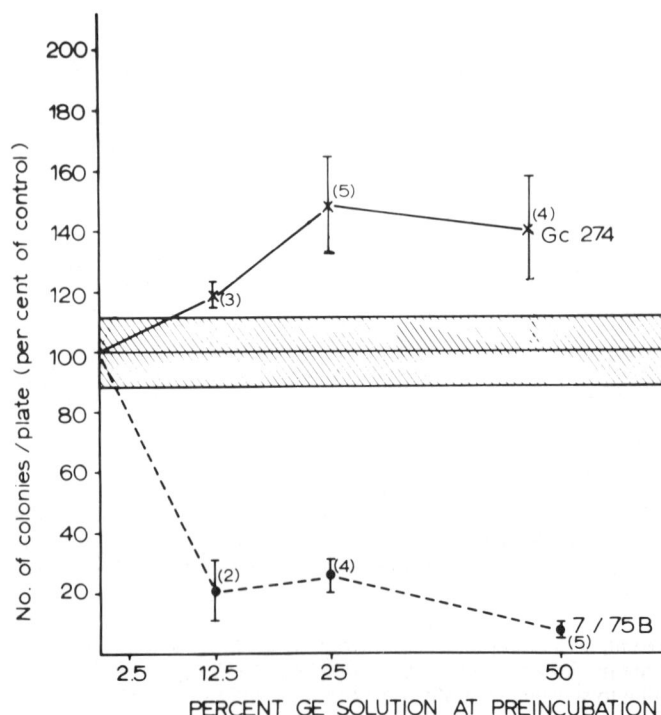

Fig. 1. Inhibitory effect of semipurified rat granulocyte extract on mouse CFU-GM in agar, and stimulatory effect of the same extract following repeated freezing and thawing (probably due to dimer formation) (from Aardal et al., 1977).

Similar findings were made by Maurer et al. (1978), using a capillary tube culture system for granulocyte colonies where the kinetics of inhibition could be directly quantitated. This was later confirmed by Schunck et al. (1979).

Apart from rat bone marrow (Paukovits, 1971), other sources such as bovine granulocytes (Rytömaa et al., 1976; Kastner and Maurer, 1980), porcine leukocytes (Kastner et al., 1984) and calf spleen (Kastner and Maurer, 1984) were found to contain the same principle. Later, Benestad and Hersleth (1984) presented convincing evidence that pure granulocyte suspensions could produce such an inhibitory factor.

Considerable efforts with purification of this factor were done by several workers, reaching the conclusion that it was a peptide of low molecular weight (Paukovits and Hinterberger, 1978; Kastner and Maurer, 1984; Balazs et al., 1980; Klupp et al., 1980).

By analogy to a general mechanism proposed by several workers (for reviews see Houck, 1976 and Iversen 1983) the active principle was termed granulocyte chalone. According to the theoretical concept such a factor should be an endogenous inhibitor of cell proliferation acting in the same tissue as it was secreted. It should be non species specific but highly tissue specific and of protein nature. Furthermore, the inhibition should be non-toxic and reversible.

In 1982 Paukovits and Laerum presented the chemical structure of a penta peptide: pyroglu-glu-asp-cys-lys-OH which had several of the properties listed above. The substance was named hemoregulatory peptide. An neutral term was choosen since its biological properties were somewhat different from those which had been claimed for a granulocyte chalone on theoretical basis.

At present it is not clear if the hemoregulatory peptide represents the active principle in all the studies on crude extracts and more or less purifed fractions cited above. The available data indicate that the peptide described by Balazs and coworkers is larger, while the peptide described by Maurer's group is very close to it. However, minor differences in chemical structures would not be detected at the level of chemical purity that has been reached so far.

THE HEMOREGULATORY PEPTIDE MONOMER-DIMER SYSTEM

Laerum and Paukovits (1982, 1984 a and b) and Foa et al. (1983) could show that the synthetic hemoregulatory peptide had strong inhibitory effects on myelopoietic cells *in vitro*, where the main effect seemed to be on myelopoietic stem cells (CFU-GM). However, in the first experiments, rather high doses were necessary to exert the effect. It then turned out that the repurification and stabilisation of the peptide after synthesis was a critical factor. When this was overcome, the peptide inhibited mouse myelopoietic cells in doses down to 10^{-13}M and human CFU-GM in doses down to 10^{-11}M (Laerum et al., 1986 and 1987). No effect was seen on granulocyte functions such as adhesion, internalisation and degradation of bacteria (Laerum et al., 1987).

A main problem both with semipurified extracts and the synthetic substance was their instability. Sometimes a strong inhibitory effect was seen, other times the inhibition was slight, and sometimes no effect or even a stimulation was seen (Aardal et al., 1977 and 1982; Lord et al., 1977). It was also found that the substance easily became oxidized, upon which a stimulatory effect was seen instead (Laerum and Paukovits, 1984a).

Recently the oxidation product of the peptide has been identified as a dimer formed by disulfide bridges in cystein, and this dimer has the opposite effects of the monomer. The dimer strongly stimulates colony formation of both human and mouse CFU-GM *in vitro* (Laerum et al., 1987) and in addition has strong stimulating effects on myelopoietic cells in mice *in vivo*.

At present, no method for the measurement of these substances in biological fluids is available. It is therefore not known to what extent they really act as regulators in the intact organism. However, it is tempting to speculate that the producer cells, the granulocytes, through their strong oxidative and reducing capacity (see e.g. Weiss et al., 1983; Klebanoff, 1977, Watanabe and Bannai, 1987), could maintain an unstable equlibrium between the monomer and dimer which causes a rapid and efficient modulation of granulopoiesis.

In this connection it is interesting that another such regulator, lactoferrin, is only active in a monomeric form, while tetramers are not inhibitory (Birgens, 1984). Similarly, a neuropeptide called head activator also looses its biological activity by dimerisation (Bodenmüller et al., 1986).

From these data a consequence would be that the producer cells of hemoregulatory peptide might be able to rapidly modulate the proliferation of their precursors, which according to the actual needs will result in a stimulation, no effect or an inhibition.

EFFECTS OF HEMOREGULATORY PEPTIDE IN VIVO

It has earlier been shown that both the presence of a high density of granulocytes (Willemze et al., 1978) as well as semipurified extracts (Perrins et al., 1980), inhibit myelopoiesis and lead to reduced granulocyte numbers.

Similarly, the hemoregulatory peptide was able to strongly depress the stem cells committed for granulopoiesis. Depending on the doses and types of administration to female C3H mice variable inhibitory effects were achieved (Laerum and Paukovits, 1984a). At most, continuous infusion by use of osmotic minipumps could in one week reduce the CFU-GM numbers in vivo to 1/4 of the normal value (Fig. 2-5). Afterwards this was reversible with an overshoot. Also the CFU-S numbers were reduced, but not to the same extent. In addition, myelopoietic cell proliferation showed a rapid decrease (Laerum and Paukovits 1984 b and c), where the main effect seemed to be a dampening of the circadian rhythmicity of myelopoietic cell proliferation. Interestingly, an immediate increase of erythropoiesis resulted, although prolonged administration of high doses of the peptide led to a secondary decrease of erythropoiesis together with decreased CFU-S numbers.

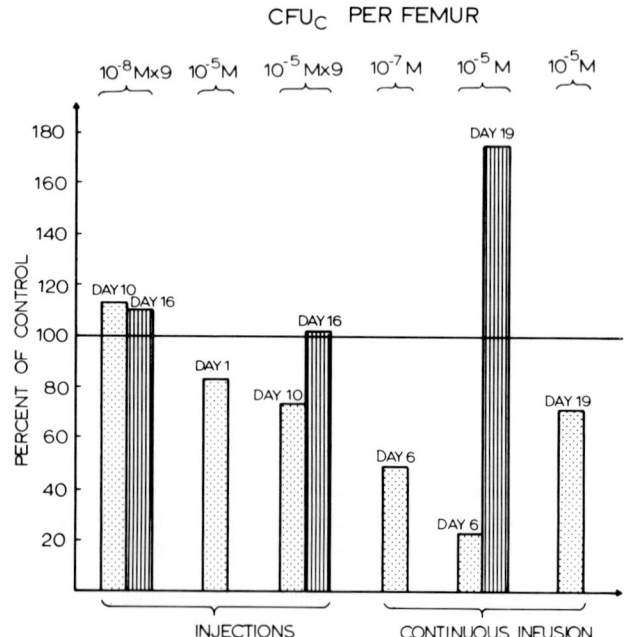

Fig. 2. Effects of different doses and types of administration (single and repeated injections and continuous infusion) on the number of CFU-GM in mouse femur, expressed as percent of controls. (From Laerum and Paukovits, 1984a).

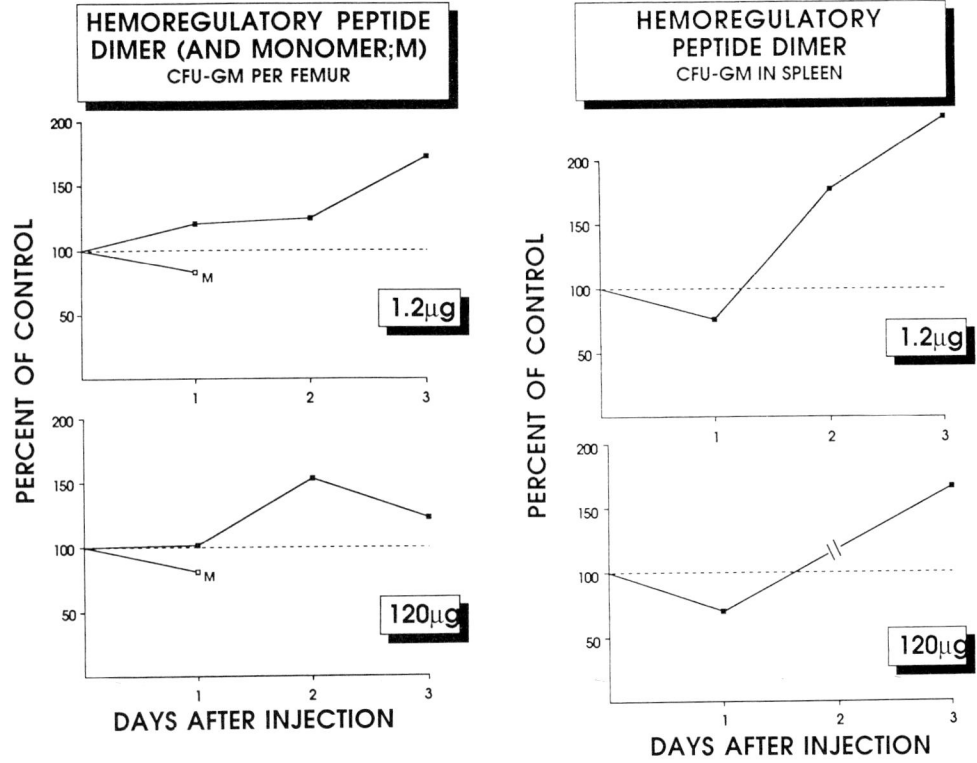

Fig. 3. Effects of a single injection of synthetic dimer on myelopoietic stem cells in mice. Effect of monomer after 1 day is shown for comparison. Strongest relative effect is after 3 days in the spleen. (From Laerum et al., in prep. 1987).

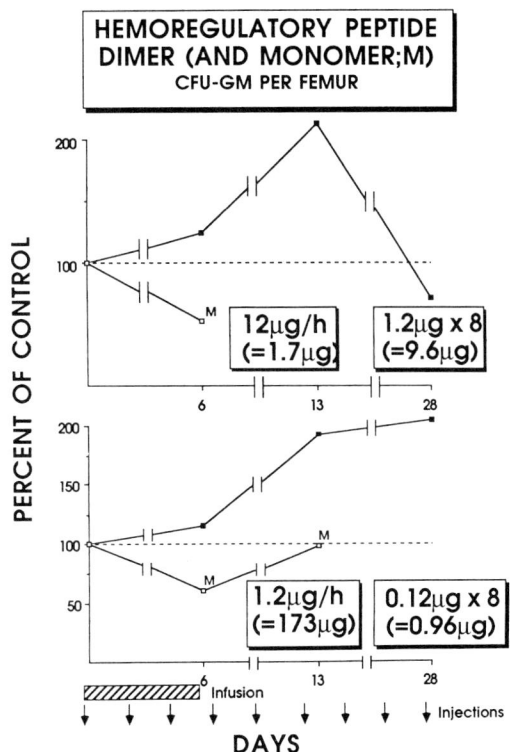

Fig. 4. Comparison of effects of dimer and monomer by continuous infusion with osmotic minipumps for 6 days (left part; 6 and 13 days), and repeated injections (right part; 28 days).

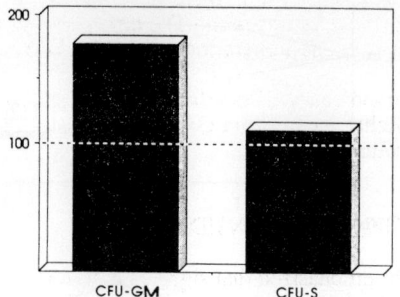

**HEMOREGULATORY PEPTIDE DIMER
CFU-GM AND CFU-S IN SPLEEN
Day 28; 8 injections (2 weekly)**

Fig. 5. Effects of repeated injections of dimer on CFU-GM and CFU-S in spleen.

As a result prolonged reduction of peripheral blood granulocyte numbers was seen (Laerum and Paukovits, 1985; Fig. 6), although a real leukopenia was never observed. This indicates that the peptide is not a cytostatic drug, but merely a physiological regulator. Interestingly, high doses and prolonged infusion of the peptide make the target cells unresponsive to further action, and sometimes even a secondary stimulation can be seen.

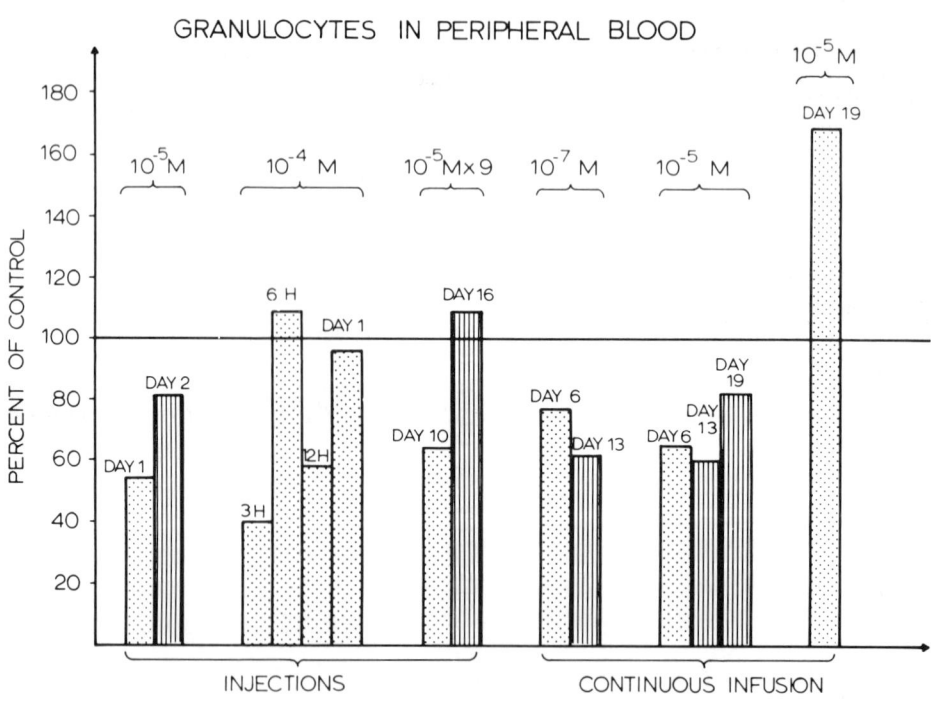

Fig. 6. Granulocyte numbers in peripheral blood of C3H-mice after different doses and types of administration (single and repeated injections and continuous infusion) expressed as percent of controls (from Laerum and Paukovits, 1985).

The dimer had roughly the opposite effects of the monomer, although granulocytosis in peripheral blood was not observed. Instead, an increase of myelopoietic cells in the bone marrow as well as lymphocytic cells was seen. The dimer also increased CFU-GM and CFU-S numbers in the spleen (Laerum et al., in preparation 1987; Fig. 3-5).

A further indication of rapidly occurring in vivo effects of the monomer is that it can protect stem cells against lethal doses of ara C in mice (Paukovits et al., 1986). This will be further dealt with elsewhere in this volume.

SOME GENERAL CONSIDERATIONS

It must be emphasized that significant biological effects of a semipurified extract or a synthetic biological substance in vitro and in vivo do not prove that they are the rate limiting regulators in the system, although it is tempting to postulate this. In order to make a final proof that the mentioned messengers really are the important regulators and not only contributing modifying factors, it is necessary to show that the concentrations of biologically active substance vary inversally with the rate of cell proliferation in the target organ. To our knowledge no such proof is available, neither for lactoferrin and the other high molecular weight messengers, nor for the hemoregulatory peptide or the putative granulocyte chalones. However, such data are urgently needed and will be one of the main goals for this research in the near future.

Furthermore, it is important that such a regulator really changes the cell proliferation as rapidly as is normally seen in the unperturbed animal. Thus, it is known that hemopoietic cell proliferation undergoes strong circadian variations, and in addition exhibit seasonal variations (for review see Laerum and Aardal, 1985). In other words, hemopoietic cell proliferation occurs in waves, by which an efficient system for maintaining the steady state is constituted. So far, there has been a tendency to neglect this fact in experimental hematology, leading to the erroneous concept that quantitative studies on hemopoiesis can be performed by only one daily observation. For further research in this field, it is therefore critical that both animals and human beings are continuously monitored when such biological regulators are tested.

Research is now on the way in our laboratories by which the cell proliferation in human bone marrow is studied through 24 hours periods, indicating that man is equally variable as the mouse (Smaaland et al., 1987).

Knowledge of these mechanisms and the availability of pure synthetic messengers give us a powerful tool for manipulating with the renewal of cells in the hemopoietic system. In this way, it should be possible both to find the time of lowest and highest proliferative activity and by use of the monomer and dimer in sequence induce waves where cytostatics can be administered at periods of minimal susceptibility to toxicity. It also opens the possibility to alter the proliferative activity of myelopoiesis in different hematological diseases, both in aplastic, hyperplastic, and possibly neoplastic conditions.

This research was supported by The Norwegian Cancer Society.

REFERENCES

Aglietta, M. Piacibello, W., Gavosto, F. (1983): Responsiveness to prostaglandin E1 of different subtype of normal and pathological committed granulomonopoietic precursors. Acta Haematol. 69, 376-381.

Aardal, N.P., Laerum, O.D., Paukovits, W.R., Maurer, H.R. (1977): Inhibition of agar colony formation. Virchows Arch. B Cell Path. 24, 27-39.

Aardal, N.P., Laerum, O.D., Paukovits, W.R. (1982): Biological properties of partially purified granulocyte extract (chalone) assayed in soft agar culture. Virchows Arch (Cell Pathol.) 38, 253-261.

Balazs, A., Sajgo, M., Klupp, T., Kemeny, A. (1980): Purification of an endopeptide to homogenity and the verification of its selective inhibitory action on myelopoid cell proliferation. Cell Biol. Int. Reports 4, 337-345.

Benestad, H.B., Rytömaa, T., Kivieniemi, K. (1973): The cell specific effect of the granulocyte chalone demonstrated with the diffusion chamber technique. Cell Tissue Kinet. 6, 147-154.

Benestad, H.B., Rytömaa, T. (1979): Granulocytic chalone inhibits rapidly proliferating committed murine progenitor cells (CFU-C). Biomedicine 31, 33-37.

Benestad, H.B., Hersleth, I.B.(1984): Production of proliferation inhibitors by mature granulocytes. Blut 48, 201-211.

Birgens, H.S. (1984): The biological significance of lactoferrin in haematology. Scand. J. Haematol. 33, 225-230.

Bodenmüller, H., Schilling, E., Zachmann, B., Schaller, H.C. (1986): The neuropeptide head activator loses its biological activity by dimerization. EMBO J. 5, 1825-1829.

Boll, I.T.M., Sterry, K., Maurer, H.R. (1979): Evidence for a rat granulocyte chalone effect on the proliferation of normal human bone marrow and of myeloid leukeemias. Acta Haematologica 6, 130-137.

Bradley, E.C., Ruscetti, F.W., Steinberg, H. et al. (1983): Inhibition of differentiation and proliferation of colony-stimulating factor-induced clonal growth of normal human marrow cells in vitro by retinoic acid. JNCI 71, 1189-1192.

Broxmeyer, H.E., DeSousa, M., Smithyman, A. et al. (1980): Specificity and modulation of the action of lactoferrin, a negative feedback regulator of myelopoiesis. Blood 55, 324-333.

Broxmeyer, H.E. (1983): Negative regulators of hematopoiesis. In Long-Term Bone Marrow Culture, eds D.G. Wright, J.S. Greenberger, pp 363-397. New York: Alan R. Liss, Inc.

Foa, P., Lombardi, L., Ciani, A. et al. (1983): A synthetic pentapeptide inhibiting normal and leukemic myelopoiesis in vitro. IRCS Med. Sci. 11, 2272-273.

Houck, J.C. (1976): Chalones. Amsterdam-Oxford: North-Holland Publ. Co.

Iversen, O.H. (1981): The chalones. In Handbook of Experimental Pharmacology, Vol. 57, ed R. Baserga, pp 491-550. Berlin-Heidelberg-New York: Springer.

Kastner, M., Maurer, H.R. (1980): Pure bovine granulocytes as a source of granulopoiesis inhibitor (Chalone). Hoppe-Seyler's Z. Physiol. Chem. 361, 197-200.

Kastner, M., Maurer, H.R. (1984): Partial purification and characterization of an endogenous granulo-monopoiesis inhibitor from calf spleen. Hoppe-Seyler's Z. Physiol. Chem. 365, 129-135.

Kastner, M., Maurer, H.R., Gerlach, U. et al. (1984): A granulopoiesis inhibitor partially purified from large-scale serum-free cultures of porcine leukocytes. Z. Naturforsch. 39c, 639-645.

Klupp, T., Balazs, A., Sajgo, M. (1980): Preparation of a target specific fraction controlling the proliferation of granulocytes. Acta Biochim. et Biophys. Acad. Sci. Hung. 15, 165-172.

Kreja, L., Hågå, P., Muller-Bérat, N. et al. (1986): Effects of a hemoregulatory peptid (HP5b) on erythroid and myelopoietic colony formation in vitro. Scand. J. Haematol. 37, 79-86.

Laerum, O.D., Maurer, H.R. (1973): Proliferation kinetics of myelopoietic cells and macrophages in diffusion chambers after treatment with granulocyte extracts (chalone). Virch. Arch. B 14, 293-305.

Laerum, O.D., Paukovits, W.R., Sletvold, O. (1986): Hemoregulatory peptide: Biological aspects. In Biological Regulation of Cell Proliferation, eds R. Baserga, P. Foa, D. Metcalf, E.E. Polli, pp 121-129. New York: Raven Press.

Laerum, O.D., Aardal, N.P. (1985): Rhythms in blood and bone marrow. In Clinical Aspects of Chronobiology, ed W.J. Rietveld, pp 85-97. Haag: CIP-Gegevens Koninklijke Bibliotheek.

Laerum, O.D., Paukovits, W.R. (1984a): Inhibitory effects of a synthetic pentapeptide on hemopoietic stem cells in vitro and in vivo. Exp. Hematol. 12, 7-17.

Laerum, O.D., Paukovits, W.R. (1984b): Modulation of murine hemopoiesis in vivo by a synthetic hemoregulatory pentapeptide (HP 5b). Differentiation 27, 106-112.

Laerum, O.D., Paukovits, W.R. (1984c): Modification of mouse hemopoietic cell proliferation in vivo by a hemoregulatory pentapeptide (HP 5b). Relation to circadian rhythms. Virchows Arch (Cell Pathol) 46, 333-348.

Laerum, O.D., Paukovits, W.R. (1985): Peripheral blood leukocyte alterations in mice induced by a hemoregulatory pentapeptide (HP 5b). Leukemia Res. 9, 1075-1084.

Laerum, O.D., Sletvold, O., Bjerknes, R., Paukovits, W.R. (1987): A synthetic hemoregulatory peptide (HP 5b) inhibits human myelopoietic colony (CFU-GM) formation but not leukocyte phagocytosis in vitro. Scand. J. Haematol., in press.

Lord, B.I., Testa, N.G., Wright, E.G., Banerjee, R.K. (1977): Lack of effect of a granulocyte proliferation inhibitor on their committed årecursor cells. Biomedicine 26, 163-169.

Muller-Bérat, C.N., Laerum, O.D., Maurer, H.R. (1973): Chalone inhibition of the committed stem cell to granulopoiesis in vitro. Abstr. VIth Meeting European Study Group for Cell Proliferation. USSR Acad. Sci.

MacVittie, T.J., McCarthy, K.F. (1974): Inhibition of granulopoiesis in diffusion chambers by a granulocyte chalone. Exp. Hematol. 2, 182-194.

Maurer, H.R., Henry, R., Maschler, R. (1978): Chalone inhibition of granulocyte colony growth in agar: Kinetic quantitation by capillary tube scanning. Cell Tissue Kinet. 11, 129-138.

Paukovits, W.R.(1971): Control of granulocyte production: Separation and chemical inhibitor (chalone). Cell Tissue Kinet. 4, 539-547.

Paukovits, W.R., Hinterberger, W. (1978): Molecular weight and some chemical properties of the granulocytic chalone. Blut 37, 7-18.

Paukovits, W.R., Laerum, O.D. (1982): Isolation and synthesis of a hemoregulatory peptide. Z. Naturforsch. 37c, 1297-1300.

Paukovits, W.R., Laerum, O.D., Guigon, M. (1986): Isolation, characterization and synthesis of a chalone-like hemoregulatory peptide. In Biological Regulation of Cell Proliferation, eds R. Baserga, P. Foa, D. Metcalf, E.E. Polli, pp 111-119. New York: Raven Press.

Pelus, L.M. (1984): Modulation of normal and leukemic human hemopoietic cell differentiation by prostaglandin E. In Icosanoids and Cancer, eds H. Thaler-Dao, A.C. de Paulet, R. Paoletti, pp 183-193. New York: Raven Press.

Perrins, D.J.D., Wiernik, G., Jones, W.A. (1980): Granulocyte chalone assayed in vivo in the mouse. Acta haemat. 64, 72-78.

Rytömaa, T., Kiviniemi, K. (1967): Regulation system of blood cell production. In Control of Cellular Growth in Adult Organisms, eds H. Their, T. Ryytömaa, pp 106-138. London and New York: Academic Press.

Rytömaa, T, Kiviniemi, K. (1968a): Control of granulocyte production. I. Chalone and antichalone, two specific humoral regulators. Cell Tissue Kinet. 1, 329-340.

Rytömaa, T., Kiviniemi, K. (1968b): Control of granulocyte production. II. Mode of action of chalone and antichalone. Cell Tissue Kinet. 1, 341-350.

Rytömaa, T., Kiviniemi, K. (1969): Chloroma regression induced by the granulocytic chalone. Nature 222, 995-996.

Rytömaa, T., Vilpo, J.A., Levanto, A., Jones, W.A. (1976): Effect of granulocyte chalone on acute and chronic granulocytic leukemia in man. Report of seven cases. Scand. J. Haematol. 27, 5-28.

Schunck, H., Schütt, M., Langen, P. (1978): Granulozytenchalon: Gewebespezifität der Wirkung in Kurz-zeitkulturen. Acta Biol. Med. Germ. 37, 593-.

Smaaland, R., Sletvold, O., Bjerknes, R., Lote, K., Laerum, O.D. (1987): Circadian variations of cell cycle distribution in human bone marrow. Abstr. XVII Internatl. Soc. Chronobiology Conference. Leiden.

Verma, D.S., Spitzer, G., Zander, A.R. et al. (1981): Human leukocyte interferon preparation-mediated block of granulopoietic differentiation in vitro. Exp. Hematol. 9, 63-76.

Watanabe, H., Bannai, S. (1987): Induction of cystine transport activity in mouse peritoneal macrophages. J. Exp. Med. 165, 628-640.

Weiss, S.J., Lampert, M.B., Test, S.T. (1983): Long-lived oxidants generated by human neutrophils: Characterization and bioactivity. Science 222, 625-628.

Willemze, R., Walker, R.I., Herion, J.C., Palmer, J.G. (1978): Marrow culture in diffusion chambers in rabbits. I. Effect of mature granulocytes on cell production. Blood 51, 21-.

Résumé

On peut distinguer parmi les inhibiteurs de la granulopoïèse, ceux qui agissent en interférant avec la production des facteurs de stimulation, tels que la lactoferrine et les prostaglandines et ceux qui agissent directement sur les cellules, tels que les interférons, l'acide rétinoïque et des protéines encore non identifiées. De plus les granulocytes semblent contenir et secréter des facteurs de faible poids moléculaire qui inhibent de manière sélective la myélopoïèse, à la fois in vitro et in vivo. Un de ces facteurs, un pentapeptide appelé peptide hémorégulateur, a été identifié chimiquement et synthétisé. Il a des effets inhibiteurs spécifiques sur les cellules myéloïdes de l'homme, du rat et de la souris. Son effet principal paraît porter sur les progéniteurs granulomacrophagiques (CFU-GM) et son administration in vivo entraine une réduction prolongée du nombre de granuleux dans le sang. Ce peptide forme facilement un dimère grâce à des ponts disulfures entre deux cystéines qui est un puissant stimulant de la myélopoïèse. Une théorie sur la régulation de la myélopoïèse fondée sur l'équilibre entre les formes monomères et dimères du peptide est proposée.

Regulatory peptides inhibiting granulopoiesis

Walter R. Paukovits[1], Ole D. Laerum[2], Johanna B. Paukovits[1], Martine Guigon[3] and Jon-Sverre Schanche[4]

[1] Inst. for Tumor Biology/Cancer Research, Univ. Vienna, Austria
[2] The Gade Institute, Dept. Pathology, Univ. Bergen, Norway
[3] Laboratoire d'Hématologie, Faculté de Médecine St-Antoine Paris, France
[4] Nycomed Research Laboratories, Oslo, Norway

ABSTRACT

Myelopoiesis is modulated in vitro and in vivo by a thiol containing hemoregulatory pentapeptide (HP), which seems to be associated with mature granulocytes. This article describes a) the purification of HP from leukocytes, b) sequence determination, c) chemical synthesis and biological properties of several analogs. In addition, we present preliminary data on HP-receptors on bone marrow cells and on the oxidative conversion of hemoregulatory peptides to stimulatory dimers by disulfide formation. We also present evidence that HP is not contained in differentiation induced leukemic granulocytes. Retinoic acid induced HL-60 cells produce a structurally unrelated inhibitory peptide.

KEY WORDS

Hemoregulatory Peptide, Leukemia, Isolation, Synthesis, Dimerization, Receptors, Structure-activity relationship.

The hemoregulatory peptide (HP) is a granulocyte associated inhibitor of myelopoietic cells which we have found initially in blood serum and in leukocyte conditioned media (Paukovits 1971, 1973) and which was subsequently isolated in pure form from rat bone marrow and/or normal human leukocytes (Paukovits & Laerum 1982, Paukovits et al. 1983).

PURIFICATION AND BIOLOGICAL PROPERTIES OF HP

The purification of HP takes advantage of the acidic nature and the thiol content of the molecule (fig.1).

Fig.1: PURIFICATION OF HEMOREGULATORY PEPTIDE

rat bone marrow conditioned medium **or** homogenate of normal human leukocytes

↓

ultrafiltration: <10 kDa

↓

adsorb to AG1x2 anion exchanger, wash, and elute with 1 M formic acid

↓

Sephadex G-10, $V_e/V_o = 1.37$

↓

Covalent binding to Thiopropyl-Sepharose, wash, displace with excess 2-mercaptoethanol

↓

high resolution anion exchange on Mono-Q with NaCl gradient

↓

HPLC on C_{18} reverse phase column with acetonitrile gradient

---

The yield is in the order of 1 µg per $5 \cdot 10^{11}$ leukocytes. Based on the protein content of the original homogenate this gives a purification factor of approximately $2 \cdot 10^9$. The HP obtained in this way is homogeneous (fig.2) in reverse phase HPLC and (after ^{3}H-carboxymethylation) in high voltage electrophoresis.

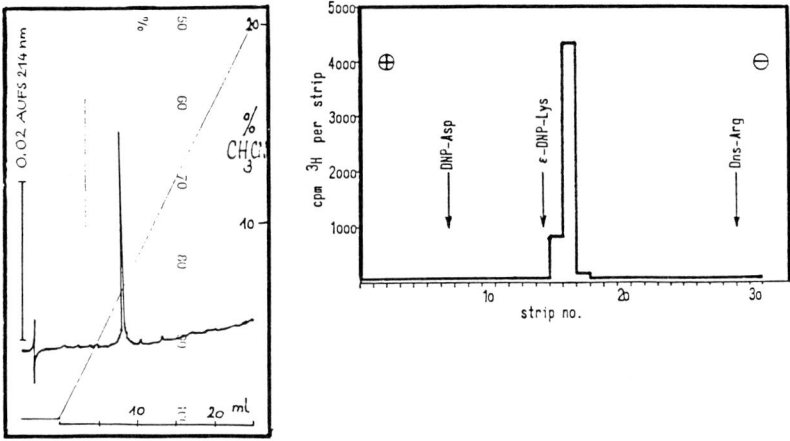

Fig.2: A......reverse phase HPLC of purified hemoregulatory peptide. Ultrasphere ODS column, 0.1 % TFA with acetonitrile gradient.
B......high voltage electrophoresis of ^{3}H-carboxymethylated HP at pH 1.9.

The purified hemoregulatory peptide is a potent inhibitor (Kreja et al., 1986) of GM-CSF stimulated CFU-GM colony formation in semisolid cultures. The dose-response curve is biphasic with an effective range from below 10^{-10} M to 10^{-6} M. At concentrations exceeding this limit the inhibition disappears. This U-shaped dose response curve can be taken as evidence that HP is not toxic to the colony forming cells. In addition it has no (or only insignificant) inhibitory effects on BFU-E and CFU-E cultures from adult mouse bone marrow and CFU-E from fetal murine liver (Kreja et al., 1986) at concentrations up to 10^{-4} M, well above the effective range for CFU-GM inhibition. The in vitro action spectrum of HP seems to be restricted to cells of the myeloid lineage. It includes also the more mature myelopoietic cells as was shown by experiments with ^{3}H-thymidine incorporation (Paukovits 1973). When administered in vivo into mice (continuous infusion for 7 days, total dose 1 µg) HP causes a reduction of the number of CFU-S (50%) and CFU-GM (25%) (Laerum & Paukovits 1984, Paukovits W.R. et al. 1986). All hematological parameters tested returned to normal some time after the end of peptide administration.

SEQUENCE DETERMINATION

HP os not inactivated by several proteolytic enzymes (Paukovits & Paukovits, 1978) but is sensitive to pyroglutamylaminopeptidase (Paukovits & Hinterberger, 1978), an enzyme which selectively removes pyroglutamyl residues from the N-terminus of proteins and peptides. The absence of a free N-terminal aminogroup precludes the application of standard sequencing procedures. Instead we had to use an indirect approach. In short HP was carboxymethylated with ^{3}H-iodoacetic acid and subjected to high voltage electrophoresis under various conditions (Paukovits & Laerum 1984). From the electrophoretic mobility of ^{3}H-CM-HP at pH 1.9 we calculated a charge of $\epsilon = +1$ and a molecular weight of ca. 650 Da. At pH 6.5 the same procedure resulted in $\epsilon = -3$ and MW = 650 Da whereas electrophoresis of native HP resulted in $\epsilon = -2$ and 600 Da. A simple calculation shows the presence of 1 (acetylatable) amino-group, 1 thiol group, and 3 carboxyl groups in the (native) HP molecule. After partial hydrolysis of ^{3}H-CM-HP we electrophoretically identified the dipeptide Cys(^{3}H-CM)-Lys. The stability of HP against trypsin (no change in biological activity and electrophoretic properties) suggests that Cys-Lys is the C-terminal end of the molecule. The combined evidence then gives as sequence of HP:

pGlu – (Asp or Glu) – (Asp or Glu) – Cys – Lys

The uncertainty at positions 2 and 3 results from the fact that by electrophoretic methods a carboxylic side chain can be identified but it is not possible to decide if this COOH belongs to Asp or Glu. Direct analysis was not possible because of the the low amount of peptide available for these experiments. In addition the strategy used, together with the inaccuracy of electrophoretic molecular weight determinations, leaves the possibility open that HP may contain additional light weight aminoacids without functional side chain groups.

LEUKEMIC GRANULOCYTES DO NOT CONTAIN HP

HL-60 cells, when induced with retinoic acid to differentiate into granulocytoid cells, release a small molecular weight inhibitor of CFU-GM into their culture medium. This inhibitory peptide is not released by uninduced HL-60 cells (table 1).

Table 1: Effects of crude (unfractionated) culture media on colony formation by CFU-GM

	colonies per 10^5 BMC		
dilutions [a]	1 : 5	1 : 10	1 : 50
fresh medium	65.3 ± 3.8	-	-
used medium of untreated HL-60 [b]	69.5 ± 6.3	57.6 ± 4.7	67.7 ± 9.0
used medium of RA-treated HL-60 [c]	27.3 ± 6.1	27.6 ± 5.8	62.0 ± 7.0

a...fresh medium was used as diluent
b...after 6 days of culture
c...after 6 days of culture in the presence of 10^{-6} M retinoic acid. RA was inactivated before assay by light exposure

The "RA-peptide" was purified (Paukovits J.B. et al. 1986) using techniques similar to those applied for isolating HP and sequenced by Chang's ultramicro technique (Chang 1983). The N-terminal sequence found (Gln-Asp-Pro-....) bears no similarity to the sequence of HP. It seems thus that the genotypically still transformed "granulocytes" obtained by differentiation induction of leukemic cells may produce regulatory factors with properties different from those of normal HP. This abnormal behavior may also have important implications for differentiation induction therapy of myeloid leukemia (Paukovits J.B. et al. 1986, Paukovits et al. 1987).

SYNTHETIS OF HEMOREGULATORY PEPTIDE(S)

The sequence found for natural HP leaves several possiblities open. Among them the four permutations of Asp and Glu in position 2 and 3 seemed to be most interesting. All these peptides were synthesized and tested for their ability to inhibit CFU-GM colony formation. Only the Glu^2-Asp^3 analog (HP5b) was found to be active (table 3).

pGlu – Glu – Asp – Cys – Lys   HP5b

It turned out that the synthesis of this peptide was not straightforward. This may account for the apparent discrepancies in the specific activities of preparations obtained by different methods (table 2). The difficulties encountered were mainly related to formation of β-bonds from Asp, nucleophilic substitution of Cys, and possibly racemization. Also the formation of dimers during handling under atmospheric conditions may contribute to low apparent specific activities (see below).

Table 2:   Biological potency of HP5b synthesized by different methods.

source	method	minimal active concentration	references
synthetic	BOC, active ester condensation in liquid phase	$10^{-9} - 10^{-10}$ M	Paukovits & Laerum, 1982
synthetic	fragment condensation	10^{-8} M	Foa et al, 1986
synthetic	FMOC, symm.anhyd. coupling, solid phase	$10^{-11} - 10^{-12}$ M	this paper
natural HP		$10^{-10} - 10^{-11}$ M	Kreja et al 1986

Fig.3 describes a synthetic approach which seems to avoid most of the above mentioned problems and consistently yields HP5b preparations of high specific activity.

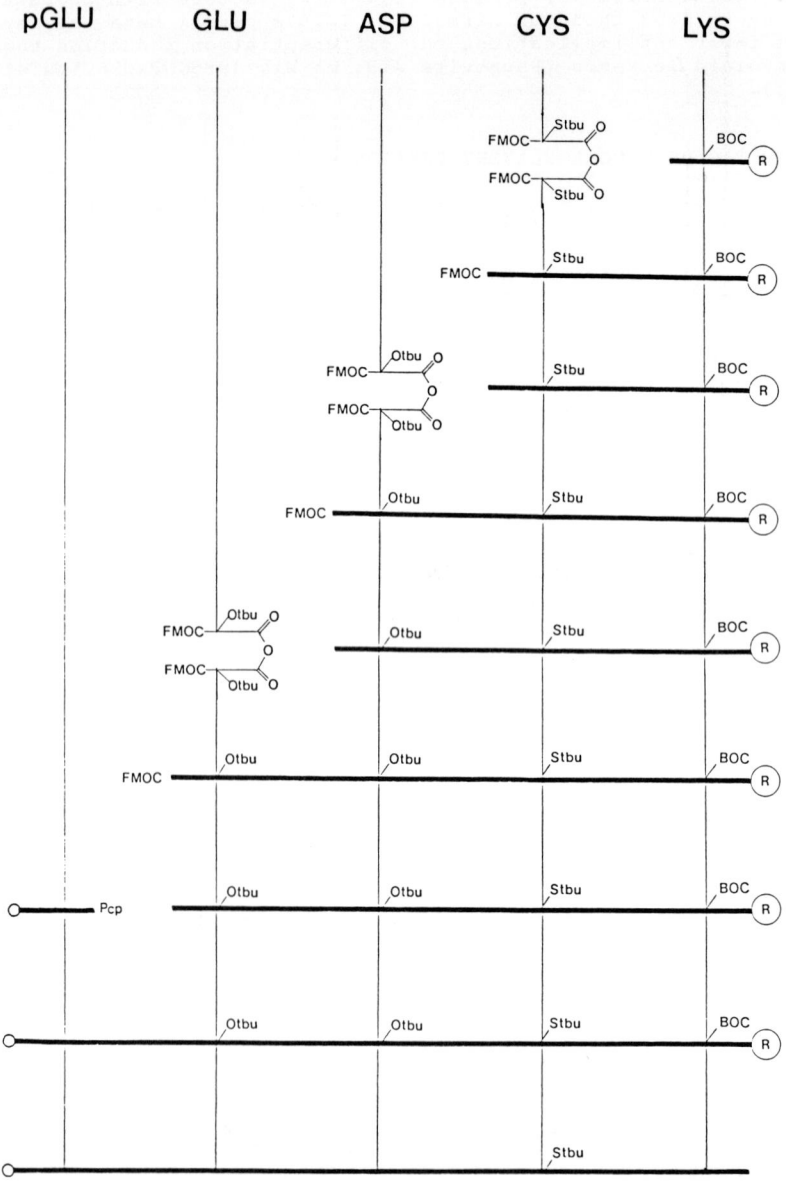

During the whole procedure, including the final removal of the acid sensitive t-butyl protecting groups the mixed disulfide protection on cysteine remains intact, preventing substitution reactions by carbocations and other reactive species. The crude synthetic product is purified by gel, anion exchange, and reverse phase chromatography and is homogeneous in RP-HPLC (fig.4A).

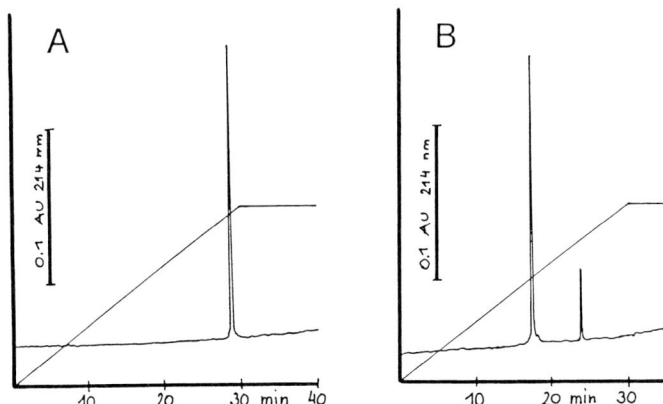

Fig.4: Reverse phase HPLC of the synthetic hemoregulatory peptide HP5b. Column: Altex Ultrasphere ODS, 0.7 ml/min, 0.1 % TFA with a gradient of acetonitrile. A...synthetic HP5b with the thiol protecting group (disulfide with t-butyl-mercaptane) still attached. B...completely deprotected HP5b. The gradient is to 20% in A and to 10 % in B.

In a last step, sometimes in situ immediately before assaying, the peptide is activated by reductive treatment with DTE. The t-butyl-mercaptan formed during this reaction is removed in vacuo. The product obtained is homogeneous in RP-HPLC (fig.4B). The second peak in this chromatogram is the dimer of HP5b, see below.

In an accompanying paper (Erikson et al, this volume) we describe another quite similar synthetic strategy. In this procedure the disulfide type protection of Cys is provided by using an appropriate cystine derivative. The end product in this case is the dimer $(HP5b)_2$ of the hemoregulatory peptide, which has interesting biological properties (see below, and Laerum et al, this volume).

HP5b is a strong inhibitor of CFU-GM colony formation in vitro as described previously (Laerum & Paukovits 1984, Paukovits W.R. et al. 1986). The dose-response curves of the synthetic (HP5b) and the natural (HP) peptide are almost identical (fig.5). HP5b is also active in vivo as a potent inhibitor of CFU-GM and CFU-S in a similar way as mentioned above for natural HP (see: Laerum et al., this volume). It also inhibits the entry of hemopoietic stem cells into the cell cycle after cytostatic treatment. This forms the basis of protection experiments against the lethality of high dose cytostatic treatment (see: Guigon et al., this volume).

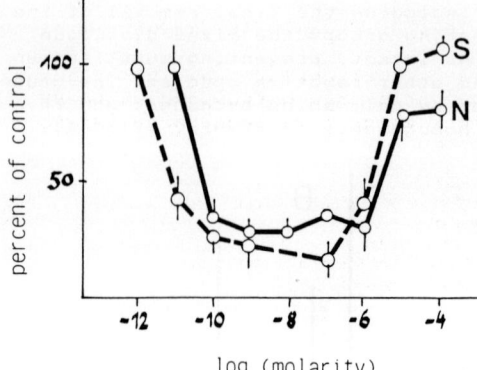

Fig. 5: Dose response curves in vitro of natural (N) and synthetic (HP5b) hemoregulatory peptides. Target: mouse CFU-GM, CSF: mouse lung conditioned medium

---

OXIDATIVE DIMERIZATION OF HP5b

HP and HP5b are both very sensitive against oxidative disulfide formation. The second HPLC peak in fig.4B represents the dimeric form of HP5b. We have described earlier (Aardal et al. 1982), that oxidation of natural HP transforms it into a stimulator of CFU-GM colony formation and the same is observed with synthetic HP5b (see: Laerum et al., this volume).

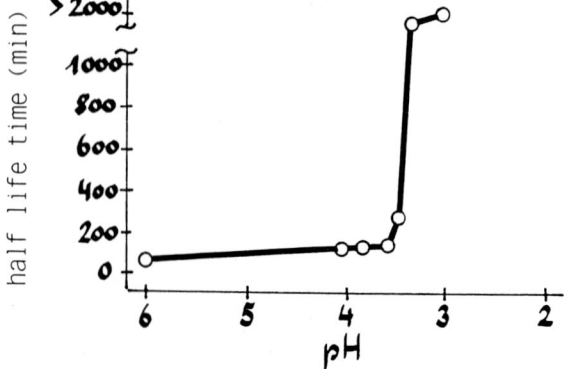

Fig.6: pH dependence of the half life time of the dimerization reaction HP5b (HP5b)$_2$. Concentrations were determined from HPLC peak areas.

The dimerization process is strongly dependent on the pH (fig.6) and is apparently catalyzed by metal ions, since the presence of small amounts of CaNa$_2$EDTA can prevent disulfide formation even at physiological pH.

HEMOREGULATORY PEPTIDES ACT THROUGH MEMBRANE RECEPTORS

Bone marrow cells remove HP from their culture medium. Fig.7 shows that the thin layer chromatographic spot corresponding to HP cannot be detected in solutions which have been incubated for 1 hour with bone marrow cells.

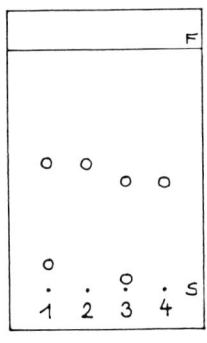

1.....dansylated
2.....incubated with bone marrow cells, centrifuged supernatant dansylated
3.....dansylaminoethylated
4.....incubated with bone marrow cells, centrifuged supernatant dansylaminoethylated

Fig.7: Bone marrow cells absorb HP from their incubation media. HP containing solutions (not completely purified were incubated with bone marrow cells. After centrifugation the supernatant was dansylated and separated by thin layer chromatography (dansylchloride reacts with aminogroups, dansylaziridine with thiols).

In an accompanying paper (Erikson et al., this volume) we describe the synthesis of ^{3}H-labeled HP5b and show that bone marrow cells, but not thymus or spleen cells, bind this radioactive peptide. This binding to bone marrow cells may be related to the presence of membrane receptors. Using the method of Goldwasser (Chang et al., 1974) we have treated bone marrow cells for 1 min at 25°C with a 0.005 percent trypsin solution (Paukovits & Paukovits, 1975). This minimal proteolytic treatment had no influence on the viability nor on the proliferative capacity of the cells. However they were no longer able to respond to the inhibitory signal mediated by HP, provided that the apparently quite rapid regeneration of surface structures was prevented by cycloheximide (fig.8).

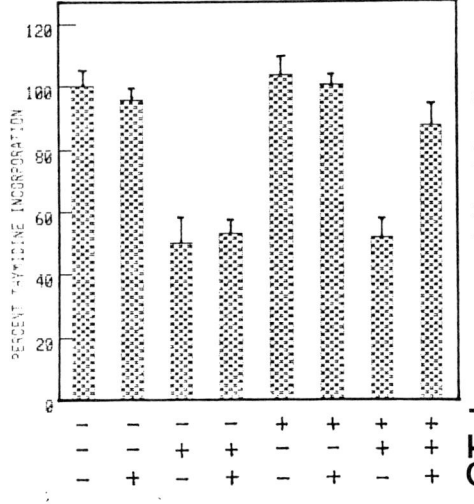

Fig.8: Trypsin treated bone marrow cells have lost their responsivity to HP when regeneration of receptors is prevented with cycloheximide. The trypsin treatment and cycloheximide concentration were chosen so that they had no influence on thymidine incorporation

REFERENCES

Aardal N.P., Laerum O.D., Paukovits W.R.
Biological properties of partially purified granulocytic extract (chalone) in soft agar.
Cell Tissue Kinet. 14: 659 (1981)

Chang J.Y., (1983)
Manual micro-sequence analysis of polypeptides using dimethylaminoazobenzene isothiocyanate.
Meth. Enzymology 91: 455-466

Chang S.C.S., Sikkema D., Goldwasser E., (1974)
Evidence for an erythropoietic receptor protein on bone marrow cells.
Biochem.Biophys.Res.Comm. 57:399-405

Foa P., Lu L., Broxmeyer H.E., Chillemi F., et al. (1986)
A synthetic pentapeptide chadidate for the role of granulocytic chalone.
In: *Biological regulation of cell proliferation*.
(R.Baserga ed.) pp.103-110, Raven Press, New York

Kreja L., Haga P., Muller-Berat N., Laerum O.D., Sletvold O., Paukovits W.R., (1986)
Effects of a hemoregulatory peptide (HP5b) on erythroid and myelopoietic colony formation in vitro.
Scand J.Hematology 37: 79-86

Laerum O.D., Paukovits W.R., (1984)
Inhibitory effects of a synthetic pentapeptide on hemopoietic stem cells in vitro and in vivo.
Exp.Hematology 12: 7-17

Paukovits W.R. (1971)
Control of Granulocyte Production: Separation and chemical identification of a specific inhibitor (chalone). *Cell Tissue Kinet* 4: 539-547

Paukovits W.R., (1973)
Granulopoiesis Inhibiting Factor (GIF): Demonstration and preliminary Chemical-Biological Characterization of a Specific Polypeptide (Chalone).
Natl.Cancer Inst. Monogr. 38: 147-155

Paukovits W.R., Hinterberger W., (1978)
Biochemical characterization of humoral factors regulating myelopoiesis.
In: *Cell Separation and Cryobiology* (H.Rainer, ed), pp.75-78, Schattauer, Stuttgart

Paukovits W.R., Laerum O.D., (1982)
Isolation and synthesis of a hemoregulatory peptide.
Z.Naturforschung 37c: 1297-1300

Paukovits W.R., Laerum O.D., (1984)
Structural investigation on a peptide regulating hemopoiesis in vitro and in vivo.
Hoppe Seyler's Z.Physiol.Chemie 365: 303-311

Paukovits W.R., Laerum O.D., Guigon M., (1986)
Isolation, characterization, and synthesis of a chalone-like hemoregulatory peptide.
In: *Biol.Regulation of Cell Prolif*. (ed: R.Baserga) pp 111-120, Raven Press

Paukovits W.R., Laerum O.D., Paukovits J.B., Hinterberger W., Rogan A:.M. (1983)
Methods for the preparation of purified granulopoiesis inhibiting factor (chalone)
Hoppe Seyler's Z.Physiol.Chemie 364: 383-396

Paukovits W.R., Paukovits J., (1975)
Mechanism of Action of Granulopoiesis Inhibiting
Factor (Chalone): I. Evidence for a Receptor Protein
on Bone Marrow Cells.
Exp.Pathol. 10:343-352

Paukovits W.R., Paukovits J.B., (1978)
Peptide nature and proteolytic sensitivity of granulo-
poiesis inhibiting factor.
IRCS Medical Science 6: 176

Paukovits J.B.,Paukovits W.R., Laerum O.D. (1986)
Identification of a regulatory peptide distinct from
normal-granulocyte-derived hemoregulotory peptide
produced by human promyelocytic HL-60 leukemia cells
after differentiation induction with retinoic acid.
Cancer Res. 46: 4444-4448

Paukovits W.R., Paukovits J.B., Laerum O.D., (1987)
Granulocytes obtained from HL-60 cells by retinoic acid
treatment behave abnormally in hemopoietic regulation -
Possible consequences for differentiation induction
treatment of leukemia.
In: J. Aarbakke (ed): The Biology and Pharmacology of
Tumor Cell Differentiation.
Humana Press, in press

Résumé

La myélopoièse est modulée in vitro et in vivo par un peptide hémorégulateur
(HP) contenant un groupement thiol qui semble être associé aux granulocytes
matures. Cet article décrit a) la purification du peptide hémorégulateur à
partir des leucocytes, b) la détermination de sa séquence, c) la synthèse
chimique et les propriétés biologiques de plusieurs analogues. En outre, nous
présentons des données préliminaires sur les récepteurs du peptide hémorégulateur sur les cellules de la moelle et sur la transformation oxydative des
peptides hémorégulateurs en dimères stimulants par formation de disulfures.
Nous démontrons également que le peptide hémorégulateur n'est pas contenu
dans les granulocytes leucémiques induits à se différencier. Les cellules
HL-60 induites par l'acide rétinoïque produisent un peptide inhibiteur de
structure différente.

Discussion

Chairpersons/*Modérateurs*: D. Metcalf (Australie)
E. Frindel (France)

B. LORD : Dr. Metcalf, you showed very dramatic increases of mast cells and megakaryocytes in the spleen with multi-CSF. Did you make equivalent measurements in the marrow as well and attempt to sum up the whole body burden? How does it compare with normal?

D. METCALF : In the mouse, there are no mast cells in the bone marrow and none appeared in mice injected with IL 3. Megakaryocyte numbers were not quantified in the bone marrow.

B. LORD : The purpose of my question is that we looked at CFU-S in mice treated with recombinant IL-3 and we also saw this quite dramatic increase of CFU-S in the spleen. In the bone marrow, there was a slight fall. Summing up the total number of CFU-S in the mouse, however, there was certainly no overall increase.

D. METCALF : That is true. One of the astonishing thing about G-CSF which nobody would have predicted from the in vitro studies is that a mouse that is injected with G-CSF has essentially no erythropoiesis in the bone marrow. All the erythropoiesis moves to the spleen so 40 to 50% of the cells in the spleen are erythroid cells with almost no erythroid cells in the bone marrow. What is mediating that enormous migration and change in the population, we don't know, but there are suggestions of similar changes seen also in response to the other CSFs. I think it's telling us that we still don't really understand very much about hematopoiesis.

G. KONWALINKA : I want to ask you on GM-CSF, did you find any burst promoting activity in this GM-CSF preparation in your experiment?

D. METCALF : If you add GM-CSF to a culture, do you get increased numbers of erythroid colonies? If it is a mouse culture, no, unless there is erythropoietin present, in which case the answer is yes. If it is a human culture, the answer is yes, there is enhancement, but we believe this is not a direct effect. Clone transfer studies suggest that for human cells, GM-CSF has an indirect effect. But this question is still in dispute.

M. TUBIANA : I have two questions for you. The first one is related to the possible effect of CSF on stem cell differentiation. You mentioned that the G-CSF, when it is given in vivo, could induce erythropoiesis in the spleen and disappearance of erythropoietic activity in the bone marrow. This would suggest that this substance may have an effect on cell differentiation. The second question is related to the possible clinical study. You mentioned that there are clinical studies which are ongoing with CSF. Could you elaborate a little on that?

D. METCALF : Well, the first question is why G-CSF, when tested in vitro, have absolutely no action on stem cells or on erythroid cells, yet have such dramatic effects in vivo? The honest answer is I don't believe anyone has an explanation. It suggests there is an indirect effect achieved by network interactions and one possible mediator molecule is IL-1. What is happening in the clinical trials, I can't tell you because we have not yet done any ourselves. GM-CSF has been tested in Boston and in Los Angeles, and I think now in North Carolina and G-CSF has certainly been tested in New-York and I suspect also in Japan. Initial reports indicate that both are able to elevate white cell levels in man but there is no experience yet on the question of whether such responses increase resistance to infections.

L. SOLBERG : My question is for anyone of the four. In the initial use of the recombinant growth factors in primates or in mouse, there have been prolonged periods of stimulation of granulopoiesis noted by continuous administration of these agents. What is the implication of this for the endogenous inhibitory systems ? Why does one not see an apparent inhibitory response to the factors in vivo ?

D. METCALF : That is a good question. It does not raise a problem for inhibitors blocking CSF production because you are injecting CSF. The ability of injected CSF to induce sustained responses does raise a problem if you think that the products of the mature cells should have direct feedback inhibitory effects on cell production. It was surprising, I think, for everyone to find that you could sustain elevated cell levels, particularly of granulocytes, for weeks in animals. As soon as you stop the CSF, the levels rapidly fall back to the normal within a few days. There is also no evidence that levels then fall below normal again raising questions about the effectiveness of inhibitors from mature cells. I think it is now appropriate for people who work with inhibitors to take CSF-injected animals and ask questions about levels of inhibitors. Are there any bad effects of excess levels of CSF? Yes there are. We have recently been looking at transgenic GM-CSF mice in which the transgene is permanently turned on. These mice develop massive numbers of peritoneal macrophages. They are blind, because their eyes are full of macrophages which destroy the retina and they die eventually at the age of two or four months with extensive muscle lesions based on macrophage activation. It is not new in biology to observe toxic effects when you have too much of any agent. Macrophages make many biologically active molecules and it will be very interesting to see in G-CSF transgenic mice whether similar lethal consequences occur. But for GM-CSF transgenic mice, there are serious diseases that develop which cause premature death.

E. FRINDEL : I just wanted to make a comment about Pr. Tubiana's question concerning the CSF having an effect on pluripotent stem cells. I don't know about men but we have shown many times that in mice, the GM-CSF has no effect on bone marrow pluripotent stem cells, either for proliferation or differentiation. Now it will be interesting to find out if there is an effect with recombinant CSFs. Maybe there was some sort of side effect or something in the molecule that does not exist in the recombinant, we have never tried to purify but the natural CSF has no effect on the pluripotent stem cells.

D. METCALF : Well, it is always important to check that the recombinant molecule has exactly the same actions as the native molecule. I, too, am very suspicious and ask all these questions. However, in general the actions of native and recombinant CSF's are the same. If you use recombinant material in vitro, there is no action on stem cells, but if you give CSF in vivo there is a possibility of interactions via regulator networks and this may be what is happening. There are reductions in the numbers of CFU-S in the marrow of mice injected with GM-CSF and there are very large rises in mice injected with G-CSF. Everyone finds the same thing, so it occurs in more than one strain of mice and we have no good explanation for the mechanisms.

O.D. LAERUM : I will try to answer your question from the inhibitor side so that you get one stimulatory answer and one inhibitory answer. People have believed in negative feedback mechanisms for regulating hematopoiesis. This is nice concept as long as you have a steady state but it cannot explain what happens in an animal during an infection : in a few hours, you certainly get a leukocytosis. If the granulocytes produce the inhibitory factor, then a leukocytosis should immediately cause inhibition of the bone marrow and it does not happen before the whole infection has come to an end, so that is contradictory. But I think in light of our data, there might be an explanation. One is that at high levels of the inhibitor, the cells become unresponsive, making a short-cut. The second possibility is that, due to the oxidative metabolism and capacity of granulocytes, it will be no problem for them to rapidly oxidise the inhibitor and make a dimer stimulator instead. Although this is purely on a theoretical basis, an inhibitory regulating mechanism of myelopoiesis must have a short-cut during an infection.

B. LORD : Some years ago, Dr. Reissmann did an experiment in which he suppressed granulopoiesis with BCNU and then restimulated it with daily injections of 19 nortestosterone decanoate. This resulted in a large overproduction of blood neutrophils which restored to normal despite the continued injections of 19 ND. He suggested this might be due to production of inhibitor and Dr. Milenkovic, working with me at the time took it up. He collected serum from mice at the peak of their granulocytosis and used it to prevent the overshoot in further series of animals. He was able to produce the same effect with our inhibitory granulocyte extract. So I think these increases produced by growth factors in vivo are self regulating but as Dr. Metcalf suggested in those experiments, the high levels were being maintained by the continuing injections of exogenous factors. In Dr Reissmann's experiments, granulocyte production was initially restimulated by the production of endogenous growth factors.

L. ROZENSZAJN : Dr. Aglietta, could you make some comments on the production of the TGF by lymphocytes and effect of this factor on T cell proliferation and/or on T cell colony formation?

M. AGLIETTA : What I discussed was published in the Journal of Experimental Medecine by Dr. Fauci and concerned IL-2 stimulated T lymphocytes. I am not aware of work on T colonies.

Brief reports
Communications brèves

Hemoregulatory peptide synthesis, purification of tritium labelled peptide and uptake of peptide in hematopoietic tissues *in vitro*

Jon Amud Eriksen, Jon-Sverre Schanche, Kjetil Hestdal*, Sten-Eirik Jakobsen*, Trygwe Tveteraas, Jon-Henrik Johansen, Walter R. Paukovits**, and Ole D. Laerum*

Nycomed, PO Box 4220 Torshov, N-0401 Oslo 4, Norway. * The Gade Institute, Department of Pathology, University of Bergen, N-5016 Haukeland Hospital, Norway. ** Institute for Tumor Biology/Cancer Research, University of Vienna, Austria

Keywords
Hemoregulatory peptide, peptide synthesis, radiolabelling, hematopoiesis.

Introduction
The hemoregulatory peptide pGlu-Glu-Asp-Cys-Lys (Paukovits and Laerum 1982), which reversibly inhibits the proliferation of bone marrow stem cells (Laerum and Paukovits 1984) has been synthesized with tritium (^{3}H) in the pGlu residue. The peptide (G-^{3}H)pGlu-Glu-Asp-Cys-Lys (I) was synthesized as shown in fig. 1. and 2. semipreparative reversed phase (C-18) HPLC was used to purify the product (fig. 3. and 4.). In vitro studies using pure I showed a time dependent linear uptake of radioactivity in bone marrow cells (BMC). Thymus cells (TH) and spleen cells (SC) showed very low or no uptake of radioactivity (fig. 5.).

I.

II.

Synthesis

The peptide was synthesized in solution as the symmetric dimer [(G-^{3}H)Gln-Glu-Asp-Cys-Lys]$_2$ (II) using either N,N-dimethylformamide (DMF) or dichloromethane (CH$_2$Cl$_2$) as solvent.

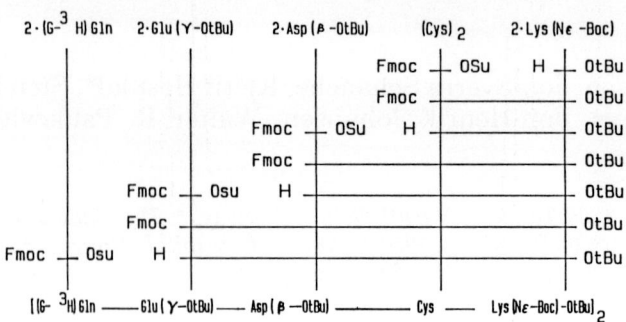

Fig. 1. Strategy for synthesis of the hemoregulatory peptide dimer.

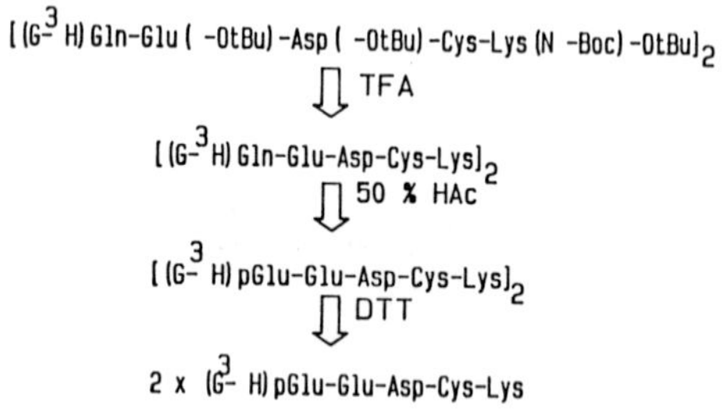

[(G-^{3}H)Gln-Glu(-OtBu)-Asp(-OtBu)-Cys-Lys(N-Boc)-OtBu]$_2$
⇩ TFA
[(G-^{3}H)Gln-Glu-Asp-Cys-Lys]$_2$
⇩ 50 % HAc
[(G-^{3}H)pGlu-Glu-Asp-Cys-Lys]$_2$
⇩ DTT
2 x (G-^{3}H)pGlu-Glu-Asp-Cys-Lys

Fig. 2. Deprotection, cyclization and reduction of the synthetic product to obtain the hemoregulatory peptide.

Hydroxysuccinimide active ester (OSu) coupling and Fmoc/Boc, t-butylester (OtBu) protecting strategy was utilized. N-terminal Fmoc protection was removed selectively after each coupling step by treatment with 20 % piperdine in either DMF or CH$_2$Cl$_2$. Trifluoroacetic acid (TFA) was used to remove Boc and OtBu protecting groups to obtain II. (G-3H)Gln was cyclized to (G-^{3}H)pGlu by heating to 50°C in 50 % aquous acetic acid (Bodanszky and Martinez 1981). Reduction of the disulfide bond by dithiothreitol (DTT) gave I (Cleland 1964).

Purification and Analysis

The radioactive monomer was purified using a semipreparative HPLC column (supelcosil LC-18, 5 t, 250 x 10 mm) coupled to a Perkin-Elmer LC4 chromatograph. The peptide was eluted (fig.3) with a gradient from 0 to 22.4 % acetonitrile in 5 mM H$_3$PO$_4$ with a duration of 30 min. The flow rate was 2 ml/min. Fractions of the eluent were collected manually.

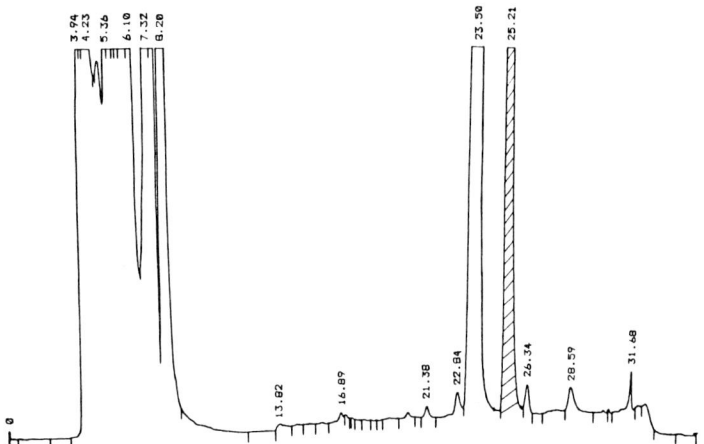

Fig.3. Purification of the labelled hemoregulatory peptide. The eluent was monitored at 214 nm, using a variable wavelength detector. The peak eluting at 25.21 min. was collected and further characterized.

Analysis was performed with a slightly different system (fig.4.). The column (supelcosil LC-18, 5 t, 250 x 4.6 mm) was eluted by a gradient from 0 to 30 % acetonitrile in 0.1 % TFA with a gradient time of 10 min. The flowrate was 2ml/min. The product was shown to be free from dimer and to be approximately 99 % pure (as compared with nonradioactive standards).

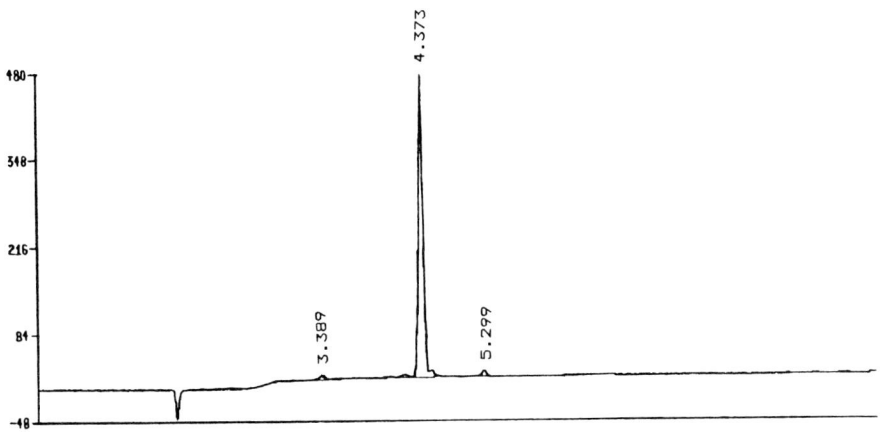

Fig.4. Analytical HPLC of the radiolabelled hemoregulatory peptide monitored at 214 nm. The peak at 4.373 min. contained all the radioactivity, and co-chromatographed with authentic standards. The other two peaks correspond to background peaks, generated by the gradient.

Uptake Studies

Cells from C3H-mice were used in these experiments. BMC were prepared by flushing femurs with physiological saline. TH and SC were prepared by tearing the respective organs with pincers. The cells (40×10^6 cells/ml) were incubated in Dulbeccos MEM, supplemented with 20 % fetal calf serum.

The cells were centrifuged and resuspended in medium containing 10^{-6} M(154 nCi/nmol) peptide. After different time intervals (fig.5.), aliquots of the cell suspension were removed, the cells were pelleted by centrifugation, and radioactivity was determined by scintillation counting. The results are presented in fig. 5.

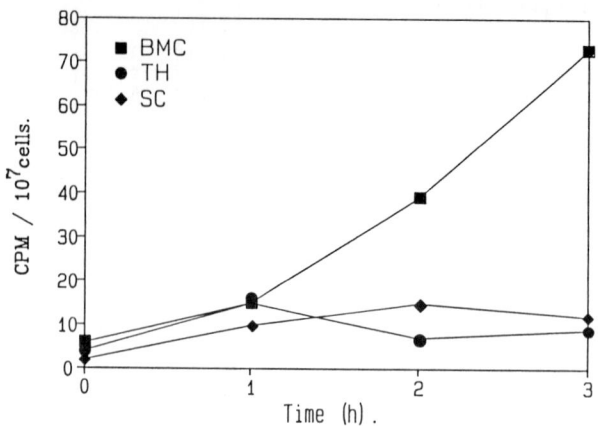

Figure 5. uptake of (G-3H)pGlu-hemoregulatory peptide by three hematopoietic tissues.

Discussion

Of the three tissues tested in these experiments, radioactive hemoregulatory peptide seems to be selectively taken up by bone marrow cells. Further experimentation with other tissues and organs are needed to ascertain this specificity. Presently we are trying to identify the cells in bone marrow responsible for the uptake, and to characterize the uptake process.

References

Bodanszky,M. & Martinez,J. (1981) Side reactions in peptide synthesis. Synthesis, 5, 333-356.

Cleland,W.W. (1964) Dithiothreitol, A new protective reagent for SH groups. Biochemistry, 3, 480-482.

Laerum,O.D. & Paukovits,W.R. (1984) Inhibitory effects of a synthetic pentapeptide on hematopoietic stem cells in vitro and in vivo. Expl. Hemat., 12, 7-17.

Paukovits,W.R. & Laerum,O.D. (1982) Isolation and synthesis of a hemoregulatory peptide. Z. Naturf, 37c, 1297-3000

A specific, low molecular mass granulopoiesis inhibitor, isolated from calf spleen

Jörg Fetsch and H. Rainer Maurer

Institut für Pharmazie der Freien Universität Berlin D-1000 Berlin 33, FRG

KEYWORDS

Granulopoiesis inhibitor, chalone, calf spleen, colony formation, peptide

INTRODUCTION

The term "CHALONE" was reintroduced by Bullough (Iversen, 1976) for substances specifically inhibiting cell proliferation by a negative feed-back-mechanism. These factors should have a cell and tissue specific but species unspecific effect. Moreover this effect should be nontoxic but reversible.
For several years we have been engaged in the isolation of a chalone-like granulopoiesis inhibitor. Kastner (1984) found some indication of such a factor in calf spleen. In continuation of this work we could recently isolate a highly potent factor, possibly a peptide, specifically inhibiting granulopoiesis in vitro at approximately 1E-9 M concentration.

METHODS

Isolation

An acetone precipitate was obtained after ethanol extraction as described by Kastner (1984). Ultrafiltration was performed in a special multi-chamber-system developed by Dr. A. Kinawi (our institute). The membranes used were XM 100 A, XM 50, YM 30, YM 10, YM 5, YM 2 and YC 05 (Amicon). Fast protein liquid chromatography (FPLC) was performed with an equipment by Pharmacia, using a Mono Q anionexchanger column, NaCl-gradient 0-0.4 M / 20 min, 0.4-1 M / 2 min, flow rate of 2 ml/min, fraction size 400 μl, detection at 254 nm. High performance liquid chromatography (HPLC) was performed with a Beckman system, using a Merck Hibar column 3x150, C18, 5 μm, eluent 0.2 M sodium acetate (pH 6.5), acetonitrile / methanol (4:1, v/v), 2.5-45% / 30 min, 0.5 ml/min, detection at 254 nm.

Biological assays

We used our T-lymphocyte and granulocyte assay in glass capillaries as described by Maurer (1976, 1977).
For the CFU-E test we used the method described by Konwalinka (1982), however we used microtiter plates instead of petri dishes.

STATISTICAL EVALUATION

We performed 3 - 4 tests in triplicates for each concentration determination. The bars represent 50% of all results of one concentration, the point being the median value. In addition the minimum and maximum value are given.

RESULTS

From calf spleen we prepared an acetone precipitate, which we dissolved in water and subjected to ultrafiltration. The fraction with a molecular mass of 0.5-1 kDa showed the most evident reduction of colony number in our granulocyte assay (GC-assay).

After FPLC (Fig. 1) we found several fractions inhibiting granulopoiesis in vitro without affecting T-lymphocyte colony growth (Fig. 2). In our GC-assay the inhibitory effect of fraction 21 (#21) of the FPLC eluate was only seen upon 1E3 to 1E4 (1000 to 10000) fold dilution of the eluate (Fig. 3). At higher dilutions the colony number reached the control level again. The result is a bell-shaped dose/response curve with a maximum inhibition at 5E-9 to 5E-10 M concentration, according to an estimation of the concentration based on the amount of material used for purification and quantitative amino acid analysis. At 1E-6 M no inhibition was observed, which precludes any cytotoxicity.

T-lymphocyte colony growth (Fig. 4) was not affected by fraction 21, as there was no reduction of colony growth over the whole concentration range tested. Similarily, in the CFU-E assay (Fig. 5) no effect was observed except a slight reduction of colony number at the highest concentration. This, however, is a 1000 fold higher concentration than that of the GC-assay.

After HPLC a single peak (22.8 min; Fig. 7) was obtained. In our GC-assay this peak again showed an inhibitory effect only in a certain dose range. The maximum inhibition was dependent on the presence of 2-mercaptoethanol (2-ME): addition of 2-ME caused a shift of the minimum of the bell-shaped curve (i.e. maximum inhibition) to lower concentrations (Fig. 6). After HCl hydrolysis this peak disappeared (Fig. 7). After derivatization with phenylisothiocyanate (PITC) and HPLC separation aspartic and glutamic acid (asp, glu) were detected among others (Fig. 8).

ACKNOWLEDGMENT

We thank Drs. W. Paukovits and O.D. Laerum for continuing counsel.

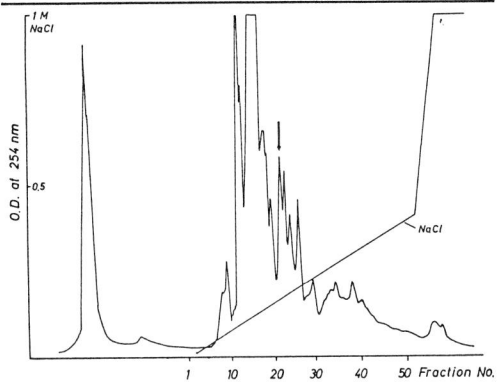

Fig. 1

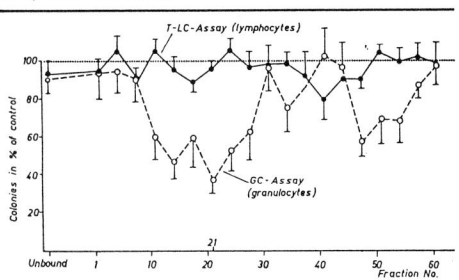

Fig. 2

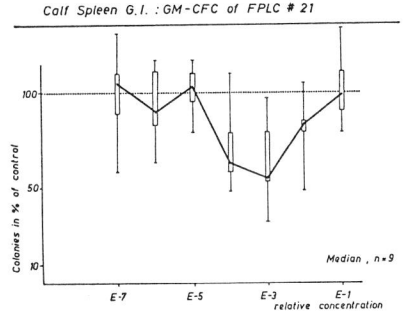

Fig. 3

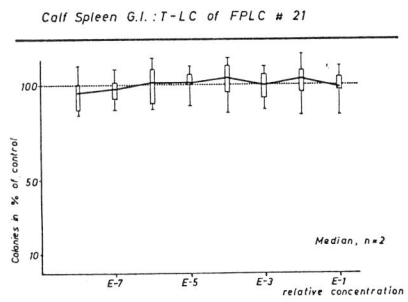

Fig. 4

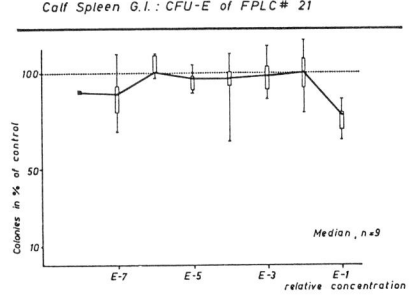

Fig. 5

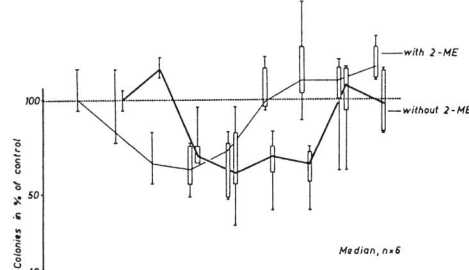

Fig. 6

57

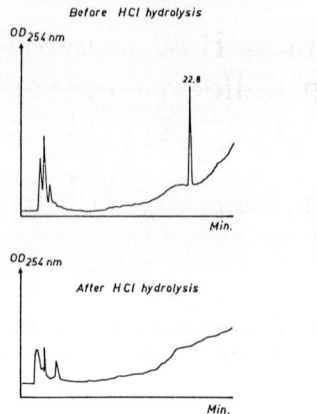

Fig. 7

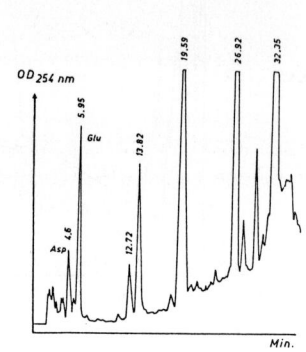

Fig. 8

REFERENCES

Iversen, O.H.(1976): The history of chalones. In Chalones, ed J.C. Houck, pp 37-69. Amsterdam: North-Holland Publishing Company

Kastner, M.(1984): Partial purification and characterization of an endogenous granulomonopoiesis inhibitor from calf spleen. Hoppe-Seyler's Z. Physiol. Chem. 365, 129-135

Konwalinka, G.(1982): A micro agar culture system for cloning human erythropoietic progenitors in vitro. Exp. Hematol. 10, 71-77

Maurer, H.R.(1976): Colony growth of mouse bone marrow cells in agar contained in glass capillaries. Blut 33, 11-22

Maurer, H.R.(1977): In vitro culture of lymphocyte colonies in agar capillary tubes after PHA-stimulation. J. Immunol. Meth. 18, 353-364

Inhibitory activity of recombinant human H-subunit ferritin on the *in vitro* growth of human granulocyte-macrophage progenitor cells

Laura Dezza, Mario Cazzola, Gaetano Bergamaschi, Carmelo Carlo Stella, Paolo Pedrazzoli, Massimo Aglietta and Edoardo Ascari

Department of Internal Medicine and Medical Therapy, University of Pavia, Pavia and Department of Biomedical Sciences and Human Oncology, University of Turin, Turin, Italy

KEY WORDS

Recombinant H-subunit ferritin, normal hemopoietic progenitor cells.

INTRODUCTION

Ferritin is a ubiquitous protein whose major function is iron storage and detoxification, although other biological functions have been suggested (Harrison et al, 1980). Ferritin is a protein shell with a molecular weight of about 500,000 Mr made up of 24 subunits. The multiple forms, or isoferritins, that occur in most human tissues appear to originate from the presence of different subunits (Worwood, 1982). Three subunit types have been described: L (light, 19,000 Mr), H (heavy, 21,000 Mr) and G (glycosylated, 23,000 Mr), which share immunological differences that are encoded on different families of genes (Cragg e coll, 1981; Watanabe and Drysdale, 1981). Tissue ferritins consist of variable proportions of H and L subunit types: H-subunit-rich isoferritins are acidic (lower pI), whereas L-subunit-rich isoferritins are more basic (higher pI).

Purified tissue H-subunit-rich isoferritins have been described from Broxmeyer et al (1981) as having inhibitory activity on the in vitro growth of human hemopoietic progenitors. These results have been obtained in our laboratory (Dezza et al, 1986) but not confirmed by others (Sala et al, 1986). More recently, Broxmeyer et al (1986) have found that recombinant human H-subunit ferritin, but not L-subunit ferritin has inhibitory activity on colony formation in vitro.

In order to clarify the effects of acidic isoferritins on human hemopoietic progenitors, we have used human recombinant homopolymers in this work.

MATERIALS AND METHODS

Recombinant human L-subunit and H-subunit ferritins were kindly provided by Dr. Paolo Arosio and Sonia Levi, University of Milan, Milan, Italy. Recombinant human H-subunit ferritin was obtained as described in detail elsewhere (Arosio et al, 1985; Levi et al, in press). The coding region of a cDNA previously identified for the H chain was inserted into the expression/modification vector pEMBLex2. Ferritin synthesis by E. coli transformed with this plasmid was induced by heating at 42°C for 4 hours. Cells were washed and disrupted by sonication; the supernatant was heated to 75°C, then centrifuged at 200,000 g for 90 minutes and loaded on a Sepharose 6B column. Recombinant human L-subunit ferritin was similarly prepared using the coding region of a cDNA previously identified for the L chain kindly provided by Dr. Jochen Salfeld, Heidelberg, West Germany. The monoclonal antibody 2A4 against human heart ferritin was developed by Cavanna et al. (1983). The monoclonal antibody L03 against human liver ferritin was developed and characterized by Luzzago et al. (1986). The two antibodies are specific for the H and L subunits respectively.

For CFU-GM assay, 1×10^5 low density normal human bone marrow cells, after Ficoll separation, were plated in a 1 ml mixture containing Iscove's Modification of Dulbecco's Medium (IMDM), 15% fetal bovine serum, 0.3% agar (Difco Laboratories, Detroit, MI, USA) and 10% conditioned medium from the 5637 cell line as a source of CSA, in the presence or absence of the various isoferritins. Cultures were incubated at 37°C in a humified atmosphere of CO_2 in air. Plates were scored for colonies with an inverted microscope after 7 and 14 days of incubation. All aggregates containing more than 40 cells were counted as colonies. In some experiments, cells were preincubated for 2 hours with ferritin, extensively washed and then plated in semisolid medium. To study the neutralization of the isoferritin-associated inhibitory activity by a specific antibody, recombinant isoferritins were preincubated at 10^{-8} M with the monoclonal antibodies L03 and 2A4 prior to the culture.

RESULTS

Recombinant isoferritins were tested for their capacity to influence colony formation of human normal hemopoietic progenitor cells. Results are reported in "Fig 1".

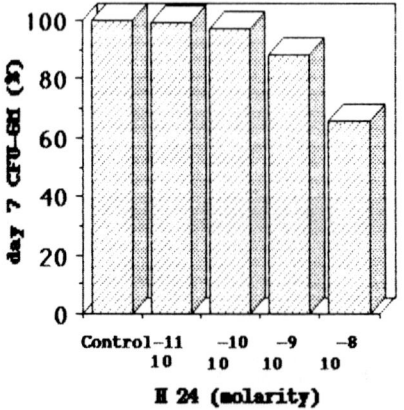

Fig. 1: Effects of various concentrations of recombinant H-subunit-rich ferritin on normal CFU-GM. Results represent the mean of 6 separate experiments. Colony growth is expressed as % of the control.

Recombinant L-subunit-rich isoferritin (L_{24}) did not show any effect on CFU-GM growth at concentrations equal to 10^{-8} M. Recombinant H-subunit-rich isoferritin (H_{24}) showed inhibitory activity at 10^{-8} M on CFU-GM (% inhibition = 36 ± 13). No consistent inhibition was obtained with concentrations lower than 10^{-9} M. The inhibition observed was removed by preincubation with the monoclonal antibody 2A4 directed against the H subunit, whereas it was unaffected by preincubation with the antibody LO3 directed against the L-subunit ferritin. When bone marrow mononuclear cells were exposed to H_{24} (10^{-8} M) for 2 hours and plated after extensive washing, no inhibition of colony formation was seen.

DISCUSSION

We previously found that some preparations of acidic isoferritins rich in H-subunits are able to inhibit normal human CFU-GM growth (Dezza et al, 1986). However, it was not clear why preparations apparently similar with respect to biochemical and immunological properties do not behave in the same way. In particular, the role of glycosylation was uncertain because in the purified preparations of ferritins from human tissues the effective degree of glycosylation is low (5%) and the H subunit is supposed to have no glycosylation site on its external surface. To clarify the effective role of acidic isoferritins in the regulation of myelopoiesis, we tested recombinant human homopolymers made up of 100% L or H subunits. Since these isoferritins had been expressed in E. Coli, they were not glycosylated. The results obtained in this work show that recombinant H-chain ferritin is capable of inhibiting colony formation by normal hemopoietic progenitor cells in vitro. Thus, it seems that the inhibitory activity of acidic isoferritins is related to the H subunit and that glycosylation is not a prerequisite for acidic isoferritin activity. The mechanism responsible for this activity remains to be clarified. However, it should be noted that a high number of molecules per cell is necessary for obtaining the inhibitory effect (10^{-8} M means 6×10^{15} molecules or 6×10^7 molecules per cell in the CFU-GM assay). Such a high number may imply a mechanism other than receptor interaction. Preliminary studies, however, have suggested that receptors for H-type ferritin exist on erythroleukemia K562 cells (Fargion et al, 1986) The recognition of this receptor suggests its possible relevance in the mechanism of action of acidic isoferritins in myelopoiesis.

REFERENCES

Arosio P., Levi S., Albertini A., Ruggeri G., Iacobello C, Luzzago A., De Simone F., Cesareni C., Cortese R. (1985):Development of monoclonal atibodies and recombinant human ferritins for structural analysis of isoferritins. In Biotechnology in Diagnostics, ed H. Koprowski, S. Ferrone, A.Albertini, p.237-244. Amsterdam: Elsevier.

Broxmeyer, H.E., Bognacki J., Dorner M.H., deSousa M.(1981): The identification of leukemia-associated inhibitory activity (LIA) as acidic isoferritins: a regulatory role for acidic isoferritins in the production of granulocytes and macrophages. J. Exp. Med., 153, 1426-1438.

Broxmeyer H.E., Lu L., Bicknell D.C., Williams D.E., Cooper S., Levi S., Salfeld J., Arosio P.(1986): The influence of purified recombinant human heavy-subunit and light-subunit ferritins on colony formation in vitro by granulocyte-macrophage and erythroid progenitor cells. Blood, 68, 1257-1263.

Cavanna F., Ruggeri G., Iacobello C., Chieregatti G., Murador E., Albertini A., Arosio P. (1983): Development of a monoclonal antibody against human heart ferritin and its application in an immunoradiometric assay. Clin. Chim. Acta 134, 347-356.

Cragg S.J., Wagstaff M., Worwood M. (1981): Detection of a glycosylated subunit in human serum ferritin. Biochem. J. 19, 565-571.

Dezza L., Cazzola M., Piacibello W., Arosio P., Levi S., Aglietta M. (1986) Effect of acidic and basic isoferritins on in vitro growth of human granulocyte-monocyte progenitors. Blood 67, 789-795.

Fargion S., Arosio P., Fracanzani A.L., Cislaghi, Levi S., Piperno A., Fiorelli G. (1986): Expression of ferritin receptor on K562 cells and specificity for H ferritin subunits. European Iron Club Meeting, Pavia September 24-27. Abstract Book p. 19.

Harrison P.M., Clegg G.A., May K. (1980): Ferritin structure and function. In Iron in Biochemistry and Medicine, 2nd edn, London: Academic Press.

Luzzago A., Arosio P., Iacobello C., Ruggeri G., Capucci L., Brocchi E., De Simone F., Gamba D., Gabri E., Levi S., Albertini A.(1986): Immunochemical characterization of human liver and heart ferritins with monoclonal antibodies. Biochim. Biophys. Acta 872, 61-71.

Sala G., Woorwood M., Jacobs A. (1986): The effect of isoferritins on granulopoiesis. Blood 67, 436-443.

Watanabe N., Drysdale J.W. (1981): Evidence for distinct mRNAs for ferritin subunits. Biochem. Biophysis. Res. Commun. 98, 507-511.

Worwood M. (1982): Ferritin in human tissues and serum. Clin. Haematol. 11, 275-307.

ACKNOWLEDGMENTS

Work supported by grant no 86.00352.44 from CNR (Consiglio Nazionale delle Ricerche), Special Project Oncology, and grants from Associazione Italiana per la Ricerca sul Cancro (AIRC).

The role of lactoferrin in regulating colony stimulating factor production

Ivan N. Rich[1] and Gunther Sawatzki[2]

[1] Dept. of Tranfusion Medicine of the University of Ulm, DRK Blutspenderzentrale, Oberer Eselsberg 10, D-7900 Ulm, FRG
[2] Research Department, Milupa AG, D-6382 Friedrichsdorf/Taunus, FRG

KEYWORDS
Lactoferrin, GM-CSF, neutrophil regulation, macrophage function.

INTRODUCTION

Besides its anti-bacterial effect, the iron-binding glycoprotein lactoferrin (Lf), has been implicated as a negative-feedback regulator for granulocyte-macrophage colony stimulating factor (GM-CSF) thereby regulating granulocyte production (Broxmeyer,1979). This postulated role for Lf does not, however, correlate with many clinical findings (Olsson et.al.1979; Chilcote et.al.1983; Venge et.al. 1984). In addition, several reports have not confirmed this reputed role for Lf (Winton et.al.1981; Galbraith,1986; Poppas et.al.1986). Species-specific Lf has recently been purified to homogeneity (Sawatzki & Kubanek, 1983) and has enabled a reevaluation of the role of Lf in GM-CSF production.

MATERIALS & METHODS

Female C57Bl/Bom (GL. Bomholtgart Ltd.,Ry,Denmark), 8-10 weeks old were used throughout. Bone marrow were prepared as described previously (Rich, 1986). Peritoneal cells were prepared by adherence and suspended in Iscove's Modified Dulbecco's Medium (IMDM).

Iron-saturated, species-specific Lf was isolated and purified from human and mouse milk and assayed by ELISA (Sawatzki & Kubanek,1983). Traces of endotoxin were removed by polymyxin B-Sepharose chromatography (Issekutz,1983).Homogeneity of the preparations was determined using SDS-PAGE, gel filtration and rocket immunoelectrophoresis.

The in vitro effect of Lf on bone marrow and adherent peritoneal cells was performed in exactly the same manner as described by Broxmeyer et.al. 1984).Endotoxin-free mouse Lf (2mg/injection) was injected into three groups of mice (see results). Assay of granulocyte-macrophage colony forming cells (GM-CFC) and GM-CSF has been previously described in detail (Rich, 1986).

RESULTS

Table 1 shows the GM-CSF content in supernatants derived from 20,000, 100,000 and 500,000 bone marrow cells incubated either alone or supplemented with endotoxin-free mouse or human Lf in the presence or absence of indomethacin. The results indicate that in nearly all cases (exception: 20,000 cells with mLf or INDO+mLf, both not significantly different from the group, cells alone), a significant increase in GM-CSF was obtained regardless of whether indomethacin was present or not. Similar results were also obtained for adherent peritoneal cells (results not shown). No carry-over effect of Lf from the preincubation to the assay system was observed.

TABLE 1. EFFECT OF LACTOFERRIN ON GM-CSF PRODUCTION BY BONE MARROW CELLS IN VITRO

GROUP	CELL CONCENTRATION		
	20,000	100,000	500,000
Cells alone	32.5+-7.5	42.5+-7.5	47.5+-2.5
Cells + INDO	70.0+-30.0	65.0+-10.0	107.5+-2.5
Cells + mLf	25.0+-5.0	55.0+-15.0	90.0+-10.0
Cells + INDO + mLf	30.0+-7.3	110.0+-20.8	140.0+-18.3
Cells + hLf	60.0+-16.3	65.0+-9.6	90.0+-5.8
Cells + INDO + hLf	90.0+-6.5	60.0+-6.3	150.0+-37.0

The experiment was performed 3 times. These results show one representative experiment. The results are given as mean GM-CFC/10^5 (performed in quadruplicate)+-1 standard deviation. INDO=1uM indomethacin.mLf=mouse lactoferrin(0.1uM).hLf=human lactoferrin (0.1uM). Cells incubated for 48 h. Supernatants added at 10% v/v to assays.

Table 2 shows the in vivo effect of Lf on GM-CFC and GM-CSF in mouse bone marrow and plasma respectively. Compared to controls, mice treated with Lf showed an increase in GM-CFC after 48h whereas GM-CSF plasma levels were increased at 12 h and decreased to normal values by 48 h. No change in GM-CFC numbers were observed in the pleen. No evidence of inhibition was observed.

DISCUSSION

Lactoferrin is present in the secondary granules of neutrophils and released from them when the cell dies or in increased amounts during a stress situation, e.g.bacterial infection. Under these circum-

TABLE 2. IN VIVO EFFECT OF ENDOTOXIN-FREE, MOUSE LACTOFERRIN ON BONE MARROW GM-CFC AND PLASMA GM-CSF

	GROUP	GM-CFC/ORGAN	PLASMA GM-CSF
Controls	1	2239 +- 165	50 +- 13
	2	2454 +- 338	35 +- 17
	3	1923 +- 174	67 +- 9
Lf-treated	1	2389 +- 291	115 +- 16
	2	2604 +- 278	73 +- 9
	3	3120 +- 246	65 +- 17

Absolute GM-CFC/organ +-1 standard deviation. Plasma GM-CSF levels (2% v/v) measured as stimulation of $GM-CFC/10^5$ +- 1 standard deviation. Lactoferrin (Lf) injected at a concentration of 2mg/injection as follows: group 1 at 12h; group 2 at 12h and 24h; group 3 at 12h, 24h and 48h. Controls received 0.25ml of 0.9% NaCl.

stances, and taking into account the short neutrophil life span, it follows that rather than an inhibition, a stimulation is necessary in order to maintain neutrophil homeostasis. Several clinical findings would correlate with such a view. The released apo-Lf binds to iron present in the plasma and thus acts as an natural anti-bacterial agent, removing iron necessary for bacterial proliferation (Van Snick et.al. 1974; Nemet & Simonovits, 1985; Sawatzki, unpublished results). The iron-bound Lf then binds to receptors on macrophages (Birgens et.al. 1983; Rich & Sawatzki, unpublished results) which are responsible for GM-CSF production (Rich, 1986).

The in vitro results reported here clearly demonstrate that endotoxin-free (Lf preparations not purified over a polymyxin B-Separose column showed the same effect as those which were purified), species-specific, iron-saturated Lf produces no inhibition of GM-CSF whatsoever even in the presence of indomethacin and regardless of whether mouse or human Lf was used. Furthermore, in most cases, the release of GM-CSF increased with increasing cell concentration in vitro, indicating that cell crowding did not play a part in this phenomenon (cf. Fletcher & Willars, 1986). Similarly, the in vivo administration of Lf, at concentrations in the same order of magnitude as that released during a bacterial infection, resulted not in a decrease but an increase in plasma GM-CSF levels within 12h leading to a significant GM-CFC increase after 48 h.

Such results can be favorably correlated with previous investigations performed by Quesenberry et.al. (1972). This and other groups correlated serum CSF levels with blood neutrophil numbers under various stress conditions and found that the serum CSF levels were inversely proportional to the neutrophil count. It is also known that the neutrophil turnover is proportional to the Lf concentration (Hansen et.al. 1984). We therefore consider release of Lf from neutrophils as a "demand signal" which is sensed by the macrophage and in turn respond by producing CSF the presence of which stimulates GM-CSF to produce sufficient neutrophils in order to maintain homeostasis.

REFERENCES

Birgens,H.S.,Hansen,N.E.,Karle,B. & Kristensen,L.O.(1983):Receptor binding of lactoferrin by human monocytes. British Journal of Haematology, 54, 383-391.

Broxmeyer,H.E. & Platzer,E.(1984):Lactoferrin acts on I-a and I-e/c antigen subpopulations of mouse peritoneal macrophages in the absence of T lymphocytes and other cell types to inhibit production of granulocyte-macrophage colony stimulatory factors in vitro. J of Immunol, 133, 306-314.

Chilcote, R.R., Rierden, W.J. & Baehner, R.L. (1983):Neutropenia, recurrent bacterial infections, and congenital deafness in patients with monocytopenia. Amer J Dis Children, 137, 964- 967.

Fletcher J. & Willars J. (1986):The role of lactoferrin released by phagocytosing neutrophils in the regulation of colony stimulating activity production by human mononuclear cells. Blood Cells 11, 447-454.

Gailraith P.R. (1986):Effects of lactoferrin on human granulopoiesis in vitro. Clin Invest Med 9, 1-5.

Gutteberg,T.J.,Haneberg,B. & Jorgensen,T.(1984):Lactoferrin in relation to acute phase proteins in sera from newborn infants with severe infections. Eur J Pediat, 142, 37-39.

Hansen,N.E.,Malmquist,J. & Thorell,J.(1975):Plasma myeloperoxidase and lactoferrin measured by radioimmunoassay: relations to neutrophil kinetics. Acta Med Scand, 198, 437-443.

Issekutz,A.C.(1983):Removal of Gram-negative endotoxin from solutions by affinity chromatography. J Immunol Meth, 61, 275-281.

Nemet,K. & Simonovits,I.(1985):The biological role of lactoferrin. Haematologia, 18, 3-12.

Poppas A, Faith M.R. & Bierman H.R. (1986):In vivo study of lactoferrin and murine rebound myelopoiesis. Am J Hematol. 22, 1-8.

Olsson,I.,Olofsson,T.,Ohlsson,K. & Gustavsson,A.(1979):Serum and plasma myeloperoxidase, elastase and lactoferrin content in acute myeloid leukaemia. Scand J Haemat, 22, 397-406.

Quesenberry,P.,Morley,A.,Stohlman,JR.,F.,Rickard,K.,Howard,D & Smith,M.(1972):Effect of endotoxin on granulopoiesis and colony-stimulating factor. New Eng J Med 286, 227-232.

Rich I.N. (1986): A role for the macrophage in normal hemopoiesis. I. Functional capacity of bone marrow-derived macrophages to release hemopoietic growth factors. Exp Hematol. 14, 738-745.

Sawatzki,G. & Kubanek, B.(1983):Isolation and ELISA of mouse and human lactoferrin, In: Structure and Function of Iron Storage and Transport Proteins (eds.: I. Urushizaki et al.):Elsevier Science Publishers B.V., p. 441-443

Van Snick,J.L., Masson,P.L. & Heremans,J.F.(1974):The involvement of lactoferrin in the hyposideremia of acute inflamation. J Exp Med, 140, 1068-1084.

Venge,P.,Foucard,T.,Henriksen,J.,Hakanssen,L. & Kreuger,A. (1984): Serum-levels of lactoferrin, lysozyme and myeloperoxidase in normal, infection-prone and leukemic children. Clin Chim Acta, 136, 121-130.

Winton,E.F.,Kinkade,JR.,J.M.,Vogler,W.R.,Parker,M.B. & Barnes, C.B. (1981):In vitro studies of lactoferrin and murine granulopoiesis. Blood, 57, 574-578.

Lactoferrin inhibits Interleukin-2 production in mixed lymphocyte culture

Kevin Slater and John Fletcher

Medical Research Centre, City Hospital, Nottingham, NG5 1PB, England

---

KEY WORDS

Lactoferrin, Interleukin 2, Mixed Lymphocyte Culture.

Lactoferrin (Lf) is contained in the secondary granules of neutrophils and is released when the cells are stimulated. A number of activities have been attributed to Lf including supression of antibody production and inhibition of Colony Stimulating Factor production.

These suggest an interaction between Lf and mononuclear cells involving control of protein synthesis; this could possibly occur by inhibition of a common intermediate. We have investigated this hypothesis using allogeneic mixed lymphocyte cultures (MLCs) which allow the measurement of growth factors in the supernatants in the absence of exogenous stimulatory substances.

We have previously reported (Slater and Fletcher, 87) that greater transcriptional activity (assessed by incorporation of tritiated uridine (^{3}H-UdR)) occurs in MLCs when the cells are crowded together in round-bottomed wells (RBWs) than when spread in flat-bottomed wells (FBWs). We have shown that a factor released from phagocytosing neutrophils inhibits the increased transcriptional activity of MLCs grown in RBWs. However, no inhibition could be achieved in FBWs. We have identified this neutrophil derived inhibitory factor as Lf and demonstrated that iron is required to activate the molecule.

The observation that Lf was only effective as an inhibitor in RBWs suggests that it acts by inhibition of a factor produced in response to cell crowding. Transfer of supernatants produced in RBWs to MLCs in FBWs increased ^{3}H-UdR incorporation into the latter.

However, with Lf present in the RBW cultures, less transcriptional activity could be detected when the crowded culture supernatants were subsequently added to FBW cultures. Thus it appears that a factor produced in response to cell crowding and able to stimulate spread cells, is inhibited by Lf.

Measurement of Interleukin-2 (Il-2) in the crowded MLC supernatants revealed that Lf inhibits the production of this lymphokine. Dose response curves for the inhibition of ^{3}H-UdR incorporation into MLCs by Lf and of Il-2 production by these MLCs were identical, with significant inhibition occurring down to 10^{-12} M Lf. We therefore suggest that Il-2 is the factor responsible for the previous observations. The inhibition of Il-2 production by Lf could also contribute to other cellular responses to Lf.

REFERENCES

Slater, K and Fletcher, J. (1987): Lactoferrin and the Mixed Lymphocyte Reaction. Blood (In Press).

Lymphocyte mediated suppression of granulopoiesis

Treen Carson Michael Morris, *Alexandra Elizabeth Irvine and *Anne French

Department of Haematology, Belfast City Hospital, Belfast, BT9 7AB, Northern Ireland.
* Northern Ireland Leukaemia Research Laboratory, Royal Victoria Hospital, Belfast, BT12 6BL, Northern Ireland

KEY WORDS

Granulocyte macrophage, colony forming cells (CFU-GM), inhibition, lymphocytes, aplastic anaemia, neutropenia.

INTRODUCTION

Since the introduction of a semi-solid agar culture system for the growth of granulocyte macrophage precursors (CFU-GM) in the mid 60's most workers have concentrated on in vitro stimulatory factors as putative regulators of granulopoiesis in vivo. Although stimulatory factors are generally more easily studied than inhibitors we have developed a technique for demonstrating the granulopoiesis regulatory activity of lymphocytes and have studied this negative regulatory factor in a number of clinical situations.

MATERIALS AND METHODS

Lymphocytes and normal bone marrow cells were prepared and co-cultured in semi-solid agar as previously described (Morris et al, 1980). Lymphocytes were routinely co-cultured with normal marrow in the upper layer at a concentration of 1×10^5 cells /ml; on each occasion lymphocytes were prepared from both normal and patient samples and paired tests performed.

RESULTS

Lymphocytes from normal individuals will inhibit the growth of CFU-GM in a dose dependent manner when co-cultured with allogeneic or autologous normal marrow cells (Morris et al, 1980). In a recent series of 57 experiments the level of inhibition produced by co-culturing 1×10^5 normal lymphocytes with 1×10^5 normal marrow cells was 43% $\pm$ 16% (Mean $\pm$ SD).

We have studied 24 patients with aplastic anaemia and co-cultured their lymphocytes with normal bone marrow cells (Morris et al, 1984; Ennis et al, 1987). In each case a paired normal control was cultured with the same normal bone marrow, the results are summarised in Fig. I. Six patients had lymphocytes which were more inhibitory than normal controls suggesting a lymphocyte mediated aetiology in their disease. In contrast 12 patients had lymphocytes which were significantly less inhibitory than normal.

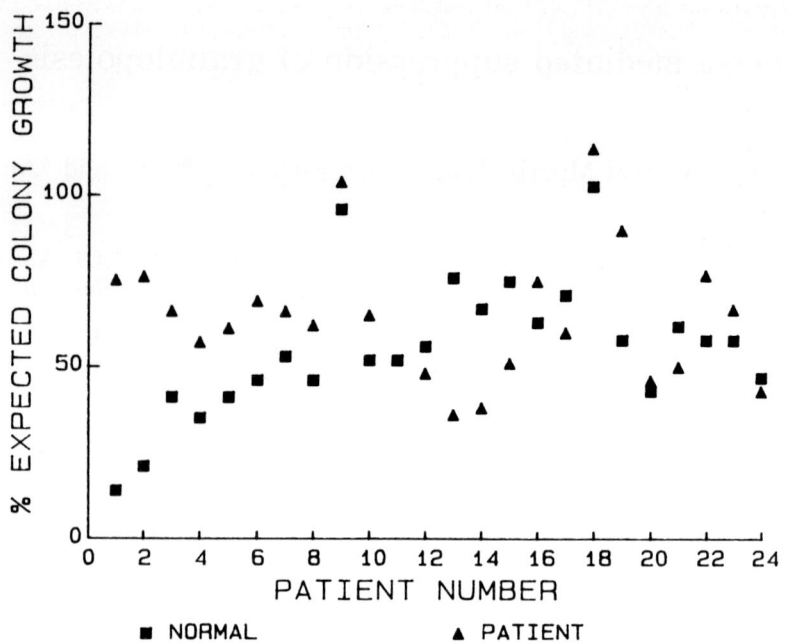

Fig. I: Shows the results of paired co-cultures of normal or aplastic patients lymphocytes with normal marrow. The results of co-culture of the normal marrow with normal lymphocytes are shown as squares while the results of co-culture of the aplastic patients lymphocytes and normal marrow are shown as triangles.

We have also studied 20 patients with neutropenia not due to aplastic anaemia. Three patients showed inhibition of normal marrow colony formation by their lymphocytes significantly greater than that produced by normal lymphocytes. When tested against their own marrow, 3 patients also showed inhibition of their own marrow cells which was greater than that produced by normal lymphocytes. However, only 2 patients had lymphocytes which inhibited both their own and normal marrow significantly. Both of these patients had clinical features of autoimmune disorders (Morris et al, submitted).

The role of lymphocytes in patients in whom neutropenia occurred as a consequence of marrow infiltration with leukaemia or other marrow replacing malignancies has also been investigated. These patients were selected because no leukaemic or other immature cells were observable by microscopy of their peripheral blood. Control cultures of these patients blood did not grow any colonies. It can be seen in Fig. II that all the patients except Patient 2 have lymphocytes which are less inhibitory than normal lymphocytes. In 5 of the patients this difference is statistically significant. In cases 4 and 6 stimulation of normal marrow growth is seen, but the reason for this is unknown.

The effect of lymphocytes from patients taking lithium in therapeutic doses on normal marrow colony formation has also been investigated (Irvine et al, 1986). In parallel with these studies we have also tested the effect of conditioned medium produced from non stimulated lymphocytes after 3 days culture from both patients on lithium and from a similar group of normal controls. Lymphocyte mediated inhibition of colony formation was abrogated in the patient group on testing both their own cells and the conditioned medium produced from their cells. Inhibition produced by conditioned medium from the normal controls was similar to that produced when lymphocytes were incorporated directly in co-culture (Fig. III). Granulocytosis as seen in patients taking lithium therapeutically may be related to the reduction in lymphocyte mediated inhibition of granulopoiesis.

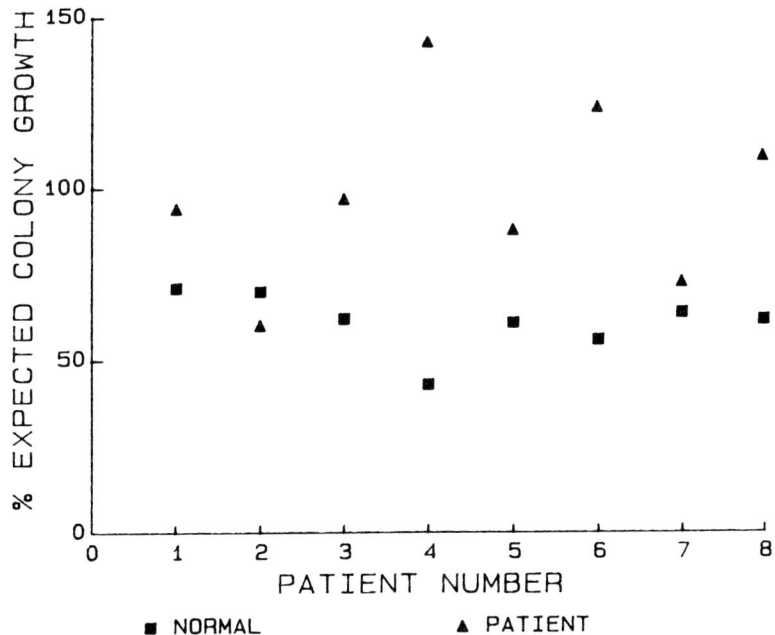

Fig. II: Shows the results of paired co-cultures of normal lymphocytes or lymphocytes from patients with neutropenia due to marrow replacing disorders. The results of co-culture of normal donors marrow with normal lymphocytes are shown as squares, while the results of co-cultures of the neutropenic patients with normal marrow are shown as triangles. Lymphocytes from Patients 3, 4, 6, 7, and 8 are significantly less inhibitory than normal controls.

DISCUSSION

We have repeatedly shown that normal lymphocytes are consistently capable of reducing the in vitro growth of CFU-GM colonies when co-cultured with normal marrow (homologous or autologous). This inhibition is highly reproducible and can be mediated by medium conditioned by non stimulated lymphocytes harvested after 3 days. We suggest that this inhibition may be of both physiological and pathological significance. In the studies of patients with neutropenia due to either aplastic anaemia or other causes a small proportion of patients (25% in the aplastic group, 10% in the neutropenic group showed inhibition which was significantly greater than that caused by normal lymphocytes. These cases with significantly increased in vitro inhibition of CFU-GM by lymphocytes may demonstrate the in vivo pathophysiology of their disease. In contrast lymphocytes from half of the patients with aplastic anaemia exerted little or no inhibitory effect on normal marrow. This may represent the abrogation of a negative controlling factor in a situation where inadequate numbers of neutrophils are being produced due to other causes. Data from the patients with marrow replacement appear to support this concept as 5 out of 8 patients had lymphocytes which were significantly less inhibitory than paired normal controls. In these patients granulopoiesis was seen to be embarrassed physically by the marrow infiltrating cells. We feel these clinical studies support or view that lymphocytes may play a physiological role in the regulation of granulopoiesis and that pathological conditions may arise as a consequence of the abnormalities of this system.

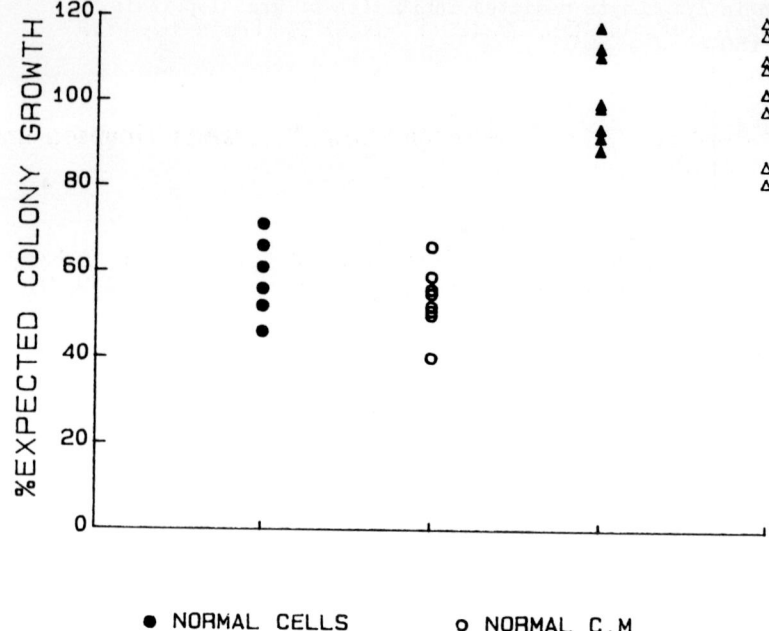

● NORMAL CELLS   ○ NORMAL C.M
▲ PATIENT CELLS   △ PATIENT C.M

Fig. III: Comparison of the effect of lymphocytes (closed symbols) and conditioned medium from unstimulated lymphocytes (open symbols) from normals (circles) and patients taking lithium (triangles). No difference is seen between the inhibition produced by lymphocytes or conditioned medium from the same group but there is a significant abrogation of lymphocyte mediated inhibition in the patients taking lithium.

REFERENCES

Ennis, K.T., Morris, T.C.M., Irvine, A.E., Alexander, H.D. (1987): T-lymphocyte colony formation by lymphocytes from patients with aplastic anaemia. Natural Immunity and Cell Growth Regulation, 6, 45-55.
Irvine, A.E., Crockard, A.D., Desai, Z.R., Ennis, K.T., Fay, A.C., Morris, T.C.M., Bridges, J.M. (1986): Lymphocytes from patients receiving lithium do not inhibit CFU-C growth. Br J Haemat. 62, 467-477.
Morris, T.C.M., Vincent, P.C., Sutherland, R., Heasey, P. (1980): Inhibition of normal human granulopoiesis in vitro by non B non T lymphocytes. Br J Haemat. 45, 541.
Morris, T.C.M., Vincent, P.C., Young, G.A.R., Sutherland, R., Forrest, P.R., Isbister, J.P. (1984): CFU-C inhibitors in aplastic anaemia. Blut, 48, 61-74.

Autoimmune neutropenia and haemolytic anaemia

Alexandra Elizabeth Irvine, * Veronica Craig, * Amanda Thomson and * Treen Carson Michael Morris

*Northern Ireland Leukaemia Research Fund, Royal Victoria Hospital, Belfast, BT12 6BL, Northern Ireland. * Department of Haematology, Belfast City Hospital, Belfast, BT9 7AB, Northern Ireland*

KEY WORDS

Neutropenia, haemolytic anaemia, CFU-GM, lymphocyte inhibitors, serum inhibitors.

INTRODUCTION

We have previously shown that normal lymphocytes inhibit normal bone marrow granulocyte macrophage (GM) colony formation (Morris et al, 1980) and this inhibition may be increased or decreased in a variety of conditions (Morris et al, 1984; Irvine et al, 1986). We describe a patient who presented with severe neutropenia and recurrent infections, who was also found to have a Coombs positive haemolytic anaemia.

MATERIALS AND METHODS

Lymphocytes and normal bone marrow cells were prepared and co-cultured in semi-solid agar as previously described (Morris et al, 1980). Lymphocytes were routinely co-cultured with normal marrow in the upper layer at a concentration of 1×10^5 cells/ml; on each occasion lymphocytes were prepared from both normal and patient samples and paired tests performed.

CASE REPORT

A 54 year old female with a 3 month history of recurrent abscesses refractory to antibiotic therapy was referred for investigation of neutropenia. She gave a history of tiredness, general malaise, fever, moderate night sweats and occasional rigors during the past 3 months and she also complained of low back pain which had been present for several years, together with chronic constipation. She had received various antibiotics and had been taking analgesics for her back pain. There were no findings of significance in her past history. On examination she was noted to be mildly icteric and anaemic with minimal pyrexia. She had marked pharyngitis and a perianal abscess. Systematic investigation was otherwise normal.

On admission her Hb was 10.1g/dl, leucocytes 2.0×10^9/l, neutrophils 0.3×10^9/l, platelets 429×10^9/l. Hb, leucocyte count and neutrophil count are shown in Fig 1

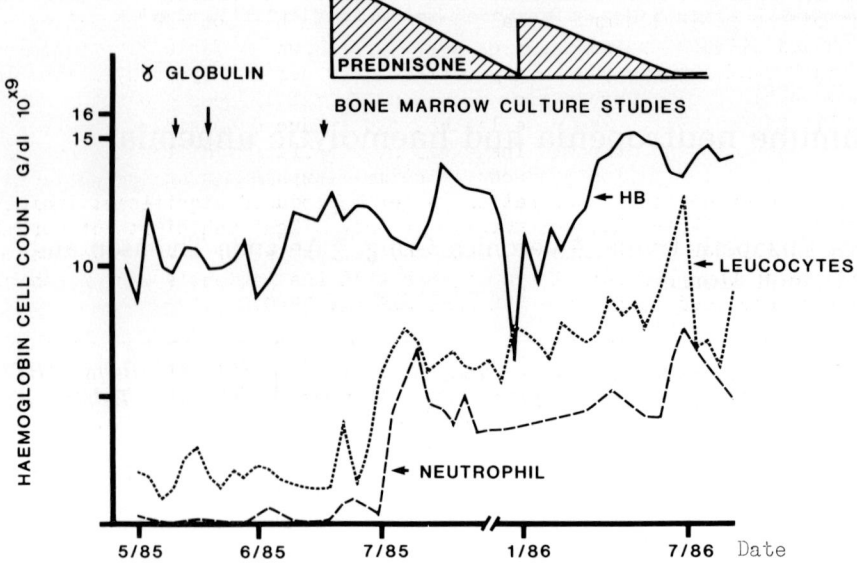

Fig. 1: Haematological Reponse to Gamma Globulin and Corticosteroids.

together with treatment details. Her platelet count remained normal throughout. Other investigations showed reticulocyte count 8%. Her Direct Coombs test was positive to IgG only. Her serum was shown to contain an auto-antibody which appeared to show no specificity but an ether eluate prepared from her cells showed an Anti E specificity. Sucrose lysis test was negative, urinary haemosiderin was not detected. The remainder of her biochemical parameters were essentially normal apart from a slight polyclonal increase in gamma globulin. Her red cell survival studies carried out later in the course of her disease showed a mean cell life of approximately 35 days.

At admission she required a blood transfusion and intravenous antibiotics. Infective lesions were slow to resolve and there was reluctance to treat her with steroids because of these lesions. Instead she was given a course of intravenous gammaglobulin 21 gm/day for 5 days with piriton and hydrocortisone cover, this can be seen in Fig. 1. There was a modest reponse to this therapy and this response was sufficient to promote resolution of her septic lesions. When her neutrophil count had dropped again it was decided to treat her with corticosteroids despite further infective complications there was a gradual improvement in her neutrophil count and this subsequently returned to normal. The steroids were gradually withdrawn but this withdrawal ended in an acute exacerbation of her haemolytic anaemia. Steroids were increased with a prompt decrease in haemolysis and were withdrawn slowly over the course of the next 6 months. During this time she maintained Hb, leucocyte and neutrophil counts at normal levels and eventually all steroid therapy was ceased without relapse of her disease. There has been no further evidence of disease activity over the following 9 months.

RESULTS

The patient's marrow grew normal numbers of colonies on both occasions cultured. At diagnosis there were 101 ± 9 colonies per 1×10^7 bone marrow cells, when grown 20 days later the number was 63 ± 7. Both these results are within our normal range 20-115.

The effect of the patient's and normal lymphocytes on her own and normal marrow CFU-GM with and without the addition of serum are shown in Table 1. It can be seen that her own CFU-GM were only modestly inhibited by her own lymphocytes and to a greater degree by normal lymphocytes, although the differences are not statistically significant. The same lymphocytes produced considerably greater inhibition of normal bone marrow. There was no significant difference between the lymphocyte mediated inhibition produced by normal lymphocytes or the patients lymphocytes. The addition of the patient's serum produced significant inhibition of her own marrow (35% inhibition P=0.01) but only slight inhibition of normal marrow (12% inhibition NS). However, when the patient's serum was added to co-culture of the patient's abnormal lymphocytes with the patient's marrow inhibition of CFU-GM was increased to 59% and 57% respectively (P=0.01).

Table 1 Study At Diagnosis	Marrow Alone	+Patient's NAM	+Normal NAM	+Patient's Serum	+Patient's NAM +Patient's Serum	+Normal NAM +Patient's Serum
Patient's Bone Marrow %Inhibition	101±1	93±19 (8)	80±5 (21)	66±4 (35)	41±6 (59)	44±10 (57)
Normal Bone Marrow %Inhibition	62±5	30±8 (52)	22±4 (64)	54±3 (12)	Not Determined	Not Determined

These experiments were repeated 20 days later after antilymphocyte globulin treatment and before steroid therapy, (Table 2). On this occasion the patient's lymphocytes exhibited greater inhibition of autologous GM colony formation than allogeneic GM colony formation (54% v 28% inhibition respectively P=0.03). Incorporating the patient's serum into co-cultures of her bone marrow and lymphocytes again caused 77% inhibition of colony formation (P=0.005), although a control culture of normal marrow with the patient's serum and lymphocytes showed only 6% inhibition.

Table 2 Study 20 Days later	Marrow Alone	+Patient's NAM	+Normal NAM	+Patient's Serum	+Patient's NAM +Patient's Serum	+Normal NAM +Patient's Serum
Patient's Bone Marrow %Inhibition	63±7	29±1 (54)	45±4 (28)	54±5 (14)	14±4 (77)	45±6 (28)
Normal Bone Marrow %Inhibition	35±4	25±3 (29)	26±7 (26)	Not Determined	Not Determined	38±9 (0)

DISCUSSION

Autoimmune disorders resulting in destruction of red blood cells or platelets or both are well recognised, but well documented cases where granulocytes are the target of the autoimmune process are only occasionally documented, although there is evidence of an autoimmune aetiology in at least some cases with pancytopenia such as aplastic anaemia (Camitta et al, 1982; Suda et al, 1981) and Felty's syndrome (Abdou, 1983). The combination of autoimmune haemolytic anaemia and neutropenia without thrombocytopenia appears to be highly unusual and we have been unable to identify cases similar to the present study.

It is therefore worthwhile to review the evidence for an autoimmune aetiology for both the haemolytic anaemia and the neutropenia. A clear cut anti erythrocyte antibody with anti E specificity was identified in the patient's serum and her red blood cells gave a positive Direct Coombs test. There was significant shortening of her mean red blood cell survival and she responded to steroid therapy. The evidence for autoimmune neutropenia comes principally from the CFU-GM studies. In the first study an inhibitor is clearly present in the patient's serum which was more active against her own CFU-GM than against allogeneic CFU-GM. Her lymphocytes were not more inhibitory than normal lymphocytes when tested against her own or normal CFU-GM. The difference in the level of inhibition of the patient's CFU-GM and the normal CFU-GM is thought to be a property of the target cell population and not the added cells (Morris et al, 1980). An additive effect was noted when the patient's serum was incorporated in co-cultures of the patient's CFU-GM with normal or autologous lymphocytes.

The second study performed after intravenous gamma globulin therapy but prior to steroid therapy shows the serum inhibitor was still present, although not so potent as the initial study. However, on this occasion the patient's lymphocytes were significantly more inhibitory than normal lymphocytes to the patients own CFU-GM but not normal CFU-GM. In addition the combination of the patient's serum and lymphocytes was highly inhibitory to her own CFU-GM but not to the normal control. A further comparison using normal lymphocytes plus the patient's serum showed a lesser degree of inhibition of the patient's CFU-GM but none of normal. Additional points in favour of an autoimmune aetiology were the transitory rise in neutrophils following intravenous gamma globulin therapy and sustained response to steroid therapy.

We have previously demonstrated cases in which lymphocyte inhibition was responsible for pancytopenia with an aplastic marrow (Morris et al, 1984; Irvine et al, 1986) or a cellular marrow (Morris et al, submitted). We have also demonstrated serum inhibitors in cases with pancytopenia and aplastic anaemia (Morris et al, 1984) or cellular marrow (Morris et al, submitted) as have others (Winkel et al, 1982). This case appears to be unique in terms of the additive effect of the lymphocytes and serum directed specifically against the patient's CFU-GM and the combination of autoimmune neutropenia with autoimmune haemolytic anaemia.

REFERENCES

Abdou, N.I. (1983):Heterogeneity of bone marrow-directed immune mechanisms in the pathogenesis of neutropenia of Felty's syndrome. Arthritis and Rheumatism, 26, 947-953.
Camitta, B.M., Storb, R., Thomas, E.D. (1982): Aplastic anaemia. N Eng J Med. 11, 645-652.
Irvine, A.E., Crockard, A.D., Desai, Z.R., Ennis, K.T., Fay, A.C., Morris, T.C.M. (1986):Lymphocytes from patients receiving lithium do not inhibit CFU-C growth. Br J Haemat. 62, 467-477.
Morris, T.C.M., Vincent, P.C., Sutherland, R., Hersey, P. (1980):Inhibition of normal human granulopoiesis in vitro by non-B non-T lymphocytes. Br J Haemat. 45, 541-550.
Morris, T.C.M., Vincent, P.C., Young, G.A.R., Sutherland, R., Forrest, P.R., Isbister, J.P. (1984):CFU-C inhibitors in aplastic anaemia. Blut, 48, 61-74.
Suda, T., Mizugachi, H., Kiura, Y., Kubota, K., Takaku, T. (1981): Suppression of in vitro granulocyte-macrophage colony formation by the peripheral mononuclear phagocytic cells of patients with idiopathic aplastic anaemia. Br J Haemat. 47, 433-442.
Winkel, C.N., Goselink, H.M., Veenhof, W.F.J., Claas, F.H.K., Jansen, J. (1982): Serum inhibitors in aplastic anaemia. Blut, 44, 193-200.

Negative regulation of erythropoiesis and megakaryopoiesis

Inhibiteurs de l'érythropoïèse et de la mégacaryopoïèse

Properties of a protein NRP that negatively regulates DNA synthesis of the early erythropoietic progenitor cells BFU-E

Arthur A. Axelrad, Hélèna Croizat, Dario del Rizzo, Denise Eskinazi, Gabriella Pezzutti, Solomon Stewart and Henk Van der Gaag

Department of Anatomy, University of Toronto, Toronto, M5S 1A8 Canada

ABSTRACT

In the bone marrow of C57BL/6 (B6) mice, most BFU-E are in a quiescent state with respect to DNA synthesis, as evidenced by their invulnerability to ^{3}H-thymidine or hydroxyurea in vivo or in vitro. Washing of normal B6 marrow cells resulted in a dramatic increase in the proportion of BFU-E that were synthesizing DNA, but had no such effect on the proportion of DNA-synthesizing CFU-S, CFU-E, or Gm-CFC. Supernatant of washed B6 marrow cells rapidly reduced the proportion of BFU-E engaged in DNA synthesis. The factor in this supernatant, whose activity could be initiated at 0°C and reversed by washing, was shown to be macromolecular (by ultrafiltration), heat stable (56°C, 30 min), and trypsin-sensitive; we have called it Negative Regulatory Protein or NRP. From the culture medium of a B6 marrow cell line, we have recovered a protein with properties indistinguishable from those of NRP. After 5 chromatographic steps, it eluted at an apparent enrichment of $>$100,000-fold. NRP bound to wheat germ lectin and could be eluted with N-acetyl glucosamine, but it did not bind to Con A or soybean lectin. Based on its behaviour on ion exchange, gel filtration, and lectin columns, we conclude that at physiological pH NRP is a neutral glycoprotein of apparent molecular weight 79,000. We provide evidence that the level of NRP in murine marrow is under genetic, developmental, and physiological control. Our data are consistent with the hypothesis that in B6 marrow, quiescence of the BFU-E with respect to DNA synthesis is not a passive state; it must be continually maintained by interactions at the surface of these progenitor cells with erythroid lineage-specific Negative Regulatory Protein molecules, which appear to be competing with lineage non-specific, but positively acting growth factors such as Il-3, Insulin-like Growth Factors, and Transferrin.

Key Words: ^{3}H-thymidine, hydroxyurea, C57BL/6, Fv-2, Interleukin-3

INTRODUCTION

History
A wise old Frenchman once gave this advice on living: 'Il faut toujours être en train de faire quelquechose d'impossible.' 'One should always be engaged in trying to do something that is impossible.' When we began the work I am about to describe to you in 1965, we had no idea how close we would come to fulfilling the advice of that wise old Frenchman.

We began the work in order to find out how it is that an allele at a single gene locus can decide whether or not a mouse will develop leukemia if infected with the Friend Leukemia Virus (FV). [Charlotte Friend, incidentally, who was a friend to many of us here, died just 3 months ago, and I would like to dedicate this paper to her].

It was in fact our desire to understand the mechanism of action of this gene, which was later shown to be Fv-2 on chromosome 9 of the mouse (Lilly, 1970), that led to the identification of, and culture assay methods for the late erythroid progenitor cell CFU-E (Stephenson et al, 1971) and the early erythroid progenitor cell BFU-E (Axelrad et al, 1974).

While trying to discover what effect FV had on the cell cycle status of the BFU-E, Dr. Shigetoh Suzuki found among his uninfected controls that in FV-resistant mice of the B6 strain of genotype $Fv-2^{rr}$, most of the BFU-E were in a quiescent, non-DNA synthesizing state, as judged by ^{3}H-thymidine suicide experiments in vivo; in our congenic FV susceptible B6.S mice of genotype $Fv-2^{ss}$ (Axelrad, et al, 1972), a high proportion of the BFU-E were engaged in DNA synthesis. And this was true not only for our mice congenic at Fv-2 but also for other congenic partner strains, as well as other inbred strains differing at Fv-2. $Fv-2^{ss}$ was associated with high percentages of BFU-E killed, $Fv-2^{rr}$ with low percentages of BFU-E killed. Our conclusion was that cycling of BFU-E is controlled at Fv-2 or a closely linked locus on Chromosome 9 (Suzuki and Axelrad, 1980).

When marrow cells were removed from B6 mice and subjected to a single wash in physiological medium their BFU-E immediately became vulnerable to ^{3}H-thymidine or hydroxyurea (HU). Since these cycle-active agents operate by totally different mechanisms but have in common the ability to kill only those cells that are engaged in DNA synthesis, we concluded that washing of B6 bone marrow cells must have removed something that prevents BFU-E from engaging in DNA synthesis.

Bone marrow cells from B6.S and B6 mice were resuspended either in medium or in their own bone marrow supernatant or in supernatant from the marrow of their congenic partners. Aliquots of the cells were subjected to the conditions of ^{3}H-thymidine suicide and then cultured for 7 days. The results showed that when washed in medium, a high proportion of BFU-E of both genotypes went into DNA synthesis. When the cells were resuspended in bone marrow supernatants from B6 mice, the BFU-E showed low percentage kills, indicating that there was something in B6 marrow supernatant that could prevent BFU-E from engaging in DNA synthesis. Moreover it could do this to BFU-E of both genotypes (Axelrad et al, 1981).

The search for what was in that B6 marrow supernatant has occupied us over the past 6 years. In the course of this work, we found that the activity in B6 marrow supernatant that reduced the proportion of DNA-synthesizing BFU-E was a trypsin-sensitive, heat-stable macromolecule (Axelrad et al, 1983). We concluded that it is a protein and named it Negative Regulatory Protein or NRP. In what follows, we shall review information on NRP that is already in the literature, add new data, and speculate on the mechanism of action and the significance of this molecule for the physiological regulation of erythropoiesis.

BIOLOGICAL PROPERTIES OF NRP

Genetic control of NRP activity
Bone marrow supernatants from B6.S mice did not reduce the proportion of BFU-E engaged in DNA synthesis. Therefore whether the BFU-E engaged in DNA synthesis or not was determined not within the BFU-E itself but by an activity present in bone marrow supernatant of B6 ($Fv-2^{rr}$) but not B6.S ($Fv-2^{ss}$) mice. Thus the Fv-2 locus (or a closely linked gene on chromosome 9) must control the activity in B6 bone marrow supernatants that shuts down DNA synthesis of the BFU-E; it does not control the response of the BFU-E to this activity. No such activity was detected in the marrow of (B6 x B6.S)F_1 hybrid mice. Thus the NRP activity in B6 marrow behaved as if it was recessively inherited (Axelrad et al, 1981b).

Reversibility of NRP action
The effect of B6 marrow supernatant on DNA synthesis of BFU-E was rapidly reversible (Axelrad et al, 1982). B6 marrow cells were maintained in B6 marrow supernatant where the percentage of BFU-E killed by ^{3}H-thymidine was kept at a low level (0-10 percent). At various times up to 3 hours, the cells were subjected to a single wash in alpha medium. We found that within the time it took to do the assay (20 minutes), the proportion of BFU-E killed had already risen to levels comparable to those of untreated control B6 marrow cells washed in alpha medium (40-50%). The level of percentage BFU-E killed by ^{3}H-thymidine that they reached after washing was not influenced by the length of time during which the cells had been maintained in B6 marrow supernatant. Thus the BFU-E behaved as if they were returning to the proportion after washing that would have been engaged in DNA synthesis had they not been prevented from doing so by the activity in B6 marrow supernatant. There was thus no evidence of accumulation of BFU-E at any point in the cell cycle during their exposure to NRP.

Specificity of NRP action
The activity in B6 marrow supernatant appeared to be specific for one particular target cell population in the hemopoietic system – the BFU-E. Washing of a marrow cell suspension had no significant effect on DNA synthesis of CFU-S, CFU-E, or GM-CFC. Only the DNA synthesis of BFU-E was affected by the washing (Axelrad et al, 1981a). We now wished to test the effect of adding B6 bone marrow supernatant to bone marrow suspensions in order to find out whether this would reduce the proportion of CFU-S already engaged in DNA synthesis. To do this it was first necessary to obtain CFU-S that were actively engaged in DNA synthesis. In normal marrow of B6 mice, very low proportions of CFU-S are in DNA synthesis (Blackett et al, 1974; Suzuki & Axelrad, 1980; Van Zant et al, 1983). A variety of attempts were made to increase this proportion. Treatment of donors with phenylhydrazine or bleeding

gave variable results, but HU 12 hours before, or total body radiation with 150 rads ^{137}Cs γ-rays 24 hours before marrow harvest proved to give high enough percentage kills with HU or ^{3}H-thymidine to be useful. We therefore carried out a series of experiments designed to test whether B6 marrow preparations known to reduce the proportion of BFU-E synthesizing DNA would also reduce the proportion of CFU-S engaged in DNA synthesis as judged by a reduction in percentage of CFU-S killed. Table 1 shows the results. B6 marrow supernatant Table 1.

B6 marrow supernatant fails to stop DNA synthesis of CFU-S*

	No. of colonies per spleen			% of
	-Hydroxyurea	+Hydroxyurea	P	CFU-S killed
Alpha medium	11.4 ± 1.14	5.8 ± 0.63	<0.001	49.1
B6 marrow supernatant	13.1 ± 0.95	5.2 ± 0.76	<0.001	60.3

*Donor CFU-S DNA synthesis was activated by total body γ-radiation (150 rads) 24 hrs before transplantation.

had no significant effect on the proportion of CFU-S killed by the cycle-active agent. Further evidence was produced by Dr. Helena Croizat who obtained from Dr. M. Guigon a sample of material known to be active in reducing the proportion of CFU-S in DNA synthesis and tried it in our system on BFU-E. It had no effect. Taken together, all these results provide strong evidence that the action of NRP is specific for the BFU-E.

Where in the BFU-E cell cycle does NRP act?
When we exposed bone marrow cells continuously to ^{3}H-thymidine (Wu, 1981) while in alpha medium we found that the BFU-E entered DNA synthesis and were killed at a rate that averaged ∼13% per hour. In the presence of B6 marrow supernatant, the proportion of BFU-E engaged in DNA synthesis was reduced, as if a number of these cells had from the start been eliminated from participation in DNA synthesis (Axelrad et al, 1983). The cells responding to the NRP activity in B6 marrow supernatant thus appeared to be marrow BFU-E in S-phase and having average cycle times of around 8 hours. Thus it is a very rapidly cycling cell, as mammalian cells go, that acts as target for the NRP in B6 bone marrow supernatants. Since the instantaneous kill by ^{3}H-thymidine or HU was 40%, S phase lasts a little over 3 hr in this cell.

The higher the concentration of NRP, the lower was the proportion of BFU-E that remained capable of undergoing DNA synthesis. Because the assay method we used to carry out this investigation provided a window through which we were able to see events that took place only during the S-phase of the cell cycle, it was impossible for us to tell whether NRP acts only on BFU-E actually engaged in DNA synthesis to arrest that process, or whether it acts on cells at all phases of the cell cycle to stop the process of cycling itself. No physiological precedent for the latter mechanism exists, to our knowledge, but there are at least 2 pharmacological precedents (Bruchovsky et al, 1967; Cormack, 1966).

When the arrest of BFU-E DNA synthesis was reversed by washing, it did not appear as if the arrest had caused accumulation of BFU-E at any single point in the cell cycle. After washing, the proportion of DNA synthesizing BFU-E promptly rose to around 40 percent irrespective of the length of time during which they had been maintained at around 0-10 percent in the presence of NRP. This fact, together with the rapidity with which DNA synthesis could be switched off and on argues that NRP does not induce a 'G_o-like state' comparable with that extensively studied in vitro in cell lines that have been deprived of serum, for example, in which the cells take many hours to reach S-phase (Pardee, 1974). BFU-E appeared to be actively cycling when freed of NRP. They thus did not seem to require 'competence factors' (Pledger et al, 1977) to activate them into cycle. Whether or not positively acting factors played any role could not be decided from these experiments.

MOLECULAR PROPERTIES OF NRP

Methodology
It is a truism to say that it is much more difficult to establish the validity of an activity that is specifically inhibitory for cell proliferation than of one that is stimulatory for cell proliferation. Many influences can impair cell proliferation in a non-specific manner, and so the onus of proof rests on those who claim specificity for any growth inhibitory activity. We have defined negative regulatory activity operationally in the present context as activity that prevents an expected reduction in number of erythropoietic bursts that would otherwise occur as a result of exposure of hemopoietic cells to a cytotoxic agent known to kill only those cells that are engaged in DNA synthesis.

Comparison of percentages of BFU-E killed are frought with danger because each percentage is not an independent value but a calculated ratio of two means, each having its own variance. Percentage data from individual assays in routine experiments employing small numbers of samples have large variances (Quesenberry and Stanley, 1980; Liestøl and Benestad, 1986). Relatively large numbers of experiments are thus required for statistical validity. These considerations render impractical the use of such an approach for following purification of a putative negative regulator. To assess negative regulatory activity, we have therefore used simple comparisons of group means for burst numbers rather than percentages of BFU-E killed.

Three criteria had to be fulfilled in order for us to accept data as reliable evidence for the presence of negative regulatory activity: 1) the mean number of bursts in the group treated with the cytotoxic agent but not with the putative negative regulator (B) must be significantly reduced as compared to that of the group treated with neither the cytotoxic agent nor the putative negative regulator (A); 2) The mean number of bursts in the group treated with both the putative negative regulator and the cytotoxic agent (D) must not be significantly different from the mean number of bursts in the control group treated with the putative negative regulator but untreated with the cytotoxic agent (C); 3) the mean number of bursts in the groups untreated by the cytotoxic agent (A) must not be significantly affected by treatment with the putative negative regulator (C).

Table 2 Example to show criteria for negative regulatory activity

	-HU		+HU	
alpha medium	(A) 30 ± 5	vs.	(B) 16 ± 2	→ significant difference
	vs.			
NRP	(C) 28 ± 3	vs.	(D) 25 ± 3	→ no significant difference
	↓			
	no significant difference			

Such an approach provides statistically reliable data but it indicates only
whether or not negative regulatory activity is present at the concentration of
the agent being assayed. It thus essentially provides an end-point dilution
method for the assay of negative regulatory activity.

When we now report mean percentages of BFU-E killed we present these data
without their standard deviations or errors, and the percentages are used, only
after the differences between experimental and control groups have been shown
not to be statistically significant, as another way of obtaining approximate
levels of NRP activity. We have defined one unit of NRP activity as the amount
required to bring about a reduction, on a linear scale, of 10% from control
(alpha medium) in the proportion of BFU-E killed by exposure to high specific
activity ^{3}H-thymidine (20 Ci/mmol, 100 uCi/ml) for 20 min or HU (14.7 mM) for
45 min at 37°C. For specific activities, protein determinations were made
either from OD_{280}, by Biorad assay, or by amino acid analysis. Negative
(alpha medium) and positive (NRP of known activity) controls were included in
each experiment, and at least 6 samples were used in each group. Inter-
observer differences have been shown not to be statistically significant.

Purification of NRP
A cell line called "B6 Pan", derived from the bone marrow cells of B6 mice, has
been found to produce considerable quantities of a material having NRP-like
activity. Experiments on this material and on that found in B6 marrow
supernatant showed that with respect to their ability to shut down DNA
synthesis of BFU-E, and to their stability to dialysis, lyophilization and
heat, both appeared to be either the same, or closely related macromolecules
(Del Rizzo et al, 1987a). Medium from B6 Pan cells served as
starting material for purification of NRP. We began with crude supernatant
which we concentrated by Amicon filtration, dialyzed against distilled water,
and then lyophilized. One gram of this material was added onto a Carboxymethyl
Affi-Gel Blue chromatographic column; phosphate buffer at pH6 was followed by a
linear NaCl gradient, and fractions were collected. Carboxymethyl is a weak
cation exchanger and binds positively charged molecules; Affi-Gel Blue binds
albumin specifically (pseudoaffinity). NRP did not bind to the Carboxymethyl
Affi-Gel Blue column but eluted with the running buffer. This material, which
we call Step I NRP, gave an enrichment of over 500-fold. The next step was gel
filtration. Step I NRP was added onto a pre-calibrated Sephadex G-100
chromatographic column. The buffer was phosphate buffered saline (PBS) at pH
7.0. Fractions were collected and the active fraction (Step II NRP) was found
to correspond to an average M_r in different experiments of $\sim$79,000. This was
confirmed on FPLC Superose 12 gel filtration.

Several of the known positive hemopoietic regulatory factors, like IL-3 and
CSF's are known to be glycoproteins. They bind to lectins and can be eluted
with specific sugars; consequently Step II NRP, after being dialyzed against
PBS, was loaded onto a Con A Sepharose affinity chromatographic column. Con A
binds mainly proteins with glucose or mannose residues. NRP was found to elute
with the unbound proteins. No NRP activity was found in association with bound
proteins eluted with alpha-D-methylmannoside. We could now obtain clean
titrations of Step III NRP. Three sequential steps of purification yielded a
preparation (Step III NRP) with specific activity of over 2200 U/mg protein,
representing an enrichment of nearly 6000-fold, but yields were small.
For the next step in purification of NRP we therefore chose anion exchange
chromatography on a Pharmacia FPLC system using a Mono-Q column on Step III
material with a 30mM Tris HCl buffer at pH 7.5. Five pooled fractions were
dialyzed against water, lyophilized and assayed for NRP activity. The results
showed that the NRP activity did not bind to the anion exchange column. The
specific activity of Step IV NRP was around 20,000 U/mg protein, representing
about 40,000-fold enrichment. Step IV material was applied to an FPLC strong

cation exchange column, Mono-S, in which the sulfate radical is the negatively charged ion. Again NRP failed to bind. At this time the specific activity of NRP Step V was in the order of 50,000 U/mg protein, representing an enrichment of 100,000-fold. The amount of protein recovered was however too low to be seen by gel electrophoresis with silver staining. NRP did not bind to the anion exchange column at pH 7.5, nor did it bind to the weak cation exchange column at pH 6, nor to the strong cation exchange column at pH 4.7. Therefore NRP seems to be a neutral protein at physiological pH.

That this protein has carbohydrate residues was shown in another experiment in which the ability of NRP to bind to Con A Sepharose, Soybean Agglutinin (SBA) Agarose, and Wheat Germ Agglutinin (WGA) Sepharose was investigated. NRP bound to WGA but not to either Con A or SBA. NRP that was bound to WGA could be eluted with 1M N acetyl glucosamine (GlcNAc). Thus NRP contains either GlcNAc or sialic acid residues (Lis and Sharon, 1977). Carbohydrate analysis of Step V NRP showed the presence of GlcNAc and GalNAc with GlcNAc representing most of the total carbohydrate measured (Sialic acid cannot be detected by the system used. On the basis of these findings we conclude that NRP is a WGA-binding glycoprotein probably containing GlcNAc residues. Using this information we have designed a simple two-step procedure for the purification of NRP that yields material suitable for biological experiments (Del Rizzo et al, 1978b).

What NRP is not
NRP is believed to be produced under the influence of the Fv-2 gene on chromosome 9 of the mouse or a gene closely linked to it. The gene for transferrin (Trf) is also located on mouse chromosome 9 in the same segment by which the congenic strains B6 and B6.S differ (Axelrad et al, 1978), and is about 4 recombination units away from the Fv-2 locus (Davisson and Roderick, 1982). The apparent M_r of NRP is around 79,000. This is also the approximate M_r of mouse transferrin (Sawatzki et al, 1981). Both are glycoproteins but both are concerned with the process of DNA synthesis. The question thus naturally arose as to whether NRP and mouse serum transferrin are the same molecule. We have investigated this question and can report that by functional, biochemical, and immunological criteria, these are different molecules.

In a serum-free medium, we have shown that Fe-saturated transferrin is absolutely essential for erythropoietic burst formation (Stewart et al, 1984). Transferrin has also been shown to operate positively in the regulation of BFU-E DNA synthesis (Shannon et al, 1986). In contrast, NRP operates negatively on BFU-E DNA synthesis.

Examination of transferrin in polyacrylamide gels of serum from B6 and B6.S strain mice and from the parent strain SIM revealed neither electrophoretic differences (all are Trf[b] (Fox, R., 1986) nor obvious quantitative differences in their serum concentration. Mouse serum transferrin is known to bind to Con A; NRP does not bind to Con A, and this fact is exploited in its purification (Del Rizzo et al, 1987a). Mouse serum transferrin did not bind to WGA (Regoeczi, E., personal communication); NRP did bind and could be eluted with GlcNAc, and this fact was also used in its purification (Del Rizzo et al, 1987b). On an FPLC Mono S anion exchange column, NRP failed to bind; transferrin did bind and it required 1 1/2 bed volumes of 1M NaCl to elute it from the column.

We have used a polyclonal goat antibody against mouse serum transferrin (Cappel) in ELISA tests to compare the content of this protein in crude supernatant taken directly from B6 bone marrow with the same supernatant after a single step of purification on CM Affi-Gel Blue chromatography. This system was shown to be capable of detecting 2 picograms of bovine and 0.01 picograms of murine serum transferrin. As shown in Table 3, transferrin was readily detectable in the crude marrow supernatant at 0.375 ug of protein, but was not

detectable at 100 times this amount of protein after 1 step of purification, while the specific activity of NRP was increased over 150-fold. The same anti-transferrin antibody was reacted with NRP and it was found not to affect its biological activity with respect to reduction in proportion of BFU-E killed by ^{3}H-thymidine. We conclude that NRP and mouse serum transferrin are distinctly different entities. Whether they possess any degree of homology will of course have to await amino acid sequence data.
Table 3.

Failure to detect transferrin in Step I NRP from B6 bone marrow supernatant by ELISA test

		µg of protein	Transferrin (ELISA test)	NRP (Units)
B6 bone marrow supernatant				
(Lot 49)	1:100	3.75	++	0.02
	1:1000	.375	+	0.002
	1:10000	.0375	−	0.0002
NRP Step I				
(Lot 49)	1:10	37.5	+	33.15
	1:100	3.75	−	3.31
	1:1000	.375	−	0.33
	1:10000	.0375	−	0.03
	1:100000	.00375	−	0.003
Bovine serum Trf		0.1	+	−
PBS		−	−	−

Opposing effects of Il-3 and NRP

Il-3 is a growth factor that acts positively and non-specifically on hemopoietic stem and progenitor cells to stimulate their proliferation (Ihle et al, 1985; Kelvin et al, 1986). NRP is a growth factor that acts negatively and specifically on early erythroid progenitor cells to prevent them from undergoing DNA synthesis. Thus at the level of the BFU-E these growth factors must have opposing effects and it should be possible to demonstrate this in vitro with respect to BFU-E DNA synthesis. To test this possibility we administered a known dose of NRP to bone marrow cell cultures and titrated the effects of Il-3 against it. The results are shown in Fig. 1. The effect of Il-3 was clearly to oppose the negative effect of NRP on the proportion of BFU-E engaged in DNA synthesis. One unit of Il-3 (5×10^6 U/mg protein) neutralized the negative effect of around 1.5 units of NRP (5×10^4 U/mg protein). Both growth factors acted within minutes on marrow BFU-E but the actions were in opposite directions. This does not necessarily imply that Il-3 and NRP compete for the same receptor. However it does imply that at the molecular level the mechanisms regulating DNA synthesis in the BFU-E are sensitive to both agents and are able to respond rapidly to both. These findings also strongly suggest that the actions of both Il-3 and NRP are directly on the BFU-E and do not appear to require the involvement of an intermediate cell or factor.

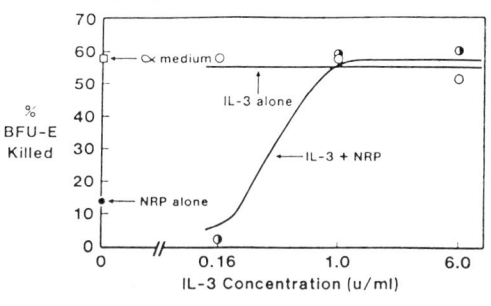
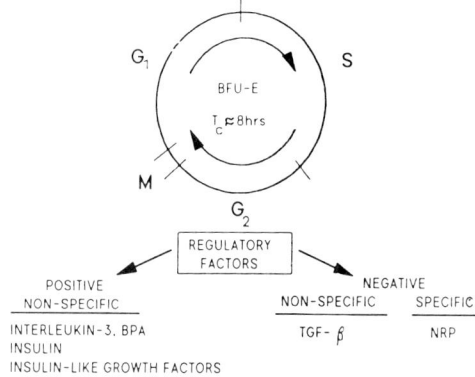

Fig. 1   Fig. 2

Molecular interactions between NRP and BFU-E
Despite the high molecular weight of NRP, interactions between NRP and BFU-E were rapid, could be initiated in the cold (Axelrad et al, 1983), and were promptly reversible by washing. They therefore seemed to take place at the surface of the BFU-E, and to represent a very low affinity type of interaction, presumably involving specific receptors for NRP.

The rapidity with which washing of the cells increased, and exposure to NRP decreased the proportion of BFU-E engaged in DNA synthesis recalls the recent findings of Davis and Czech (1986). They showed that several growth factors known to be potent mitogens of BALB/c 3T3 fibroblasts, PDGF, IGF-1 and EGF, could increase the expression, without new synthesis, of transferrin receptors at the surface of these cells within 5 minutes. Transferrin receptors represent the normal route of entry into cells of the iron required for DNA synthesis (via ribonucleotide reductase). Thus it is possible that NRP brings about a decrease in the number of transferrin receptors at the surface of the BFU-E; this in turn could be responsible for the rapid reduction in the proportion of BFU-E undergoing DNA synthesis (Compare Shannon et al, 1986). In line with such a mechanism, and on the same time scale, is our finding that iron-saturated transferrin in the medium permitted a high proportion of BFU-E to undergo DNA synthesis, but apotransferrin substantially reduced the proportion of DNA-synthesizing BFU-E.

PHYSIOLOGICAL PROPERTIES OF NRP

During the purification of NRP it became apparent from calculations of recovery that our B6 Pan cell culture medium was contaminated with factors that masked the negative effect of NRP on DNA synthesis of BFU-E. Comparison of total activity of NRP recovered after CM-Affi-Gel Blue chromatography (Step I) with total activity of crude NRP from B6 Pan cell culture medium yielded efficiencies of recovery greater than 100% (Del Rizzo et al, 1987a). The same phenomenon was seen with NRP recovered from B6 bone marrow supernatant after CM-Affi-Gel Blue chromatography. This implies that we were not merely isolating NRP from a variety of contaminating proteins, but that we were also removing positively acting factors that prevented the full expression of NRP. Both FCS and bone marrow are known sources of factors such BPA (Iscove, 1978; Porter and Ogawa, 1982) which can increase the proportion of BFU-E engaged in DNA synthesis. Removal of positively acting factors allowed us to estimate the actual amount of NRP in normal B6 bone marrow. Each B6 mouse was thus found to have an average of around 10U of NRP in his 2 femurs.

Variation in NRP levels under different physiological conditions

Effect of age
We could not find NRP in infant B6 mice whose hemopoietic systems were still developing (Axelrad et al, 1981); it was first detectable in young adults. NRP production or activity thus appeared to be developmentally regulated.

Effect of bleeding
If the NRP in B6 marrow were a physiological negative regulator of BFU-E DNA synthesis, then its concentration in the marrow of B6 mice should be affected by bleeding. To test this prediction, we subjected young adult B6 mice to one, two or three bleedings by orbital vein and assayed supernatants of marrow from these mice for their negative regulatory activity by measuring the effect of marrow supernatants from these mice on the proportion of their own BFU-E engaged in DNA synthesis compared with alpha medium as control on the same cells. The results are shown in Table 4. They demonstrate that repeated bleeding rendered NRP undetectable in B6 marrow. These data are consistent with the hypothesis that NRP, which is capable of turning off DNA synthesis of BFU-E in vitro, also acts in vivo.

Table 4.

EFFECT OF BLEEDING ON CONCENTRATION OF NEGATIVE REGULATOR IN B6 MARROW

Bleeding	Suspended in	NO. OF BURSTS/ 10^5 CELLS		Percent of BFU-E killed
		dThd	^{3}HdThd	
no bleeding controls	own supernatant	30.6 ± 1.2	23.1 ± 0.9	24.5
	alpha	36.0 ± 2.6	15.8 ± 1.9	56.1
24 hrs after 1st bleeding	own supernatant	28.5 ± 1.6	16.6 ± 1.4	41.7
	alpha	24.1 ± 2.3	6.5 ± 0.9	70.5
24 hrs after 2nd bleeding	own supernatant	16.6 ± 0.6	9.8 ± 0.8	40.9
	alpha	13.0 ± 1.0	7.3 ± 1.2	43.8
24 hrs after 3rd bleeding	own supernatant	27.0 ± 1.7	9.0 ± 1.4	66.6
	alpha	23.3 ± 1.8	8.2 ± 2.5	63.5

Effect of time of day
It is known that B6 mice are hematologically normal (Suzuki and Axelrad, 1980). If their BFU-E are permanently shut down by NRP, how could these animals produce red blood cells on a day-to-day basis? We wondered whether the levels of negative regulatory activity in B6 marrow were in fact constant throughout the day. We therefore investigated the level of this activity in bone marrow supernatants as a function of time of day in B6 and B6.S mice (Axelrad, Croizat and Eskinazi, unpublished experiments). Samples of marrow were taken at 4-hour intervals around the clock and the supernatants were assayed on washed B6 bone marrow cells with the ^{3}H-thymidine suicide method.

We found that 1) overall, B6 marrow supernatants showed lower percentages of BFU-E killed and therefore higher levels of negative regulatory activity than B6.S marrow supernatants. 2) The levels of negative regulatory activity were not constant but varied in what appeared to be a circadian rhythm in both strains. 3) The peak in negative regulatory activity in B6 marrow was between 10 a.m. and 2 p.m. It was as if the brake was off at night and BFU-E would then be capable of synthesizing the DNA necessary for producing the precursors of red blood cells. In fact it was fortunate that we had chosen to work in the morning for these experiments. Had we worked at 6 p.m. or 2 a.m. instead we would never have seen the phenomenon of negative regulation of DNA synthesis in BFU-E. These findings strongly indicate, along with the evidence already presented, that the mouse uses this mechanism physiologically to regulate negatively, rapidly, and reversibly, the proliferative behaviour of his early erythroid progenitor cells. The state of quiescence with respect to DNA synthesis of the BFU-E may therefore not be a passive process, but one requiring ongoing interactions of BFU-E with NRP molecules at the cell surface.

Sharkis et al (1974) have documented a circadian variation in erythropoiesis in the B6 mouse as judged by mitotic indices of erythroid cells in smears of marrow and spleen suspensions. They found reciprocal variations in erythropoietic activity of marrow and spleen, and postulated an on-off mechanism of control, involving both organs. Our data are at present limited to marrow, but they are consistent with the findings of Sharkis et al., and could provide a mechanism for the observed circadian variation in erythropoiesis in these animals.

Our findings strongly indicate, along with the evidence already presented, that the mouse uses the NRP mechanism physiologically to regulate negatively, rapidly, and reversibly, the prolferative behaviour of his early erythroid progenitor cells. The state of quiescence with respect to DNA synthesis of the BFU-E may therefore not be a passive process, but one requiring ongoing interactions of BFU-E with NRP molecules at the cell surface.

A model for the regulation of hemopoietic stem and progenitor cell proliferation.
Our own data and those of others seem to fall naturally into a hierarchical growth regulatory scheme in which the proliferation of each stem and early progenitor cell in the hemopoietic system is separately controlled, and each responds to both negative and positive regulatory signals. All positively acting regulatory factors so far investigated that affect hemopoietic stem and early progenitor cells appear to act non-specifically with respect to cell lineage. This is perhaps not surprising since the stem and earliest progenitor cells themselves are not lineage restricted.

At the level of the BFU-E, as shown in the present work (Fig. 2), negative regulation of DNA synthesis exerted by NRP was cell lineage-specific. TGF-β, a lineage-nonspecific growth factor (Sporn et al, 1986), in our system also acted negatively on BFU-E DNA synthesis, but it could be shown on other cells to be distinct from NRP. Positive regulation of BFU-E proliferation was cell lineage non-specific and could be exerted by a variety of agents. Among these are Il-3 (Fig. 1), BPA (which may not be different from Il-3 in this regard) (Porter and Ogawa, 1983), Insulin and Insulin-like Growth Factors I (Akahane et al, 1987) and II (Dainiak and Kreczko, 1985), and transferrin (Stewart et al, 1984; Shannon et al, 1986; Axelrad et al, unpublished experiments).
When the multipotential stem cell is released from its normal proliferative restraint, it must give rise to progenitor cells belonging to all of the hemopoietic lineages and each capable of extensive proliferation. Selection of a single lineage for amplification from among these many lineages may well occur by stopping the proliferation of all progenitors but one. This could be accomplished by a series of negative regulators, each of which would act physiologically and specifically on one of these restricted progenitor cells to turn off its DNA synthesis. It seems safe to predict, at the present time, that more negative regulators will be discovered.

ACKNOWLEDGMENTS

This work was supported by the National Cancer Institute, the Medical Research Council, and the Leukemia Research Fund of Canada and the Fraternal Order of Eagles.

REFERENCES

Akahane, K., Tojo, A., Urabe, A., and Takaku, F. Pure erythropoietic colony and burst formation in serum-free culture and their enhancement by Insulin-like Growth Factor I. Exp. Hematol. In press.

Axelrad, A.A., Ware, M., and Van der Gaag, H.C. (1972). Host cell susceptibility and resistance to murine leukemia viruses and their genetic control. In RNA Viruses and Host Genome in Oncogenesis. eds P. Emmelot and P. Bentvelzen, pp 239-254. Amsterdam: North Holland Publishing Co.

Axelrad, A.A., McLeod, D.L., Shreeve, M.M., Heath, D.S. (1974): Properties of cells that produce erythrocytic colonies in plasma culture. In Robinson WA (ed) Proceedings of the second international workshop on hemopoiesis in culture. pp 226-237. New York: Grune and Stratton.

Axelrad, A.A., Suzuki, S., Van der Gaag, H., Clarke, B.J. and McLeod, D.L. (1978): Friend virus and malignant erythropoiesis. A genetic analysis. In Hematopoietic Cell Differentiation, eds D.W. Golde, K.J. Cline, D. Metcalf, and C.F. Fox, pp 69-90. New York: Academic Press

Axelrad, A.A., Croizat, H.J., and Eskinazi, D. (1981a): A washable macromolecule from $Fv-2^{rr}$ marrow negatively regulates DNA synthesis in erythropoietic progenitor cells BFU-E. Cell 26, 233-244.

Axelrad, A.A., Croizat, H., and Eskinazi, D. (1981b): Gene control of progenitor cell proliferation during erythropoietic differentiation. In Hemoglobins in Development and Differentiation. eds G. Stamatoyannopoulos, and G. Neinhuis, pp 45-55. New York: Alan R. Liss.

Axelrad, A.A., Croizat, H., Eskinazi, D., Stewart, S., Vaithilingam, D. and Van der Gaag (1982): Gene controlled negative regulation of DNA synthesis in erythropoietic progenitor cells. J. Cell. Physiol. Suppl. 1, 165-173.

Axelrad, A.A., Croizat, H., Eskinazi, D., Stewart, S., Vaithilingam, D. and Van der Gaag, H.C. (1983): Genetic regulation of DNA synthesis in early erythropoietic progenitor cells: Mechanisms and consequences for polycythemia. In Haemopoietic Stem Cells, Alfred Benzon Symposium 18, eds Sv-Aa. Killmann, E.P. Cronkite and C.N. Muller-Berat, pp. 234-251, Copenhagen: Munksgaard.

Blackett, N.M., Millard, R.E. and Belcher, H.M. (1974): Thymidine suicide in vivo and in vitro of spleen colony forming and agar colony forming cells of mouse bone marrow. Cell Tissue Kinet. 7, 309-318.

Bruchovsky, N., Mak, S. and Till, J.E. (1967): Effects of -phenethyl alcohol on mouse L cells in suspension culture. II Effects on the cell division cycle. Mol. Pharmacol. 3, 133-141.

Cormack, D.H. (1966): Site of action of ribonuclease during its inhibition of egg cleavage. Nature, 209, 1364.

Dainiak, N. and Kreczko, S. (1985): Interactions of insulin, insulin-like growth factor II, and platelet-derived growth factor in erythropoietic culture. J. Clin. Inv. 76, 1237-1242.

Davis, R.J. and Czech, M.P. (1986): Regulation of transferrin receptor expression at the cell surface by insulin-like growth factors, epidermal growth factor and platelet-derived growth factor. EMBO J. 5, 653-658.

Davisson, M.T. and Roderick, T.H. (1982): Gene-mapping in the mouse (abst.) Cytogenet. Cell Genet. 32, 263.

Del Rizzo, D.F., Eskinazi, D., and Axelrad, A.A. (1987a): Negative regulation of DNA synthesis in early erythropoietic progenitor cells (BFU-E) by a protein purified from medium of a C57BL/6 mouse marrow cell line. Submitted.

Del Rizzo, D.F., Eskinazi, D., and Axelrad, A.A. (1987b): Negative regulator of DNA synthesis in BFU-E is a wheat germ lectin-binding glycoprotein. Submitted.

Fox, R.(1986): The Jackson Laboratory, Bar Harbor, Maine. Personal communication.

Ihle, J.N., Weinstein, Y., Keller, J., Henderson, L. and Palaszyn, E. (1985): Interleukin-3. Meth. Enzymol. 116, 540-552.

Iscove, N.N. (1978): Erythropoietin-independent stimulation of early erythropoiesis in adult marrow cultures by conditioned media from lectin-stimulated mouse spleen cells. In <u>Hematopoietic Cell Differentiation</u> ed D.W. Golde, M.J. Cline, D. Metcalf, and C.F. Fox, pp 36-52. New York: Academic Press.

Kelvin, D.J., Chance, S., Shreeve, M., Axelrad, A.A., Connolly, J.A., and McLeod, D. (1986): Interleukin-3 and cell cycle progression. <u>J. Cell. Physiol.</u> 127, 403-409.

Liestol, K. and Benestad, H.B. (1986): How should we estimate confidence intervals for a ratio, for example, between counts of colony-forming cells? <u>Exp. Hematol.</u> 14, 187-191.

Lilly, F. (1970): <u>Fv-2</u>: identification and location of a second gene governing the spleen focus response to Friend Leukemia Virus in mice. <u>J. Natl. Cancer Inst.</u> 45, 163-169.

Lis, H. and Sharon, N. (1977): Lectins: Their chemistry and application to immunology. In <u>The Antigens Vol</u> 4, ed M. Sela, Ch.7, pp 429-529. New York: Academic Press.

Pardee, A.B. (1974): A restriction point for control of normal animal cell proliferation. <u>Proc. Natl. Acad. Sci. USA</u> 71, 1286-1290.

Pledger, W.J., Stiles, C.D., Antoniades, H.N., and Scher, C.D. (1977): Induction of DNA synthesis inj BALB/c 3T3 cells by serum components: Reevaluation of the commitment process. <u>Proc. Natl. Acad. Sci. USA</u> 74, 4481-4485.

Porter, P.N. and Ogawa, M. (1982): Characterization of human erythroid burst-promoting activity derived from bone marrow conditioned media. <u>Blood</u> 59, 1207-1212.

Porter, P.N. and Ogawa, M. (1983): Erythroid burst promoting activity (BPA). In <u>Current Concepts in Erythropoiesis</u>, pp. 81-98. New York: John Wiley and Sons.

Quesenberry, P.J. and Stanley, K. (1980): A statistical analysis of murine stem cell suicide techniques. <u>Blood</u> 56, 1000-1005.

Regoeczi, E. (1987): McMaster Univ., Hamilton, Canada, Personal communication.

Sawatzki, G., Anselstetter, V., and Kubanek, B. (1981): Isolation of mouse transferrin using salting-out chromatography on sepharose CL-6B. <u>Biochem. Biophys. Acta</u> 667, 132-138.

Shannon, K.M., Larrick, J.W., Fulcher, S.A., Burch, M.B., Pacely, J., Davis, J.C. and Ring. D.B. (1986): Selective inhibition of the growth of human erythroid bursts by monoclonal antibodies against transferrin or the transferrin receptor. <u>Blood</u> 67, 1631-1638.

Sharkis,S.J., Palmer, J.D., Goodenough, J., LoBue, J. and Gordon, A.S. (1974): Daily variations of marrow and splenic erythropoiesis, pinna epidermal cell mitosis and physical activity in C57BL/6J mice. <u>Cell Tissue Kinet.</u> 7, 381-387.

Sporn, M.B., Roberts, A.B., Wakefield, L.M. and Assoian, R.K. (1986): Transforming Growth Factor-β: Biological function and chemical structure. <u>Science</u> 223, 532-534.

Stephenson, J.R., Axelrad, A.A., and McLeod, D.L. (1971): Induction of colonies of hemoglobin-synthesizing cells by erythropoietin in vitro. <u>Proc. Natl. Acad. Sci. USA</u> 68, 1542-1546.

Stewart, S., Zhu B.D., and Axelrad A. (1984): A "serum-free" medium for the production of erythropoietic bursts by murine bone marrow cells. <u>Exp. Hematol.</u> 12, 309.

Suzuki, S. and Axelrad, A.A. (1980): <u>Fv-2</u> locus controls the proportion of erythropoietic progenitor cells (BFU-E) synthesizing DNA in normal mice. <u>Cell</u> 19, 225-236.

Van Zant, G., Eldridge, P.W., Behringer, R.R. and Dewey, M.J. (1983): Genetic control of hematopoietic kinetics revealed by analyses of allophenic mice and stem cell suicide. <u>Cell</u> 35, 639-645.

Wu, A.M. (1981): A method for measuring the generation time and length of DNA synthesizing phase of clonogenic cells in a heterogeneous population. Cell Tissue Kinet. 14, 39-52.

Résumé

Dans la moelle des souris C57BL/6 (B6), la plupart des BFU-E ne sont pas en situation de prolifération, comme le montre leur résistance à la thymidine tritiée ou à l'hydroxyurée in vivo ou in vitro. En étudiant l'influence du gène de résistance du virus érythroleucémique de Friend sur l'état du cycle des BFU-E, nous avons trouvé que le lavage de cellules médullaires normales de souris B6 (Fv-2rr) provoquait une augmentation brutale de la proportion des BFU-E en phase de synthèse du DNA. Le lavage par contre n'avait aucun effet sur la proportion de CFU-S, CFU-E ou CFU-GM en phase de synthèse de DNA de la même moelle. Le surnageant des cellules médullaires B6 ainsi lavées réduisait rapidement la proportion de BFU-E en cycle. Le facteur, présent dans le surnageant, dont l'activité pouvait être déclenchée à 0° et supprimée par le lavage est une macromolécule, stable à la chaleur (30 minutes à 56° C) et sensible à la trypsine. Nous l'avons appelé la protéine de régulation négative ou NRP. Nous avons trouvé une protéine dont les propriétés ne sont pas distinguables de celles de NRP à partir d'un milieu de culture d'une lignée cellulaire dérivée des cellules médullaires B6 (B6 Pan). Cette protéine ne se fixe pas à une colonne de Carboxy-Méthyl-Affi-Gel Blue (échangeuse de cations faibles - pseudo-affinité) à pH 6.0 mais est éluée avec le tampon, enrichi 500 fois. Sur Sephadex G-100 et par gel filtration sur Superose 12 en FPLC, elle a un poids moléculaire apparent de 79.000. En chromatographie sur Sepharose A concanavaline (affinité de lectine), elle s'élue avec les protéines non fixées. Elle se fixe à la lectine de germe de blé et peut être éluée avec la N-acétyl glucosamine, ce qui traduit sa nature d'hydrate de carbone mais elle ne se fixe pas à la lectine de soja. NRP ne se fixe pas à une colonne mono Q (échangeuse d'anions forts) en FPLC à pH 7.4 ; elle ne se fixe pas non plus à une colonne mono S (échangeuse de cations forts) à pH 7.4. Après 5 étapes de chromatographie, elle est éluée avec le tampon initial avec un enrichissement apparent de 100 000. Nous avons conclu, sur ce comportement en colonnes échangeuses d'ions et de lectine, qu'au pH physiologique, la NRP est une glycoprotéine neutre. La proportion de BFU-E en cycle est à peu près inversement reliée à la concentration de NRP dans la moelle. Peu ou pas d'activité de type NRP n'a été trouvée dans la moelle de souris B6-S congénique avec B6 au niveau du locus Fv-2 ni dans la moelle de souris B6 nouveaux nés, ni dans la moelle de souris B6 adulte soumises de manière répétée à des saignées, toutes situations au cours desquelles les BFU-E sont en grande activité de prolifération. De plus nous avons montré que le niveau de NRP dans la moelle varie avec le nycthémère, étant le plus bas la nuit. L'interleukine 3, connue comme un facteur de croissance non spécifique de lignée agissant sur une grande variété de cellules souches et de progéniteurs hématopoïétiques, et la NRP, la protéine de régulation négative spcéfique de la lignée érythroïde, ont des actions opposées sur la proportion de BFU-E en phase de synthèse d'ADN, déterminée par la technique de suicide par la thymidine tritiée. Nos données sont en faveur de l'hypothèse que dans la moelle B6, la quiescence des BFU-E n'est pas un état passif ; il doit être maintenu sans cesse par l'interaction à la surface de ces progéniteurs érythroïdes de molécules de NRP spécifique de la lignée érythroïde avec des facteurs de croissance non spécifiques de lignée mais ayant une activité de stimulation comme IL3, la transferrine et les facteurs Insulin like.

Suppression of human erythropoiesis by inactivation of topoisomerases

Nicholas Dainiak *, Sandra Kreczko * and Phyllis R. Strauss +

*Departments of Medicine and Biomedical Research, St. Elizabeth's Hospital and Tufts University and + Department of Biology, Northeastern University, Boston, MA, USA

ABSTRACT

The importance of higher-order chromatin structure to DNA replication and transcription in eukaryotic cells has recently been emphasized. Since DNA is both helical and supercoiled, topological changes in DNA are an important part of replication and transcription. Topoisomerases have the capacity to relieve the torsional stress associated with DNA function by catalyzing the reversible nicking of DNA strands. We have investigated the effects of inactivation of topoisomerases on erythroid differentiation in culture. When added to human bone marrow cultures, camptothecin and teniposide, inhibitors of type I and type II topoisomerases, respectively, suppress the proliferation of CFU-E and BFU-E progenitor cells. Suppression is directly related to the time of inhibitor addition to culture and to inhibitor concentration. Furthermore, camptothecin and teniposide alter the release of detergent-soluble DNA (a topoisomerase-dependent phenomenon) from nucleated erythroblasts. Analysis of colony size and colony hemoglobinization reveals that topoisomerase inhibitors also impair terminal maturation of human erythroid cells in culture. These results suggest that enzymes crucial to the topological integrity of DNA are critical to normal erythroid differentiation.

KEYWORDS

Erythroid differentiation, DNA replication, detergent-soluble DNA, topoisomerases.

INTRODUCTION

Recently, a large number of regulatory proteins have been identified as important to differentiation of hematopoietic cells (Whetton and Dexter, 1986). These include factors which may have specificity for hematopoietic cells such as erythropoietin, granulocyte/macrophage-colony stimulating factor, burst promoting activity, Interleukin-III, pluripoietin, and multipotential hematopoietic stimulating activity. Other hematopoietic regulatory factors whose target cells are not limited to the hematopoietic system include platelet-derived growth factor, insulin-like growth factors I and II, prostaglandins, insulin, thyroid hormones, androgens and others (Dainiak, 1985). In addition, surface-derived activities that appear to act via close apposition to target hematopoietic progenitor cells or by cell-cell contacts have also been described (Axelrad et al, 1981; Dainiak and Cohen,

1982). Little information is available concerning intracellular events initiated by such factors. Furthermore, virtually no information is available regarding regulatory proteins that interact with DNA directly and hematopoietic growth control. In eukaryotes, it is possible that proteins that bind to DNA might change the superhelicity of DNA and thereby alter transcription and/or replication.

Here, we have investigated the importance of higher-order DNA structure to erythroid proliferation and differentiation. Since DNA molecules are compacted into chromatin in a highly controlled fashion, we have asked whether interference with enzymes that are crucial to release of torsional stress during DNA replication has an effect on erythroid progenitor cell growth and differentiation. The enzymes selected for study, topoisomerases, interconvert various topological isomers of DNA without altering nucleotide sequence (Wang, 1985; Wasserman and Cozzarelli, 1986). They accomplish this by reversible covalent insertion into the DNA molecule. In the case of topoisomerase II, after a pass-through reaction occurs, the enzyme reseals the break and departs. Type I topoisomerases catalyze single-strand DNA breaks, while Type II topoisomerases catalyze double-strand cutting. Accordingly, as daughter strands form during replication, the tension generated ahead of and behind replicating regions is relieved (see Figure 1).

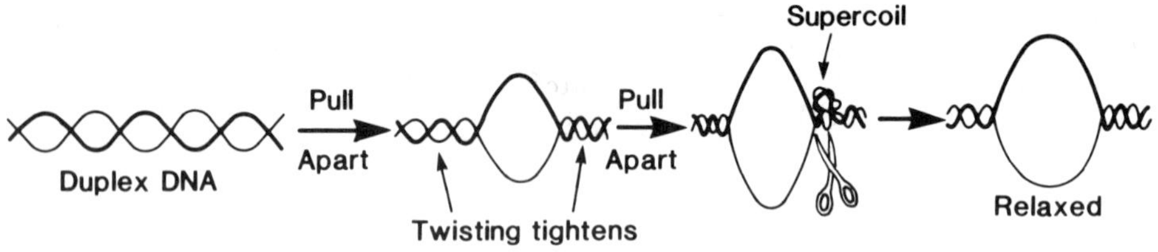

Figure 1: When DNA strands become separated in preparation for replication or transcription, twisting increases in adjacent regions. Twisting would lead to the formation of supercoils except that topoisomerases (denoted as scissors) relieve the tension. Type I topoisomerases which catalyze nicking of the single stranded DNA are often associated with actively transcribed genes. In contrast, Type II topoisomerases cut both DNA strands so that neighboring regions of the helix can pass through the cut ends. They are associated with both replication and transcription (insert citation).

METHODS AND RESULTS

To probe the importance of topoisomerases to erythroid progenitor cell proliferation, we employed specific inhibitors of the enzymes. Camptothecin (lactone form) and teniposide inhibit Type I and Type II topoisomerases, respectively. We exposed human and murine hematopoietic cells to these specific enzyme inhibitors for various periods of time. We observed that short-term incubation with topoisomerase inhibitors alters release of DNA into the detergent-soluble fraction, while long-term incubation blocks proliferation and maturation of erythroid progenitor cells.

Hematopoietic cells release detergent-soluble DNA.

Detergent-soluble DNA (DS-DNA) is a rapidly labeled fraction of DNA which separates from chromatin upon disturbance of cell architecture (Strauss et al, 1984; 1987) We have previously shown that topoisomerase inhibition prevents the release of DS-DNA (Zhang et al, 1986). To determine whether DS-DNA is present in normal human marrow cells, we analyzed DNA from the detergent-soluble fraction of erythroblasts isolated from human sternal bone marrow or spleens of phenylhydrazine-treated mice. Cells were incubated in the presence or absence of inhibitor for 30 min and tritiated thymidine was added. After 2 hrs the cells were washed free of exogenous tritiated thymidine with buffer and lysed with Nonidet P-40. DNA was resolved on native 1.2% agarose and the distribution of incorporated radiolabel was determined by fluorography.

We observed that murine erythroblasts, human lymphocytes, and human erythroblasts separated by density centrifugation released a small amount of DNA into the detergent soluble fraction. A ladder of size classes was observed with the smallest size class being 190 base pairs. Exposure of murine cells to teniposide resulted in reduction of the amount of DNA released into the detergent-soluble fraction while exposure to camptothecin resulted in increased release of DNA into the soluble fraction. Furthermore, topoisomerase inhibitors also diminished the incorporation of tritiated thymidine by up to 80 to 90%.

Camptothecin and teniposide inhibit erythroid differentiation.

Human marrow light-density mononuclear cells were cultured under serum-free conditions as previously described (Dainiak et al, 1985b). Cultures contained IMDM, iron-saturated transferrin, albumin, erythropoietin and lymphocyte-conditioned medium. Camptothecin or teniposide was added at various times (Day 0 thru Day 9) and at various concentrations (0-100 uM). As shown in Figure 2, there was a dramatic reduction in the number of CFU-E-derived colonies formed in cultures containing teniposide. This was evident at each marrow cell seeding concentration tested. The colony inhibitory effect was directly related to time of inhibitor addition. When added at the onset of incubation (Day 0), or on Days 1 and 2 of culture, virtually no CFU-E-derived colonies formed. When added on Days 3, 4, and 5, suppression was observed at a progressively lower level (see Figure 2). However, when added to culture on Day 6, no inhibition was observed.

These effects may be explained by what is known about the time course of events that occur in bone marrow culture under normal circumstances. Presumably, the CFU-E-derived colonies had already formed by Day 6 in culture. Furthermore, since colony formation occurs in a semi-synchronous pattern, the intermediate responses seen when inhibitor was added on Days 3 thru 5 might be anticipated. A similar pattern of response was observed for BFU-E derived colony formation, except that no erythroid bursts formed in cultures containing inhibitor added on Days 0 thru 6. Intermediate burst levels were observed at each marrow seeding density when inhibitors were added on Days 7 thru 9 (data not shown). Again, this is consistent with the expected pattern of colony and burst appearance in culture.

We next determined the influence of topoisomerase inactivation on hemoglobinization of erythroid colonies. Human marrow cells were cultured as above in the presence and absence of teniposide plus camptothecin added at Days 0 thru 9. Cultures were maintained for 12 days and the mean colony size was determined. As shown in Table I, mean colony size was significantly reduced by early addition of inhibitor (Days 1 thru 5). In addition, virtually none of the "colonies" formed contained hemoglobin, as assessed by staining with benzidine. Furthermore, cultures established with inhibitor at the onset of incubation contained singly dispersed cells and no colonies (see Table I). The data are consistent with the hypothesis that topoisomerases appear at detectable levels primarily in proliferating cells (Heck and Earnshaw, 1986).

The majority of colonies formed in cultures plated with inhibitor added on Days 3 thru 5 appeared to be composed of benzidine-negative or weakly benzidine-positive erythroblasts. To determine whether the cells composing these colonies were erythroid in nature, we performed immmunofluorescence staining with an antispectrin antibody. The antibody was determined to be specific for erythroid spectrin. In contrast to freshly isolated peripheral blood lymphocytes and to granulocytes and macrophages composing colonies derived from CFU-G/M progenitors, cells composing colonies formed in the presence of erythropoietin plus topoisomerase inhibitor stained intensely with the antibody. A similar degree of antibody staining was observed with colonies formed in cultures containing erythropoietin alone (data not shown). Therefore, the data suggest that in addition to inhibition of cell proliferation, topoisomerase inhibitors retard or block terminal maturation of colony erythroblasts.

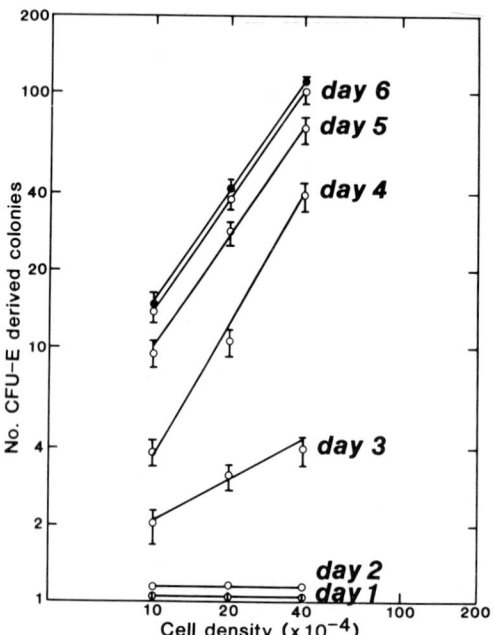

Figure 2: Serum-free human marrow cultures were established in the presence and absence of 100 uM teniposide added on the indicated day of incubation. While cell viability is unchanged (greater than 95% trypan blue exclusion after 12 days of incubation), there is a significant reduction in colony formation in cultures containing inhibitor added on Days 1 through 5. A similar effect was observed in cultures containing camptothecin. Suppression of colony formation was directly related to inhibitor concentration with half maximal suppression observed at approximately 20 uM of inhibitor.

Table I–Influence of Topo Inhibitors on Hemoglobinization

Day of Addition	Mean Colony Size (Per Cells)	% Benzidine Positive
0	0	0
1	2.0	0
2	2.8	0
3	3.1	0
4	3.6	0
5	5.8	0
6	8.9	10
7	14.5	19
8	28.1	21
9	45.5	54
No Inhibitor	69.6	93

Human marrow cells were cultured in the presence of 100 uM teniposide plus 100 uM camptothecin added on the indicated day of incubation. Colonies appearing after 12 days of culture were scored and the data was analyzed. Note that both colony size and percentage of benzidine-positive colonies are sharply reduced in the presence of the inhibitors.

CONCLUSIONS

Our findings indicate that mammalian erythroid cells contain detergent-soluble DNA which is associated with cell proliferation and which arises through the actions of topoisomerases. Inhibition of these enzymes with camptothecin or teniposide alters the release of detergent-soluble DNA. Inactivation of topoisomerases also abruptly halts the division of human erythroblasts and hematopoietic progenitor cells in serum-free culture. Furthermore, topoisomerase inhibition arrests terminal maturation of human erythroid cells which are found to react with antibodies against erythroid spectrin. This suggests that topoisomerases play an important role in proliferation and differentiation of erythroid cells. The data presented raise potentially intriguing questions regarding the importance of these enzymes in erythroid cell development. Moreover, since an agent known to be induced during hematopoietic differentiation, protein kinase C, directly phosphorylates topoisomerase II (Sahyoun et al, 1986), they also raise the possibility that topoisomerases may be involved with differentiation events initiated by soluble and cell surface-associated hematopoietic growth regulatory proteins.

REFERENCES

Axelrad, A.A., H. Croizat and D. Eskinazi (1981): A washable macromolecule from $Fv2^{rr}$ marrow negatively regulates DNA synthesis in erythropoietic progenitor cells BFU-E. Cell 26, 233-244.

Dainiak, N. and C.M. Cohen (1982): Surface membrane vesicles from mononuclear cells stimulate erythroid stem cells to proliferate in culture. Blood 60, 583-594.

Dainiak, N. (1985): Hematopoietic Stem Cell Physiology, eds Cronkite, E.P., Dainiak, N., McCaffrey, R., Palek, J. and Quesenberry, P., pp 59-76, New York, Alan R. Liss.

Dainiak, N., S. Kreczko, A. Cohen, R. Pannell and J. Lawler (1985b): Primary human marrow cultures for erythroid bursts in a serum-substituted system. Exp Hematol 13, 1073-1079.

Heck, M.M.S. and W.C. Earnshaw (1986): Topoisomerase II: A specific marker for cell proliferation. Jour Cell Biol 103, 2569-2581.

Sahyoun, N., M. Wolf, J. Besterman, T-S Hsieh, M. Sander, H. LeVine, K-J Chang and P. Cuatrecasas (1986): Protein kinase C phosphorylates topoisomerase II: Topoisomerase activation and its possible role in phorbol ester-induced differentiation of HL-60 cells. Proc Natl Acad Sci USA 83, 1603-1607.

Strauss, P.R., A.T. Andrutis, S.R. Leong, S. Nickeson and E. Supple (1984): Characterization of rapidly labeled detergent soluble DNA in murine splenocytes Biochemistry (Wash) 23, 915-921.

Strauss, P.R., S.C. Christensen, and E.P. Palome (1987): Detergent soluble DNA can be released from eukaryotic cells by using a variety of disruptive agents Biochem Biophys Res Commun. Manuscript submitted.

Wang, J.C. (1985): DNA topoisomerases. Annu Rev Biochem 54, 665-697.

Wasserman, S.A. and N.R. Cozzarelli (1986): Biochemical topology: Applications to DNA recombination and replication. Science 232, 951-960.

Whetton, A.D. and T.M. Dexter (1986): Hematopoietic growth factors. TIBS 11, 207-211.

Zhang, H., S.C. Mui, J.C. Todt and P.R. Strauss (1986): Role for topoisomerases in the release of DNA into the detergent-soluble fraction of eukaryotic cells. Proc Natl Acad Sci USA 83, 5871-5874.

Résumé

L'importance de la structure de la chromatine pour la replication de l'ADN et la transcription dans les cellules eucaryotes est un fait maintenant bien établi. La structure helicoîdale et très enroulée de l'ADN conduit à des changements topologiques au cours de ces phénomènes. Les topoisomérases sont des enzymes qui diminuent la force de torsion qui s'exerce sur l'ADN en catalysant l'interruption réversible des brins d'ADN. Les auteurs ont ainsi étudié les effets de l'inactivation des topoisomérases sur la différentiation érythroîde en culture. L'adjonction à des cultures de moelle osseuse de camptothecine et de teniposide, inhibiteurs respectifs des topoisomérases de type I et II, empêche la prolifération des progéniteurs érythroîdes CFU-E et BFU-E. Ce phénomène dépend directement du moment de l'adjonction à la culture et de la concentration de l'inhibiteur. Ces 2 inhibiteurs de plus, modifient le relargage d'ADN soluble par le détergent (phénomène dépendant de topoisomérase) des érythroblastes nucléés qui sont fraichement isolés à partir de rates de souris ou de moelle humaine. Les inhibiteurs des topoisomérases modifient aussi la maturation terminale des cellules érythroîdes en culture. Ces résultats suggèrent que ces enzymes indispensables à l'integrité de la topologie de l'ADN jouent un rôle important dans la différentiation érythroîde normale.

In vitro inhibition of human megakaryocyte colony formation by platelet products : its relationship to TGF-β

William Vainchenker, Maria-Térèsa Mitjavila, Giovanna Vinci, Nelly Kieffer, Jean-Luc Villeval, Annie Henri and Janine Breton-Gorius

INSERM U91, Hôpital Henri Mondor, 94010 Créteil, France

ABSTRACT

The goals of this work were first to demonstrate that the inhibitor of human megakaryocyte (MK) colony formation present in the serum was a platelet product, second to characterize this product more precisely. In this purpose, the capacities of whole blood serum (WBS) and of platelet-poor plasma derived serum (PDS) to sustain MK colony formation were compared. The number of MK colonies grown in PDS was always higher than when their respective WBS was used. Increasing amounts of WBS inhibited MK colony formation. Crude platelet extracts or platelet secretory products also inhibited MK colony formation in a dose dependent manner (50% inhibition for 1.10^8 platelets) whatever the culture conditions. However, they also inhibited erythroid colony formation in the same range of concentration whereas they stimulated growth of day 7 CFU-GM. The(s) inhibitor(s) was located in the platelet alpha granules since platelets from gray platelet syndromes elicited a mild inhibition (0 to 20%). The fact that the inhibitory activity of platelet extracts on erythroid and MK clonal assays was a 10 to 50 fold enhanced by acidic treatment stressed on TGF-β. Homogenous TGF-β had exactly the same effects as the crude platelet products on hematopoietic progenitor growth, especially it was able to totally suppress MK colony formation at a 1 ng/ml concentration. Therefore this study indicates that the serum inhibitor of "in vitro" megakaryopoiesis is indeed of platelet origin; it is likely to correspond to TGF-β which is a molecule implicated in regulation of cell proliferation with a wide range of activities.

Key words: TGFβ, CFU-MK, serum, platelet, inhibitor

INTRODUCTION

Hematopoiesis has been demonstrated to be positively regulated by several growth factors; most of them being now purified to homogeneity (Metcalf, 1986). In addition, increasing reports have underlined a possible autocrine negative regulation by the mature cells of each hematopoietic cell line (Broxmeyer, 1986). This negative regulation could permit a better adjustement of the hematopoietic cell production. "In vivo" and "in vitro" regulation of human megakaryopoiesis remains mostly unknown. It seems that the positive regulation involves two kinds of factors, one called MK-CSF which could be identical to interleukin 3 as in the mouse (Ihle et al, 1983; Quesenberry et al, 1985; Williams et al, 1985), the other MK-potentiator could be the "in vitro" analogous of thrombopoietin (Williams et al, 1982). Only this last factor may be regulated by the platelet demand. The MK colony assays in human have pointed out the crucial role of the serum. Indeed, it has been demonstrated by several investigators that human serum was able to inhibit MK colony formation (Vainchenker et al, 1982) and that only plasma (a biologic product devoid of platelet factors) (Kimura et al, 1984; Messner et al, 1982; Solberg et al, 1985) could optimally sustain MK colony formation. All these findings suggest that during clot formation platelets may release (a) factor(s) which can negatively regulate "in vitro" megakaryopoiesis (Kimura et al, 1984; Solberg et al, 1985; Vinci et al, 1983). The present study was performed with this goal. We could demonstrate that platelet contain factor(s) present in their alpha granules capable of inhibiting MK colony formation. However, this inhibition is not specific involving the erythroid lineage and is mainly ascribed to a single factor called type beta transforming growth factor (TGF-β).

MATERIAL AND METHODS

Preparations of whole blood serum and platelet-poor plasma-derived serum.

The method of Vogel et al (1978) was used to prepare whole blood serum (WBS) and platelet-poor plasma-derived serum (PDS). Briefly, fresh blood from normal donors was collected in plastic tubes containing 0,38% sodium citrate (W/V) and was subsequently divided into two equal portions. The first portion was clotted by addition of 1 M $CaCl_2$ solution (1/70 v/v), followed by incubation for 2 h at 37° C. WBS was obtained by a centrifugation. PDS was prepared by centrifugating the second portion before clotting by addition of 1 M $CaCl_2$ (1/50 ; v/v) in glass tubes. WBS and PDS were overnight dialyzed against Iscove's modification Dulbecco's medium (IMDM).

Preparations of crude platelet extracts and of acid-soluble platelets extracts

1 unit of fresh platelets from normal donors was obtained from the Centre Départemental de Transfusion Sanguine du Val de Marne (Créteil, France). The platelet pellet was then suspended in PBS-EDTA and counted. Platelets were resuspended at the concentration of 1.10^{10} platelets/ml in IMDM. After five cycles of freezing and thawing, samples were centrifuged. A part of the supernatant was used in culture without subsequent treatment. The other part was either dialyzed overnight against PBS or heated for 10 min. in a boiling bath or treated by trypsin.

Platelets of two patients with a gray platelet syndrome (Raccuglia, 1971) were also obtained and were proceeded as above except that their concentration was adjusted to a 1.10^9 cells/ml.

In order to prepare acid soluble platelet extracts, supernatant of platelet extracts from normal subjects was dialyzed 24 h at 4° C against two changes of 1 M acetic acid. After centrifugation, the supernatant was recovered and lyophilized. Lyophilized material was dissolved in Hank's medium and the pH was adjusted to 7.2 by addition of NaOH.

Preparation and purification of platelet secretory products

Human platelets (1.10^9) were incubated 2 hours at 37° C in IMDM in the presence of purified thrombin (2,5 µg/ml, Sigma, St Louis, Missouri). The supernatant was removed and stored at - 30° C. As a control, thrombin was directly included in the culture medium at the same concentration.

Homogenous platelet factor 4, semi-purified β thromboglobulin, semi-purified platelet-derived growth factor, generous gifts of Dr. L. Tranqui (CEA, Grenoble, France) and homogenous native type β tumor growth factor (TGF-β) a generous gift of Dr. A Roberts (NIH, Bethesda, Maryland) were also tested.

- Hematopoietic clonal assays

  Bone marrow cells were aspirated from donors undergoing bone marrow transplantation. Light-density mononuclear cells were obtained by Ficoll metrizoate density centrifugation (Lymphoprep, Nyegaard, Norways, d:1.077). Adherent cells were usually removed by a 2 hour incubation of the cells in petri dishes containing IMDM, 10% fetal calf serum. Non adherent cells were recovered. Cultures were performed either by the plasma clot or methylcellulose techniques.

  - The plasma clot technique had the advantage for CFU-MK assay to permit an easy quantification of colonies by fluorescent labeling. Cells were plated at $2.5\ 10^5$ cells per ml in the presence of 2.5% PHA-leucocyte conditioned medium in triplicate cultures.

  - Methylcellulose cultures were used for the CFU-MK, CFU-E, BFU-E and CFU-GM assays. Stimulating factors for erythroid progenitors were 5% supernatant from the Mo cell line plus 1 IU/ml semi-purified Epo preparations and supernatant from the 5637 cell line for the CFU-GM assays. Cells were plated at $2.5\ 10^4$ to 5.10^4 cells per ml.

Colonies were scored under an inverted microscope for methylcellulose cultures at day 7 for CFU-E and day 7 CFU-GM and at day 14 for BFU-E, day 14 CFU-GM and CFU-MK. In the plasma clot technique, cultures were directly labeled by a platelet GP IIIa (C 17) or a platelet GP IIb-IIIa (J15) MoAb, binding was revealed by a fluorescein conjugated goat Fab'2 against mouse Ig (Bioart, Meudon, France). Colonies were scored under a fluorescent microscope at a 150 x magnification at day 12 of culture.

RESULTS and DISCUSSION

- Effects of WBS and PDS

WBS and PDS from 12 normal subjects were compared upon their ability to support MK colony formation. At a 10% concentration, WBS sustained 20%-50% less MK colonies than their respective PDS (number of MK colonies from $17-80/10^5$ cells). Increasing the amounts of WBS from 2.5% to 20% enhanced the inhibition of MK colony formation (up to 40%); in contrast PDS in this range of concentration did not inhibit this assay **(Fig. 1)**.

These data show that WBS has the same inhibitory effect as previously reported for human serum (Vainchenker et al, 1982; Solberg et al, 1985); however PDS does not contain this activity suggesting that inhibition is directly correlated to the presence of a platelet factor. Therefore we directly tested platelet products.

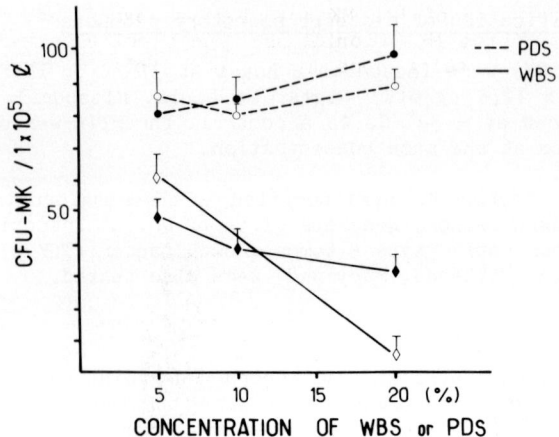

Fig. 1: Effect of increasing amounts of WBS and PDS. Increasing amounts of WBS diminished the number of colonies. Therefore, this result demonstrates that differences between WBS and PDS are not related to a MK-CSF present in PDS and absent from WBS, but are due to the presence of an inhibitor in WBS.

- Inhibition of megakaryocyte colony formation by crude platelet extracts and platelet secretory products

Crude platelet extracts and platelet secretory products inhibited MK colony formation. A 50% inhibition was obtained for the equivalent of 1.10^8 or $2-5.10^7$ platelets for platelet extracts and platelet secretory products respectively. Platelet extracts from 1.10^8 platelets corresponded to 30-50 µg of proteins. A complete inhibition could be obtained for $1\ 10^9$ platelets (**Fig. 2**).

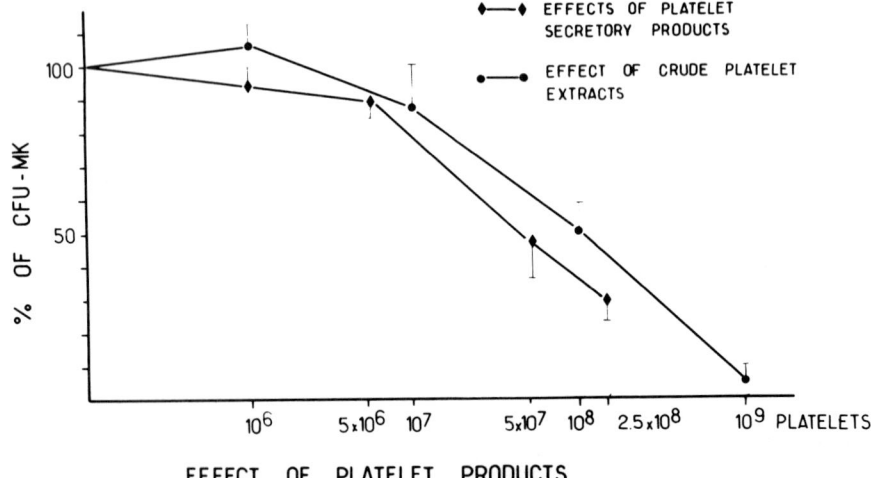

Fig. 2: Effect of crude platelet extracts and platelet secretory products on CFU-MK colony formation from normal marrow cells. Crude platelet extracts and platelet secretory products were able to inhibit in a dose dependent manner the CFU-MK growth. Platelet secretory products were slightly more efficient than crude platelet extracts in the induction of the inhibition. Cultures were performed in plasma clot in the presence of PDS.

In addition, when the size of the MK colonies was evaluated, platelet products diminished the size of the MK colonies and large colonies were rare. A large number of megakaryocytic cells were also found isolated or in clusters of 2 cells (**Fig. 3**).

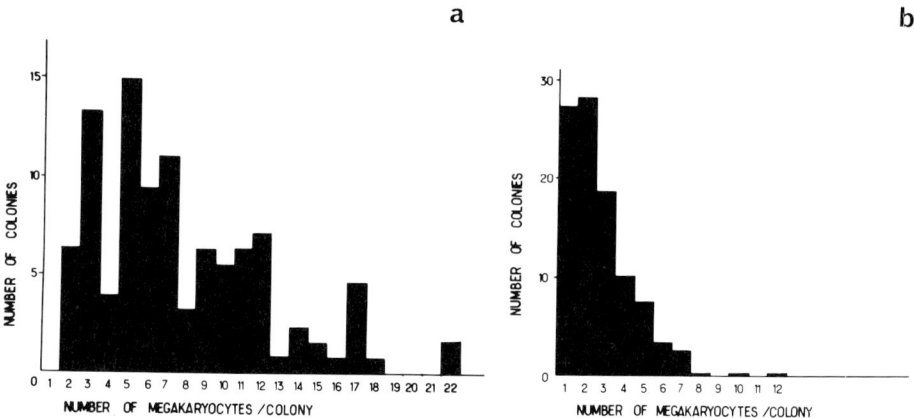

Fig. 3: Size of the megakaryocytic colonies in the presence or the absence of crude platelet extracts (30 µg). The size of the colonies was evaluated in plasma clot cultures which were labeled by the C17 MoAb. Each individual megakaryocytic cells or clusters were scored, and the number of cells composing each cluster was evaluated. The curve was established from 200 "colonies".
a: in the absence of crude platelet extracts
b: in the presence of crude platelet extracts

Therefore, these results demonstrate that the inhibition mediated by WBS is the consequence of the release of platelet products since 100 µl-200 µl/ml serum elicited a similar inhibition as their equivalent in platelet products. We subsequently tested the specificity of this inhibition.

- Inhibition of erythroid colony formation by platelet products

Crude platelet extracts or platelet secretory products were tested in other hematopoietic assays. Both of them had no inhibitory activity on the CFU-GM assay, but inhibited CFU-E colony formation in PDS, or "serum free" cultures. This inhibition was dose related, in the same range as the CFU-MK assay. Results on BFU-E colonies were not clear cut.

This inhibition observed for erythroid colony formation is quite surprising for the two following reasons:

- previous studies have reported that PDS in contrast to WBS cannot efficiently sustain CFU-E colony formation and that platelet secretory products or platelet extracts stimulate erythroid colony formation (Dainiak et al, 1983). This stimulating activity of WBS was ascribed to platelet-derived growth factor.

- In the present clonal assays, we did not find any differences between PDS and WBS for erythroid colony formation. However, recently we have found that plating efficiency for CFU-E was constantly higher in "serum free" than in serum cultures.

Therefore, we can hypothesize that human serum contains both an inhibitory activity of platelet origin and a stimulatory activity able to partially conterbalance this inhibitory activity. Potentiators of CFU-E growth have been recently described in normal murine serum (Arnaud and Blanchet, 1986).

- <u>Inhibitor(s) are stored in the platelet alpha-granules</u>

The presence of the inhibitor in the supernatant of platelets activated by thrombin strongly suggests that this inhibitor is located in platelet alpha granules. Indeed, thrombin induces secretion of the different alpha granule proteins such as platelet factor 4, β thromboglobulin, thrombospondin and fibrinogen which become detectable in the supernatant. However, thrombin has many other effects on the platelet functions especially on the secretion of all types of granules.

Therefore, we tested whether crude platelet extracts from 2 gray platelet syndrome patients could mediate an inhibition of MK colony formation. This syndrome is characterized by a congenital defect in platelet alpha granules; however recent studies have shown that the defect in the platelet alpha granule proteins may not be complete. No inhibition was detected by platelets from one patient, whereas a moderate inhibition was mediated by platelets of the other patient. In this last patient, some abnormal platelet granules containing residual amounts of thrombospondin and vWF were observed whereas in the other patient these proteins were undetectable (Cramer et al, 1985).

- <u>Partial characterization of the inhibitor</u>

In a first time we tested purified or semi-purified proteins contained in the alpha granules, neither platelet-derived growth factor, nor platelet factor 4 nor β thromboglobulin nor fibrinogen at physiologic concentration i.e. equivalent to 100 μl serum could inhibit MK colony formation. Therefore we tried to better characterize biochemically this inhibitor.

The inhibitory activity was inactivated by trypsin treatment but not by boiling. Strickingly the inhibitory effect contained in platelet extracts both on CFU-E and CFU-MK growth was enhanced when acid soluble platelet extracts were tested. A quasi total suppression of these two assays were observed for 10 μg of platelet acidic proteins. Therefore acidic treatment elicited a 10 to 50 fold greater inhibition than the crude platelet extracts. Surprisingly acid platelet extracts enhanced plating efficiency of day 7 CFU-GM but had no effect on day 14 CFU-GM.

This preliminary characterization (resistance to boiling and enhancement of the inhibitory activity by acid treatment) prompted us to test TGF-β a factor recently characterized and present in the platelet alpha granules (Assoian et al, 1983). Indeed TGF-β has a molecular weight of 25 000; it is present in platelets under a partially inactive form which is activated by acidic treatment; in addition TGF-β has been described to be the serum inhibitor of several culture systems (Childs et al, 1982).

- <u>Effects of TGF-β</u> (**Fig. 4**)

Homogenous native TGF-β was subsequently tested from 10 pg/ml to 1 ng/ml (optimal concentrations in other assays) in plasma clot cultures containing PDS for CFU-MK and in "serum free" methylcellulose cultures in the presence of homogenous Epo and GM-CSF for CFU-E and BFU-E. CFU-MK growth was nearly completely inhibited at 1 ng/ml, 50% inhibition was observed for 100 pg/ml. A similar result was observed for the CFU-E assay with the exception that only in one of 4 experiments a complete inhibition was observed at 1 ng/ml; in the three other experiments the maximum inhibition was 70% of the control. Day 12 BFU-E growth was similarly inhibited; but blood BFU-E growth was only slightly inhibited (10 to 20%). TGF-β in the same range of concentration enhanced the growth of day 7 CFU-GM stimulated by 5637 supernatant (30 to 40%) but has no effect on day 14 CFU-GM growth.

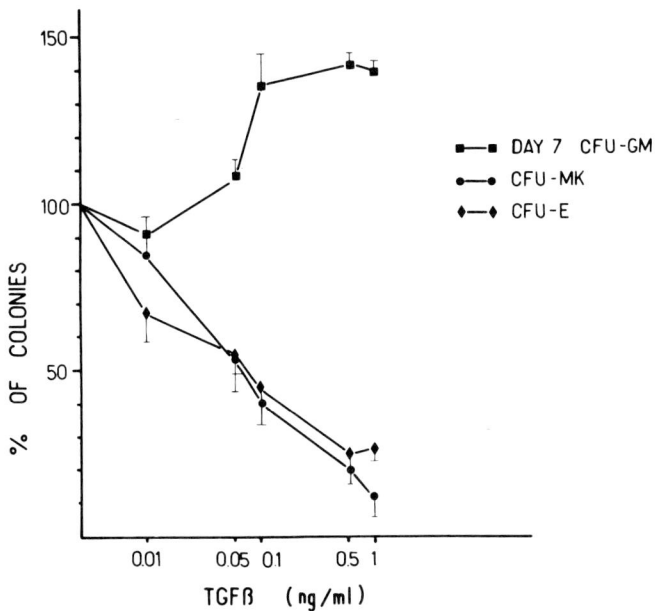

Fig. 4: Effects of homogenous TGF-β on the growth of CFU-MK, CFU-E and day 7 CFU-GM. CFU-MK were grown in plasma clot culture whereas CFU-E and day 7 CFU-GM were grown by the methylcellulose technique. CFU-MK and CFU-E growths were similarly inhibited whereas day 7 CFU-GM growth was enhanced.

Therefore, this study shows that a unique homogenous factor has the same and complex effect than the crude platelet extracts or platelet secretory products on hemopoietic progenitor growth. It is able to suppress CFU-E and CFU-MK growth but to enhance day 7 CFU-GM growth. TGF-β has been demonstrated to be a bifunctional regulator depending upon the culture systems. It can stimulate or inhibit several types of culture (Roberts et al, 1985). Moreover, for the same target cells, TGF-β may stimulate or inhibit colony formation dependent upon the growth factors used. Especially, TGF-β inhibits the proliferation of T cells, B cells and the function of NK cells (Kehrl et al, 1986; Rook et al, 1986) all cells germane to the hematopoietic cells.

Our study does not totally eliminate that other inhibitors of MK colony formation may exist in platelet products, but their possible action is totally masked by that of TGF-β or may be synergystic with TGF-β.

Finally, it remains speculative whether TGF-β may be implicated in the normal regulation of megakaryopoiesis "in vivo". Indeed, its action is too broad to permit a selective regulation of megakaryopoiesis and its adjustment to platelet demand. In contrast, it may play an important role in some congenital or malignant diseases of the MK in which MK alpha granule content is secreted directly in the marrow potentially able to result in a direct inhibition of normal megakaryopoiesis.

REFERENCES

Arnaud, S., and Blanchet, J.P. (1986): Mouse serum enables CFU-E to grow under physiologic concentration of erythropoietin in vitro. Exp. Hematol. 14:143-146.

Assoian, R.K., Komoriya, A., Meyers, C.A., Miller, D.M., and Sporn, M.B. (1983): Transforming growth factor β in human platelets. Identification of a major storage site purification and characterization. J. Biol. Chem. 258, 7155-7160.

Broxmeyer, H.E. (1986): Biomolecule-cell interactions and the regulation of myelopoiesis. Int. J. Cell Clon. 4, 378-405.

Childs, CB, Proper, J.A., Tucker, R.F. and Moses, H.L. (1982): Serum contains a platelet-derived transforming growth factor. Proc. Natl. Acad. Sci. USA 79, 5312-5316.

Cramer, E.M., Vainchenker, W., Vinci G., Guichard, J. and Breton-Gorius, J. (1985): Gray platelet syndrome: Immunoelectron microscopic localization of fibrinogen and von Willebrand factor in platelets and megakaryocytes. Blood, 66, 1309-1316.

Dainiak N, Davies, G., Kalmanti, M., Lawler, J., and Kulkarni, V. (1983): Platelet-derived growth factor promotes proliferation of erythropoietic progenitor cells in vitro. J. Clin. Invest., 71, 1206-1214.

Ihle, J.N., Keller, J., Oroszalan, S., Henderson, .E., Copeland, T.D., Fitch, F, Prystowsky, M.B., Goldwasser, E., Schrader, J.W., Palaszynski, E., Dy M., and Lebel, B. (1983): Biologic properties of homogenous interleukin 3. Demonstration of WEHI-3 growth factor activity, must cell growth factor activity, P cell-stimulating activity and histamine-producing cells-stimulating activity. J. Immun. 131, 282-287.

Kehrl, J.H., Wakefield, L.M., Roberts, A.B., Jakowlew, S., Alvarez-Mon, M., Derynck, R., Sporn, M.B., and Fauci, A. (1986): Production of transforming growth factor β by human T lymphocytes and its potential role in the regulation of T cell growth. J. Exp. Med. 163, 1037-1050.

Kimura, H., Burstein, S.A., Thorning, D., Powell, J.S., Harker, L.A., Fialkow, P.J., and Adamson, J.W. (1984): Human megakaryocytic progenitors (CFU-M) assaying in methylcellulose. Physical characteristics and requirements for growth. J. Cell Physiol. 118, 87-96.

Messner, H.A., Jamal, N., Izaguirre, C. (1982): The growth of large megkaryocyte colonies from human bone marrow. J. Cell. Phys. Suppl. 1, 45-51.

Metcalf, D. (1986): The molecular biology and functions of the granulocyte-macrophage colony-stimulating factors. Blood, 67, 257-267.

Quesenberry, P.J., Ihle, J.N., and McGrath, H.E. (1985): The effect of interleukin-3 and GM-CSA-2 on megakaryocyte and myeloid clonal colony formation. Blood, 65, 214-217.

Raccuglia, G. (1971): Gray platelet syndrome: a variety of qualitative platelet disorder. Am. J. Med. 51, 818-828.

Roberts, A.B;, Anzano, M.A., Wakefield, L.M., Roche, N.S., Stern, D., and Sporn, M. (1985): Type β transforming growth factor: A bifunctional regulator of cellular growth. Proc. Natl. Acad. Sci. USA 82, 119-123.

Rook, A.H., Kehrl, J.H., Wakefield, L.M., Roberts, A.B., Sporn, M.B., Burlington, D.B., Lane, H.C., and Fauci, A.S. Effects of transforming growth factor β on the functions of natural killer cells: Depressed cytolytic activity and blunting of interferon responsiveness. J. Immun. 136, 3916-3920.

Solberg, L.A., Jamal, N., and Messner, H.A. (1985): Characterization of human megakaryocytic colony formation in human plasma. J. Cell. Physiol. 124, 67-74.

Vainchenker, W., Chapman, J., Deschamps, J.F., Vinci, G., Bouguet, J., Titeux, M., and Breton-Gorius, J. (1982): Normal human serum contains a factor(s) capable of inhibiting megakaryocyte colony formation. Exp. Hemat. 10, 650-660.

Vinci, G., Vainchenker, W., Tabilio, A., Testa, U., Villeval, J.L., and Breton-Gorius, J. (1983): In vitro inhibition of human megakaryocyte colony formation by platelets. Blood, 62, 148a.

Vogel, A., Raines, E., Kariya, B., Rivest, M., Ross, B. (1978): Coordinate control of 3T3 cell proliferation by platelet-derived growth factor and plasma components. Proc. Natl. Acad. Sci. USA, 75, 2810-2814.

Williams, N., Eger, R.R., Jackson, H.M. and Nelson, D.J. (1982): Two factor requirement for murine megakaryocyte colony formation. J. Cell Phys. 110, 101-104.

Williams, N., Sparrow, R., Gil, K., Yasmeen, D. and McNiece, I. (1985): Murine megakaryocyte colony stimulating factor: its relationship to interleukin 3. Leuk. Res. 9, 1487-1496.

Résumé

Les buts de ce travail ont été d'abord la démonstration de l'origine plaquettaire de l'inhibiteur sérique de la mégacaryopoïèse "in vitro" puis sa caractérisation précise. Dans un premier temps, les capacités du sérum dérivé de plasma riche en plaquettes (SDPRP) et de sérum dérivé de plasma pauvre en plaquettes (SDPP) à supporter la croissance des colonies de mégacaryocytes humains ont été comparées. Le SDPP permet l'obtention d'un plus grand nombre de colonies de mégacaryocytes et de plus grande taille que le SDPRP du même sujet. Des quantités croissantes de SDPRP inhibent de manière dose dépendante la mégacaryopoïèse. Les extraits plaquettaires bruts ou les produits de sécrétion de plaquettes activées par la thrombine inhibent également ces cultures, à des concentrations en plaquettes équivalentes à celles du sérum. Mais cette inhibition n'est pas spécifique car elle intéresse également la formation des colonies érythroïdes; par contre la formation des colonies dérivées des CFU-GM du 7ème jour est stimulée par les produits plaquettaires. Cet inhibiteur(s) a été localisé dans les alpha granules plaquettaires car les plaquettes de patients atteints du syndrome des plaquettes grises n'entrainent soit aucune inhibition soit une inhibition modérée de la formation des colonies mégacaryocytaires. Finalement, les propriétés biochimiques de cet inhibiteur et surtout le fait que l'activité inhibitrice soit multipliée par un facteur de 10 à 50 lorsque les extraits plaquettaires sont obtenus en milieu acide ont suggéré que cet inhibiteur pouvait être le TGF-β. Ce facteur purifié à homogénéité est capable à la dose de 1 ng/ml d'inhiber totalement la formation des colonies mégacaryocytaires et érythroïdes "in vitro" tout en stimulant la pousse des CFU-GM du 7ème jour. Ce travail démontre donc que l'inhibiteur sérique de la mégacaryopoïèse "in vitro" est un facteur d'origine plaquettaire, le TGF-β, mais il ne s'agit pas d'un inhibiteur de la prolifération cellulaire spécifique de la lignée plaquettaire.

Transforming growth factor-β inhibits colony formation from human megakaryocytic, erythroid, and multipotent stem cells

Lawrence A. Solberg Jr*, Ronald F. Tucker **, Barbara W. Grant **, Kenneth G. Mann ** and Harold L. Moses ***

* Hematology Research Section, Mayo Clinic and Foundation, Rochester, Minnesota, 55905, USA
** Departments of Biochemistry and Internal Medicine (BWG), University of Vermont, Burlington, Vermont 05405, USA
*** Department of Cell Biology, Vanderbilt University School of Medicine, Nashville, Tennessee, 37232, USA

ABSTRACT

Transforming growth factor-β (TGF-β) is a polypeptide growth factor found in platelets, lymphocytes, and other normal and neoplastic cells. TGF-β is multifunctional and can stimulate differentiation or proliferation of cells, or inhibit proliferation of cells. This work shows that TGF-β potently inhibits human megakaryopoiesis in vitro. TGF-β also inhibits formation of erythroid and multilineage colonies in vitro. TGF-β does not inhibit granulocytic macrophage colony formation. TGF-β may play a role, in vivo, of inhibiting hematopoietic stem cell proliferation at the sites of wounds, or of providing autoinhibition of megakaryopoiesis.

KEY WORDS

Transforming growth factor-β, megakaryocytic colony formation, erythroid colony formation, multilineage colony formation, glycoprotein IIb/IIIa.

INTRODUCTION

Transforming-growth factor-β (TGF-β) is an important regulatory polypeptide synthesized by a variety of normal and neoplastic cells (Moses, 1985; Sporn, 1986). TGF-β is particularly remarkable as both a growth inhibitor and a growth stimulator depending upon the target-cell and the presence of additional interacting growth factors such as platelet-derived growth factor (PDGF) (Moses, 1985).

Platelets are a rich source of TGF-β (Childs, 1982) and platelets release TGF-β during degranulation (Assoian, 1986). Observers have reported that rat platelets contain a masking protein for TGF-β (Nakamura, 1986) and that human platelets release TGF-β in a biologically inactive complex (Wakefield, 1987) (Moses, 1987) that can be activated by proteases. The family of growth factors released by platelets at wounds probably participate in regulating the complex cellular processes that occur during wound repair (Sporn, 1986b).

Human megakaryocytic progenitor cells (CFU-M) have been observed to be inhibited by serum (Messner, 1982; Vainchenker, 1982) and platelet-rich plasma (Solberg, 1985). Because TGF-β is present in serum (Childs, 1982), we hypothesized that TGF-β might be an inhibitor of megakaryopoiesis in vitro.

METHODS

Transforming Growth Factor-β

Drs. R. Tucker and H. Moses provided TGF-β purified from human platelets by published methods (Tucker, 1984). TGF-β purified from human platelets was also obtained from R and D Systems, Inc. (Minneapolis, Minnesota, Sigma Chemical Co., St. Louis, Missouri). Stock solutions of TGF-β were prepared in 4 mM HCl with 1 mg/ml bovine serum albumin.

Hematopoietic Stem Cell Assays

Bone marrow cells were obtained from consenting normal subjects and plasma from consenting patients by plasmapheresis after approval by the Mayo Clinic Institutional Review Board.

The inhibitory effects of TGF-β on colony formation were tested in a culture system optimized for the growth of megakaryocytic (CFU-M), erythroid (BFU-E), and multilineage (CFU-GEMM) colonies (Messner, 1982) (Solberg, 1985). The culture system contained 30% plasma from a patient with aplastic anemia, 0.9% methylcellulose (Dow Chemical Co. Midland, Michigan), 5×10^{-5}M 2-mercaptoethanol (Sigma, St. Louis, Missouri), 10% conditioned medium from human peripheral blood leukocytes stimulated by phytohemagglutinin (Wellcome, Dartford, England) and 200,000 light-density bone marrow cells per ml in Iscove's Modified Dulbecco's Medium (Gibco, Grand Island, New York).

The effects of TGF-β on colonies containing granulocytes and macrophages was tested on 200,000 light-density bone marrow cells per ml in the same culture system described above except that 30% fetal calf serum (Hyclone Laboratories, Logan, Utah) was substituted for human plasma.

The stimulatory effects of TGF-β on megakaryocytic colony formation was tested by substituting a mixture of 20% fetal calf serum and 10% aplastic anemia plasma for the 30% aplastic anemia plasma. This results in growth conditions that support megakaryocytic colony formation, but allow further stimulation.

When indicated, adherent cells and T cells were depleted by published techniques (Messner, 1982). The effectiveness of macrophage-monocyte depletion was tested with non-specific esterase staining and of T-cell depletion by flow cytometry using the Leu-5 anti sheep erythrocyte receptor antibody.

Cultures were incubated for 12-14 days in an humidified incubator at 37°C with 5% CO_2 and air. Colonies were scored by inverted microscopy.

According to the experimental design, timing of addition of TGF-β to cultures, or the type of responding cell populariion were varied. Descriptions of these experimental procedures will be presented in the section on Results.

Radioimmunoassay

The effects of TGF-β were also tested in a radioimmunoassay which measures the rate of generation of cellular-bound glycoprotein IIb/IIIa in liquid suspension cultures. Details of the procedure are in press (Grant, 1987). Briefly, cells are grown at 200,000 - 500,000 cells/ml in multiple wells with the same culture conditions as described for the stem cell assays, above, except methylcellulose is absent. Standard curves of known numbers of platelets are run with each assay. At the conclusion of the assay period, cells or platelets are pelleted and washed and reacted with 10 ng of ^{125}-I labelled monoclonal antibody to GP IIb/IIIa (HP1-1D). After 60 min of incubation at 4°, chilled buffer is added containing 4×10^6 GP IIb/IIIa negative carrier cells (murine EL-4) and mixed. The cells are washed and counted in a gamma counter. With each assay, one set of target cells is incubated first with excess unlabelled HP1-1D to be certain that only specific binding is being measured.

Immunocytochemistry
Immunocytochemical techniques were used to identify types of cell present in liquid suspension culture or retrieved from culture dishes. Cytochemical stains included chloroacetate esterase for granulocytes and non-specific esterase for monocytes and macrophages. Immunocytochemical staining with alkaline phosphatase conugated antibodies was used to identify megakaryocytes (GP IIb/IIIa expression) and T lymphocytes (Leu 5). Details of the staining procedures have been described (Li, 1984).

Effects of TGF-β on DNA Synthesis in Megakaryocytes
In 2 experiments, the proportion of megakaryocytes synthesizing DNA in the presence or absence of 50-100 ng/ml TGF-β was analyzed. On day 10 of culture, the plates were labeled with final concentrations of 8 μM 5-bromo-2-deoxyuridine (Brd-Urd) and 0.8 μM 5-bromo-2-deoxyuridine (Fld-Urd). The Fld-Urd was added to inhibit the thymidine scavenger pathway so that Brd-Urd is preferentially incorporated into newly synthesized DNA. After 18 hours of incubation with Brd-Urd, individual megakaryocytic colonies were picked with a finely drawn Pasteur pipette and transferred to wells of methanol washed, teflon-masked microscope slides (Cel-Line Associates, Newfield, NJ). After rapid air drying, slides were fixed for 30 minutes in 100% methanol and then stored at $-70°C$ for later analysis by immunofluorescent microscopy. Preliminary experiments showed that there was insignificant cell loss when colonies were rapidly air-dryed and methanol fixed.

Methanol-fixed slides were incubated at 15 minutes in SPBS, then each colony was stained for 30 minutes with 10 μl of an appropriately diluted murine monoclonal antibody (BU-1) reactive with Brd-Urd (Jenkins, 1984). The slides were then washed with PBS-BSA, incubated with 10% normal goat serum (NGS) in PBS for 10 minutes, washed with PBS, and then incubated with appropriately diluted rhodamine-conjugated goat anti-mouse IgG (Cappell), for 30 minutes. After washing with PBS and PBS-BSA the slides were stained for 30 minutes with a fluorescein-conjugated murine monoclonal antibody which recognizes the platelet glycoprotein IIb/IIIa complex (HP1-1D).

Stained slides were stored at $4°C$ until completion of examination and photographic recording, which were accomplished using a Zeiss Standard 18 microscope suitably equipped for phase contrast and for two-color (fluorescein and rhodamine) epifluorescence microscopy and photography. Ektachrome ASA 400 photographic film (Eastman Kodak, Rochester, New York) was used to record color images. Identical photographic exposures, development and printing times were used to depict immune and nonimmune (control) immunofluorescence. Films were processed by the Section of Photographic Services, Mayo Clinic.

RESULTS

Effects of TGF-β on Colony Formation
TGF-β was added, over a five-log concentration range, to cultures optimized for maximum growth of CFU-M, BFU-E, CFU-GEMM, and CFU-GM derived colonies. Table 1 shows the dose-response data. The estimated concentration (ID_{50}) of TGF-β producing half-maximal inhibition was for CFU-M 2.2 ng/ml (88 pM), for CFU-GEMM 1.9 ng/ml (76 pM), and for BFU-E 9.0 ng/ml (360 pM). Over this same concentration range, no significant inhibition of granulocytic macrophage colony formation was observed.

Table 1. The effects of TGF-β on hematopoietic colony formation

Colonies per 200,000 cells
(Means ± SEM)

concentration TGF-β, ng/ml	CFU-M	CFU-GEMM	BFU-E	CFU-GM
0	60 ± 40	10 ± 6	417 ± 268	82 ± 31
0.001	55 ± 8	8 ± 1	480 ± 91	-
0.01	71 ± 17	-	505 ± 91	-
0.1	72 ± 5	9 ± 1	421 ± 93	-
1.0	53 ± 3	9 ± 1	404 ± 10	-
3.0	23 ± 1	-	372 ± 22	-
5.0	22 ± 4	3 ± 2	-	-
10.0	11 ± 1	2 ± 2	192 ± 42	-
20.0	4 ± 1	-	167 ± 33	-
50.0	2 ± 1	0	104 ± 14	85 ± 32

ALL n = 3 except at 50 ng/ml n = 6 for CFU-M and CFU-GM

It was possible, that under conditions deliberately set to maximally stimulate CFU-M, and thus test for inhibition caused by TGF-β, a stimulating effect by TGF-β might be masked. This was tested for CFU-M by culturing 200,000 marrow cells in a mixture of 20% fetal calf serum and 10% plasma from a patient with aplastic anemia. No PHA-LCM was added to the culture system. Under these conditions growth of megakaryocytic colonies is about one-half to two-thirds that seen under fully stimulated conditions. The average megakaryocytic colony formation from 3 normal donors without added TGF-β was 32 ± 5 colonies under these suboptimal conditions. Addition of 0.1 ng/ml TGF-β reduced colonies to 28 ± 5; 1.0 ng/ml to 10 ± 3; and 10 ng/ml to 9 ± 4 (means ± SEM). Over these concentration ranges, 4-400 pM, TGF-β does not stimulate megakaryocytic colony formation.

Influence of Accessory Cells
Bone marrow cells from 5 normal donors were prepared to make light-density bone marrow cells (LDBMC), non-adherent LDBMC, and T cell depleted, NALDBMC. The inhibitory effect of 50 ng/ml TGF-β was then tested on each of these cell preparations. Differential cell counts after staining with non-specific esterase and flow cytometry of T cells labelled with Leu-5 were used to assess the effectiveness of monocytic and T cells. Both types of cells were reduced by 90%. Table 2 shows that removal of 90% of adherent cells and of T cells did not reduce the capacity of TGF-β to inhibit megakaryocytic colony formation.

Table 2. Influence of accessory cells on TGF-β inhibition

Target cells	+/- TGF-β (50 ng/ml)	Colonies per 200,000 cells Means ± SEM; n=5		
		CFU-M	CFU-GEMM	BFU-E
Light-density (LDBMC)	−	77 ± 25	25 ± 10	286 ± 89
	+	13 ± 6	0	38 ± 6
Non-adherent (LDBMC)	−	69 ± 17	23 ± 3	344 ± 116
	+	3 ± 1	0	23 ± 10
T cell depleted (NA-LDBMC)	−	113 ± 28	14 ± 3	285 ± 23
	+	17 ± 11	0	94 ± 58

Time Course of the Effects of TGF-β

To test for the time course and reversibility of inhibition of colony formation caused by TGF-β, bone marrow cells were incubated with or without 50 ng/ml TGF-β for 10 min, 24 or 72 hr. The cells then were thoroughly washed, dispersed into a single-cell suspension, and plated in methylcellulose under optimum conditions for colony formation. Colonies were enumerated 10 days after plating the cells.

TGF-β rapidly caused 60% of its inhibition of megakaryocytic colonies with an exposure time as short as 20 min. This extent of inhibition was not increased by exposure times of cells to TGF-β of up to 72 hrs. Similar inhibition was noted for CFU-GEMM and BFU-E derived colonies. CFU-GM derived colonies appeared to be increased by short-term preincubation with TGF-β. These data are shown in Table 3.

Table 3. Time course of inhibition of colonies by TGF-β

Length of pre-incubation	+ or − TGF-β 50 ng/ml	Colony Counts Means ± SEM, n=3			
		CFU-GM	BFU-E	CFU-M	CFU-GEMM
10 min	+	112 ± 42	192 ± 59	9 ± 2	4 ± 1
	−	52 ± 4	306 ± 14	22 ± 10	11 ± 5
24 hr	+	68 ± 8	149 ± 40	9 ± 5	2 ± 1
	−	44 ± 13	198 ± 32	23 ± 9	8 ± 4
72 hr	+	19 ± 6	177 ± 26	15 ± 9	2 ± 1
	−	30 ± 7	256 ± 58	43 ± 12	13 ± 6

Effects of TGF-β on the Cellular Expression of GP IIb/IIIa

The recognition and enumeration of a colony in the human plasma-methylcellulose system requires that megakaryocytic cells within colonies be large and exhibit characteristic optical properties by inverted microscopy. This recognition is dependent upon empirical optimization of growth conditions for proliferation of CFU-M and cytoplasmic maturation of megakaryocytes. This optimization is insured by using plasma from patients with aplastic anemia and batches of PHA-LCM screened and selected for optimum growth promoting properties.

The failure to detect a colony after exposure to TGF-β could merely reflect an interference with cytoplasmic maturation such that the colonies could not be recognized as being megakaryocytic. To test for inhibition of megakaryopoiesis independently of colony recognition, log-dose response studies were undertaken in the radioimmunoassay.

Four log-dose response curves were measured by Dr. Barbara Grant in Burlington, Vermont, and 3 in the author's laboratory. Pooled data from both laboratories are shown in Table 4. Confirmation of the inhibition of megakaryocytes in the liquid suspension culture was done by performing cell counts and differential stains with anti-glycoprotein IIb/IIIa antibody. As may be seen, in the radioimmunoassay the dose producing 50% inhibition of GP IIb/IIIa expression is approximately 0.3 ng/ml (12 pM). TGF-β potently inhibits the cellular expression of GP IIb/IIIa.

Table 4. Inhibition by TGF-β of cellular expression of GP IIb/IIIa

Concentration, TGF-β ng/ml	Number of Experiments	^{121}I - HP1-1D Bound CPM (mean ± SEM)
0	7	13,695 ± 3953
0.003	3	10,255 ± 1859
0.01	5	9,666 ± 1347
0.03	4	11,663 ± 3484
0.1	6	10,329 ± 2299
0.3	6	6,761 ± 934
1.0	7	5,478 ± 728
3.0	7	3,996 ± 833
10.0	6	3,082 ± 727
30.0	2	4,471 ± 191
50.0	3	3,037 ± 499

Influence of Initial Concentration of Cells

In the colony assays, the initial concentration of cells plated is 200,000 per ml. With the radioimmunoassay, initial concentrations of cells ranging from

200,000 to 500,000 are commonly used. We observed that the inhibitory effect of TGF-β in the radioimmunoassay is less, and more variable, if the initial cell concentration is greater than 200,000 cells. For example, in experiments with 2 different normal donor cells, inhibition noted by 50 ng/ml TGF-β in an initial cell density of 500,000 cells/ml averaged 22% with a range of 3 to 54%. When less than or equal to 200,000 cells were placed into culture, the average inhibition was 40% and the range 28 to 62%.

Time course of TGF-β Inhibition in Radioimmunoassay
In two experiments, the inhibition of cellular GP IIb/IIIa expression caused by the delayed addition of TGF-β was measured. Maximal inhibition was defined as the inhibition caused by TGF-β, 50 ng/ml, being present for the full 12 days of liquid suspension culture. As Table 5 indicates, the addition of TGF-β even on day 12 did inhibit GP IIb/IIIa expression by 14 to 50%.

Table 5. Effects of delayed addition of TGF-β into radioimmunoassay

Donor	0	2	4	6	8	10	12
		Per cent of maximal inhibition					
1	100	56	42	31	40	14	0
2	100	42	67	59	86	50	0

Effects of TGF-β on the Synthesis of DNA by CFU-M Derived Megakaryocytes
Megakaryocytic colonies grown in the presence of bromodeoxyuridine with or without 50 ng/ml TGF-β were picked and double stained to assess GP IIb/IIIa expression and bromodeoxyuridine incorporation. Of 55 control colonies analyzed, the mean number of megakaryocytes per colony in S phase was 15.5 ± 1.8. For 45 colonies analyzed after exposure to TGF-β, 16 ± 2.3 megakaryocytes per colony were in S phase. TGF-β did not inhibit DNA synthesis in megakaryocytes.

DISCUSSION

We find that homodimeric TGF-β, prepared from human platelets, potently inhibits colony formation from human CFU-M, BFU-E, and CFU-GEMM. This inhibition occurs in a culture system containing plasma from a patient with aplastic anemia and PHA-LCM. The inhibition by TGF-β has been confirmed with bone marrow cells from approximately 25 normal subjects and with plasmas from 2 patients with aplastic anemia. TGF-β does not inhibit granulocytic-macrophage colony formation at doses ranging up to 50 ng/ml.

In addition, TGF-β inhibits the cellular expression of GP IIb/IIIa in a radioimmunoassay that does not depend upon colony recognition by inverted microscopy. Immunocytochemical analysis of cells in the liquid suspension culture confirm the marked reduction in numbers of megakaryocytes.

The concentration of added TGF-β producing half-maximal inhibition of colony formation is particularly remarkable. For example, for inhibition of megakaryopoiesis the ID_{50} is 12 pM for the radioimmunoassay and 80 pM for the megakaryocytic colony assay. The high-affinity receptors that have been isolated for TGF-β have a K_d of 50 pM (Fanger, 1985). Therefore, whatever the mechanism by which TGF-β inhibits colony formation, the response at some point probably is mediated through high-affinity receptors for TGF-β.

Removal of over 90% of adherent cells and T cells does not change the inhibition of CFU-M, BFU-E, or CFU-GEMM derived colonies. Nevertheless, some involvement of a very minor cell population of adherent or T cells can not be eliminated by our studies. Monocytes and macrophages are regenerated with time in the culture system.

The inhibition of megakaryocytic, erythroid, and multilineage colonies can occur very rapidly with only 10 min of incubation with 50 ng/ml TGF-β causing a 60% reduction in colony formation 12 days later. Granulocytic-macrophage colonies, in contrast, may actually be increased by short-term exposures to TGF-β of 20 min or 24 hrs. TGF-β is not a toxin for all stem cells.

Our observations do not allow us to infer that the inhibition of megakaryopoiesis occurs as a result of a direct interaction of TGF-β with stem cells and/or megakaryocytes. The effects might be mediated through accessory cells, or by changing synthesis of macromolecules in the culture system. TGF-β has many potential effects on cells that might be expected to be present in our culture system. TGF-β stimulates the production of fibronectin and collagen in fibroblasts (Ignotz, 1986); enhances production of plasminogen activator inhibitor (Laiho, 1986); and can inhibit production of tumor necrosis factor-α and gamma interferon from mononuclear cells (Palladino, 1987). Hunt (1987) has reported that TGF-β can stimulate monocytes to release a growth factor for fibroblasts. TGF-β also inhibits IL-2 stimulated T cell proliferation and may be an autoinhibitory lymphokine (Kehrl, 1986).

Our observations suggest that TGF-β does inhibit both megakaryocytic stem cell proliferation as well as megakaryocytic maturation. The marked reduction in detectable megakaryocytic colonies; the quantitative decrease in GP IIb/IIIa expression and recognizable megakaryocytes in the liquid suspension cultures; and the significant inhibition produced by brief exposure to TGF-β of marrow cells indicate that TGF-β inhibits megakaryocytic stem cells. The reduction in GP IIb/IIIa cellular expression produced by adding TGF-β into liquid suspension cultures as late as day 10 of 12 days of culture suggests that TGF-β may reduce maturation of individual megakaryocytes.

Burstein (1986) has reported that TGF-β potently inhibits maturation (cell size) of individual megakaryocytes stimulated by interleukin-3 in serum free culture. He also observed an inhibition of murine megakaryocytic colony formation.

It is of interest that exposure to TGF-β for 24 hrs did not inhibit DNA synthesis in CFU-M derived megakaryocytes. It is possible that an inhibitory effect of TGF-β on cytoplasmic maturation might not interfere with nuclear cell cycles.

It is important to emphasize that we have no information that the inhibitory effects noted in these _in vitro_ culture systems have any physiological relevance _in vivo_. The myriad of cell types and cellular responses to TGF-β will make the ultimate clarification of the role of TGF-β difficult. Moreover, it is becoming clear that TGF-β secreted from cells or platelets can be in biologically inactive forms (Wakefield, 1987) and that local, microenvironmental activation of TGF-β may be a significant process _in vitro_ or _in vivo_.

Nevertheless, the potency of the inhibitory effect of TGF-β on hematopoietic colony formation is remarkable and deserves extensive investigation. It is possible that TGF-β released by platelets at wounds may suppress the proliferation of circulating CFU-M, BFU-E, and CFU-GEMM stem cells while allowing the amplification of granulocytic ormonocytic cells that might be needed in an inflammatory response. TGF-β must also be evaluated as a potential molecule that inhibits megakaryopoiesis, in vivo, in a negative feedback loop.

REFERENCES

Assoian, R.K., Sporn, M.B.(1986): Type β transforming growth factor in human platelets: release during platelet degranulation and action on vascular smooth muscle cells. J. Cell. Biol. 102, 1217-1223.

Cheifetz, S., Weatherbee, J.A., Tsang, M. L.-S., Anderson, J.K., Mole, J.E., Lucas, R., Massague, J.(1987): The transforming growth factor-β system, a complex pattern of cross-reactive ligands and receptors. Cell 48, 409-415.

Childs, C.J., Proper, J.A., Tucker, R.F., and Moses, H.L.(1982): Serum contains a platelet-derived transforming growth factor. Proc. Natl. Acad. Sci. USA 79, 5312-5316.

Fanger, B.O., Wakefield, L.M., Sporn, M.B.(1985): Structure and properties of the cellular receptor for transforming growth factor type β. Biochemistry 25, 3083-3091.

Gewirtz, A.M., Sacchetti, M.K., Bien, R., and Barry W.E.(1986): Cell-mediated suppression of megakaryocytopoiesis in acquired amegakaryocytic thrombocytopenic purpura. Blood 68, no. 3, 619-616.

Hunt, D., Wakefield, L., McCartney-Francis, N., Wahl, L., Roberts, A., Sporn, M. and Wahl, S.(1987): Transforming growth factor beta (TGF-β-induced monocyte chemotaxis and growth factor production. Fed. Proc. 46, no. 3, 1036 (abstract).

Ignotz, R.A., Massague, J.(1986): Transforming growth factor-β stimulates the expression of fibronectin and collagen and their incorporation into the extra- cellular matrix. J. Biol. Chem. 261, no. 9, 4337-4345.

Ishibashi, T., Jirik, F., Sorge, J., Burstein, S.A.(1986): Transforming growth factor-β (TGF-β) is a potent direct inhibitor of murine megakaryocyte (MK) growth in vitro. Blood 68, (suppl. 1) pp. 167a (abstract).

Jenkins, R.B., Gonchoroff, N.J., Oles, K.J., Nichols, W.L., and Solberg, L.A. (1984): Analysis of DNA synthesis in human megakaryocytic colonies using a monoclonal antibody reactive with bromodeoxyuridine. Blood 64 (suppl 1), 122a (abstract).

Kehrl, J.H., Wakefield, L.M., Roberts, A.B., Jakowlew, S., Alvarez-Mon, M., Derynck, R., Sporn, M.B., and Fauci, A.S.(1986): Production of transforming growth factor β by human T lymphocytes and its potential role in the regulation of T cell growth. J. Exp. Med. 163, 1037-1050.

Keski-Oja, J., Leof, E.B., Lyons, R.M., Coffey, R.J., Jr., and Moses, H.L.(1987): Transforming growth factors and control of neoplastic cell growth. J. Cell. Biochem. 33, 95-107.

Li C.Y.(1984): Immunocytochemical techniques for identifying leukemias. Mayo Clin. Proc. 59, 185-188.

Massague, J.(1985): Transforming growth factor-β modulates the high-affinity receptors for epidermal growth factor and transforming growth factor-α. J. Cell. Biol. 100, 1508-1514.

Messner, H.A., Jamal, N., and Izaguirre, C.(1982): The growth of large megakaryocyte colonies from human bone marrow. J. Cell. Physiol. (suppl 1), 45-51.

Moses, H.L., Tucker, R.F., Leof, E.B., Coffey, R.J., Jr., Hapler, J., and Shipley, G.D.(1985): Type-β transforming growth factor is a growth stimulator and a growth inhibitor. Cancer Cells 3, 65-71.

Nakamura, T., Kitazawa, T., and Ichihara, A.(1986): Partial purification and characterization of masking protein for β-type transforming growth factor from rat platelets. Biochem. Biophys. Res. Commun. 141, no. 1, 176-184.

Grant, B.W., Nichols, W.L., Solberg, L.A., Yachimiak, D.J., Mann, K.G.(1987): Quantitation of human in vitro megakaryocytopoiesis by radioimmunoassay. Blood, in press.

Solberg, L.A., Jamal, N. and Messner, H.A.(1985): Characterization of human megakaryocytic colony formation in human plasma. J. Cell. Physiol. 124, 67-74.

Sporn, M.B., Roberts, A.B., Wakefeld, L.M., Assoian, R.K.(1986a): Transforming growth factor-β: biological function and chemical structure. Science 233, 532-534.

Sporn, M.B., and Roberts, A.B.(1986b): Peptide growth factors and inflammation, tissue repair, and cancer. J. Clin. Invest. 78, 329-332.

Palladino, M.A., Czarniecki, C.W., Chiu, H.H., McCabe, S.M., Figari, I.S., and Ammann, A.J.(1987): Regulation of cytokine production and class II antigen expression by transforming growth factor-beta. J. Cell. Biochem. (suppl 11a), 10 (abstract).

Tucker, R.F., Branum, E.L., Shipley, G.D., Ryan, R.J., and Moses, H.L.(1984): Specific binding to cultured cells of ^{125}I-labeled type β transforming growth factor from human platelets.

Vainchenker, W., Chapman, J., Deschamps, J.F., Vinci, G., Bouguet, J., Titeux, M., and Breton-Gorius, J.(1982): Normal human serum contains a factor(s) capable of inhibiting megakaryocyte colony formation. Exp. Hematol. 10, 650-660.

Wakefield, L.M., Smith, D.M., Flanders, K.C., Sporn, M.B.(1987): Characterization of a latent form of transforming growth factor-β secreted by human platelets. J. Cell. Biochem. (suppl 11a), 46.

Résumé

Le Transforming growth factor (TGF-β) est un facteur de croissance polypeptidique que l'on trouve dans les plaquettes, les lymphocytes et d'autres cellules normales et malignes. Le TGF-beta a plusieurs fonctions : il peut stimuler la différenciation et la prolifération cellulaires ou à l'inverse être un inhibiteur de la prolifération. Dans cet article, les auteurs montrent que le TGF-beta est un puissant inhibiteur de la mégacaryopoïèse chez l'homme in vitro. Le TGF-beta inhibe aussi la formation de colonies érythroïdes et mixtes in vitro alors qu'il n'a pas d'effet sur la formation de colonies granulo-macrophagiques.

On peut ainsi émettre l'hypothèse que le TGF-beta peut intervenir in vivo en bloquant la prolifération des cellules souches hématopoïétiques au niveau des blessures, ou en favorisant le contrôle de la mégacaryopoïèse.

Discussion

Chairpersons/*Modérateurs*: N. Young (États-Unis)
B. Varet (France)

N. YOUNG : Dr Dainiak, I was interested in the suppression of globin production that is unusual for other DNA inhibitors, like folic acid and vitamin B12. Have you any thought as to why that occurs?

N. DAINIAK : The enzymes are active in transcription as well as in replication. Some of them have direct effects on activated genes. Furthermore, not only does enzyme inhibition impair hemoglobinization, but also globin DNA that is released in the detergent soluble form is inhibited. The latter effect is one of those curious findings that we expect to pursue.

N. GORIN : Dr Dainiak, I suppose that the topoisomerases are not restricted to the BFU-E. Could you comment on the other progenitors, CFU-GM or CFU-MK?

N. DAINIAK : We haven't really assayed for the effects in a systematic manner, although proliferation of other progenitor cell types, including the CFU-GEMM, appears to be affected. Furthermore, we haven't looked carefully at whether the effects are exerted at an accessory cell versus the progenitor cell level. Therefore, it is likely that other cells are affected as well. Any cells that we put into the cultures are exposed to the enzyme inhibitors and there could well be that the real effects we are looking at are mediated by other cells. Certainly, other progenitor cell types may well be affected.

V. PAVLOVIC-KENTERA : I would like to ask Dr Axelrad a question about the BFU-E state in different physiological conditions. You have only mentioned what happens to BFU-E after bleeding, I mean to this protein that negatively regulates BFU-E. I would be interested particularly to know whether you have done some experiments on polycythemic animals. The reason why I am asking this question is that we did find a difference between the number of cells in the S-phase in mice that were made polycythemic by chronic hypoxia, where the number of BFU-E suicide was very low as compared to the BFU-E in post-transfusion polycythemia in mice and the whole situation changed after the injection of erythropoietin into those mice. So I was wondering whether you have some data on polycythemic mice, with hypoxia-induced or with transfusion-induced erythrocytosis.

A. AXELRAD : We do not have data on polycythemia brought about by transfusion, but this work was really started to try to understand how Friend virus works, so we have a little bit of data on Friend virus-induced polycythemia. And it looks as if the negative regulator does not work on those cells.

M. AGLIETTA : First a short comment. I am glad to see that in three different groups there are just the same results on TGF beta on hematopoiesis. Second is a question to either Dr Solberg or Dr Vainchenker, whether they have any data on the production of TGF beta by megakaryocytes in essential polycythemia or other myeloproliferative disorders and whether these diseases respond to the inhibition of TGF beta or not?

W. VAINCHENKER : I have no answer, but what we have done a few years ago was to test normal serum on CFU-MK growth from CML patients and we found that normal serum was able also to inhibit this growth. So I think in most CML, TGF beta will inhibit the growth of CFU-MK, but I don't know in thrombocythemia.

B. LORD : Dr Axelrad, I am interested in the cell kinetics of your system. Could you tell me what part of the cycle do you think your inhibitor acts on?

A. AXELRAD : I think it acts in S-phase. If you try to visualize it elsewhere than in S, the time required to enter S would not fit with the rapidity with which we get the effect.

B. LORD : In fact, what you are doing is to apply your inhibitor and every single cell within S-phase immediately switches off. Is it suspended at that stage?

A. AXELRAD : It seems to be suspending in that stage and if you wash away the NRP, you return to the same original high level of percent kills, independently of the time that the cells spent in the presence of NRP.

B. LORD : What is happening to the cells outside the S-phase? Do they continue to progress around the cycle, coming to a full stop at some point?

A. AXELRAD : I showed a slide in which there was a diminution in the proportion of BFU-E killed in the presence of NRP. The lower curve represented the remainder of the cells. In tritiated thymidine, they continued to increase in percent kill, linearly with time, as if new cells were continually entering DNA synthesis ; only those that were initially in DNA synthesis were protected by the negative regulator.

B. LORD : If you keep them inhibited, let's say for six hours or so, this means that you should be getting an accumulation of cells ready to go into DNA synthesis. Within a short time of restimulation by washing or whatsoever, do you get virtually 100 % of the cells in DNA synthesis?

A. AXELRAD : No, we could not keep them in as long as six hours, but we did it for as long as three hours and we could not see any accumulation. The proportion killed does not seem to change after they are kept for various periods of time in the Negative Regulatory Protein, and then washed.

G. KONWALINKA : It is interesting that TGF inhibits the pluripotent stem cell and the bone marrow BFU-E but does not inhibit apparently the peripheral blood BFU-E. Have you any comment about this?

W. VAINCHENKER : In our experiments, we do not try to inhibit pluripotent stem cells. We have used TGF beta only to one ng per ml on BFU-E growth and Dr Solberg has used it till about 10ng per ml. So I think that the only differences are in the dose response curve. So as we have used it at a lower concentration, it only means that BFU-E are less sensitive to the inhibiting effect of TGF beta than CFU-E.

J. CAEN : It is quite well known that TGF beta does influence the receptor of other growth factors, and for instance, the platelet derived growth factor (PDGF) which acts on the platelet membrane and inhibits for instance the phosphoinositide metabolism and the platelet activation. My question is : have you tried, either Dr Vainchenker or Dr Solberg, whether or not the TGF beta does interfere with the effect of the PDGF on the megakaryocyte and on the platelet function as well?

L. SOLBERG : I can answer very simply, Dr Caen, we have not done that.

W. VAINCHENKER : We have tested both molecules together to see if there were synergistic in their inhibition. We did not find that the addition of PDGF to TGF beta increases this inhibition in one experiment.

L. SOLBERG : May I make one comment about TGF beta and BFU-E? The high affinity receptor for transforming growth factor beta have Kd of around 50pM. In our work, we find the dose producing 50 % inhibition of megakaryopoiesis is 12-50 pM. But consistently, we have seen for BFU-E values for ID 50 up between 300 and 400pM. So I think there is some difference in the dose response curve of BFU-E versus CFU-M and CFU-GEMM.

D. ROODMAN : Dr Solberg, TGF beta is a potent stimulator of fibroblast proliferation and we have tested it in long-term marrow cultures using different assays systems, besides hematopoiesis. We found that 1ng of TGF beta stimulated overgrowth of fibroblasts of these cultures. I notice you didn't see many fibroblasts. Is that a unique feature of your assay system or have you depleted fibroblast progenitors prior to assay?

L. SOLBERG : In my experiments, cells are grown in plasma from patients with aplastic anemia. These plasmas do not usually support the growth of adherent cells, in liquid suspension culture, as does fetal calf serum. From our earlier conversation, I understand you grow your fibroblasts in fetal calf serum. Therefore, I suspect the reason you see a stimulation of fibroblasts by TGF beta, and I do not, is because of the differences in the mixture of growth factors, or other properties of our two culture systems.

M.M. LOMBARD : In relation with Dr Lord's question to Dr Axelrad, my question to Dr Solberg is about what do we know of the precise cell cycle step at which TGF beta acts, both for its positive, stimulating, and its inhibiting effect?

L. SOLBERG : We do not yet know for normal hematopoietic stem cells, what phase of the cell cycle is affected by TGF beta. In other cell types, that has been looked at more extensively on cell lines, for example, Dr Moses's group looked at fibroblast cell lines and in that case TGF beta will actually induce the synthesis of PDGF, which then is secreted and autostimulates cells. But for hematopoietic cells, cell-cycle related effects have not been looked at. I have looked at DNA synthesis in megakaryocytes, it is interesting that in those megakaryocytes that survive in the presence of TGF beta, their DNA synthesis as measured by bromodioxyuridine incorporation is not inhibited.

A. AXELRAD : We have looked at the effect of TGF beta on DNA synthesis of BFU-E as judged by percent kills, and the results show a linear diminution in percent kill with increase in concentration of TGF beta on a short time scale. This is within minutes. TGF beta also competes with Interleukin-3 in the same way.

N. DAINIAK : I have a question for you. Could you comment or speculate on the mechanism by which these inhibitory factors may be released from the cell surface? Is it simply a matter of physical disruption of molecules at the surface or could the active factor be enzymatically cleaved? Or, is it possible that release of the active factor can be altered by glycosidases?

A. AXELRAD : As you know, we have not done anything systematic on that. Initially from the marrow, we got NRP in a single wash in the cold, so it is hard to believe that an enzymatic action was involved in cutting it from the surface. It is either not attached at all, but just active on the surface of the BFU-E or it is very loosely attached to the surface and easily washed off.

O.D. LAERUM : I have a general comment and then a specific question to Dr Axelrad. My general comment is the following : I think when we are looking at stimulating or at inhibitory factors it is also important to look at the time factor. If we look at hematopoiesis in vivo, it changes quite rapidly. There can be a difference of cell proliferation of up 50 to 80 % within a few hours and such circadian variations in hemopoiesis are also documented in man. Something must actively regulate these waves of activity, which also occur at the stem cell level. Other regulatory mechanisms act more slowly by increasing or decreasing the general level of activity. I would like to have comments on all the different regulators we have heard of today, at what level they act and the time factor. For example, in our experiments with the hemoregulatory peptide, we find both a rapid effect and a more long lasting effect. Then concerning your experiments, Dr Axelrad, I was quite impressed by the quick action. But are your systems only based on suicide techniques? Can you exclude the possibility that the protein affects the membrane fluidity or permeability, so that there might be a transport effect instead of a direct effect on the DNA synthesis?

A. AXELRAD : No, I don't think that can be ruled out yet. But I think it is very unlikely because NRP reduces the percentage of BFU-E killed by either tritiatedthymidine or hydroxyurea. These are very different kinds of molecules. Also the effect of NRP is opposed by Interleukin-3 in a dose-dependent manner, another very different kind of molecule. A transport effect would have to influence all these molecules in the same way.

A.C. RICHES : Do you have an anti-NRP so that you can inhibit your inhibitor at all and the second question is, do you know from what cell the NRP is being produced in the bone marrow?

A. AXELRAD : As for the first question, I wish we had an anti-NRP but we have so little of the semipurified material, we have zero of purified material for immunization. With the material that was 100,000-fold enriched, we cannot see a band on a gel stained with silver. We are trying to immunize in vitro because there isn't anywhere near enough to give it to an animal. Your second question was which cell produces it. We don't really know much about the NRP-producing marrow cell that we have got growing in culture as a continuous line : we know that it is a hypotetraploid line, that it has three copies of chromosome 9, it is partially adherent, and it is not able to make erythroid colonies. We, of course, started with the hypothesis that if a cell late in the differentiation sequence could make NRP, if there were enough of those, it could send a signal back to the early part of the system to say turn off, that would have been very neat. It doesn't seem to be like that. It is a semiadherent cell that was got originally by adherence. We have not done yet a systematic phenotype of that cell.

H. BROXMEYER : What is the relationship between the apotransferrin and the Negative Regulatory Protein and could they be the same?

A. AXELRAD : That was a question that worried us earlier on and in fact, Dr Mel Greaves in London and I have a bet on, that Negative Regulatory Protein is (that is his side of the bet), and is not (that is my side of the bet) the same molecule as transferrin or apotransferrin. We have looked in a number of ways at possible differences between them. Antibodies made in goat directed against mouse serum transferrin do not affect the Negative Regulatory Protein biological activity. It doesn't show by Elisa test after purification. Initially, transferrin is clearly present in the crude material,but as one purifies, it disappears, while the specific activity of the Negative Regulatory Protein goes up 150 times in one step. That is on material from bone marrow taken directly from the animal. We have looked at the behaviour of transferrin and NRP on ion exchange and on wheat germ lectin columns and it is quite different. We thus have a whole series of pieces of evidence now that these are not the same molecule.

H. BROXMEYER : Can the iron-saturated transferrin block the action of the Negative Regulatory Protein?

A. AXELRAD : We get a normal level of percent kill in the presence of iron-saturated transferrin but a diminution in the level of percent kill with apotransferrin. We have not done competition experiments between iron-saturated transferrin and NRP. One of the reasons also that we thought maybe NRP and transferrin were the same molecule is that the difference between B6 and B6-S mouse strains on chromosome 9 represents a sequence of about ten recombination units ; the transferrin gene and transferrin receptor gene are right there, near where FV-2 is, and we believe that FV-2 controls the production of NRP. Also the molecular weight is approximately the same for the two. But the other evidence clearly shows, I think, that the two are not identical.

B. VARET : It is interesting to see that erythropoetin is able to stimulate both erythroid colonies and megakaryocytic colonies. Do you think it is possible that there is a competition between TGF beta and erythropoetin on erythropoietin receptors?

W. VAINCHENKER : I think it is a good question, especially for erythroid lineage, since TGF beta has a main action on CFU-E and a low action on BFU-E. Therefore, TGF beta may downregulate the Epo-receptor.

Brief reports

Communications brèves

Inhibitors of erythropoiesis in patients with chronic renal failure

Vera Pavlović-Kentera, Ljubica Djukanović, Lidija Biljanović-Paunović, Nevenka Stojanović and Pavle Milenković

Institute for Medical Research, 11001 Beograd, POBox 721, Yugoslavia

KEYWORDS

Inhibitors of erythropoiesis, renal failure, anemia, hemodialysis

INTRODUCTION

Anemia with hypoproliferative marrow is a frequent complication in patients with chronic renal failure. In the pathogenesis of this anemia, inadequate erythropoietin (Ep) production is a major cause. The role of inhibitors of erythropoiesis as a contributing factor, although demonstrated in different test systems, is questioned. We present here the results of a study of inhibitors in the sera of patients with chronic renal failure undergoing maintenance hemodialysis.

PATIENTS STUDIED AND METHODS

Thirty five patients were dialysed three times a week for five hours using a dialyser with a cuprophane membrane. The etiology of the end stage renal disease was: glomerulonephritis (7), pyelonephritis (10), polycystic kidney disease (8), Balkan nephropathy (5) and other (5). The patients were on dialysis treatment for one to 16 years. Blood samples were obtained immediately before and after regular hemodialysis. Serum samples were kept frozen at -20°C until tested. All the patients were HBS negative and had given their consent before the study.

Serum inhibitors were tested <u>in vitro</u> (Wallner, 1978) in a mouse bone marrow CFU-E assay in methylcellulose (Biljanović-Paunović, 1985). Plasma from mice previously irradiated and injected with phenylhydrazine was the source of Ep. Mouse bone marrow CFU-GM (Jovčić, 1986) with phytohemagglutinin stimulated human leucocyte conditioned medium (PHLCM) as a source of CSF (Aye, 1974) was used to test inhibitor cell specificity. Sterile filtered sera from the patients or normal human sera (10%) was added to the culture mixture at the beginning of incubation.

Inhibitors were tested in polycythemic mice (Lindemann, 1969) in vivo by injecting 0.5 ml of patients' sera 3 hours before Ep injection. Mice were given 0.5 uCu of ^{59}Fe 48 hours later and after another 48 hours ^{59}Fe incorporation was measured.

RESULTS

Inhibition of CFU-E colony formation was seen in all but one patient. In Fig. 1 the hematocrit in the patients studied is presented as a function of inhibition of CFU-E colony formation.

In Fig. 2 the inhibition of CFU-E colony formation with predialysis and postdialysis serum samples of different patients is portrayed. The inhibition was not significantly different, indicating that CFU-E inhibitors were not lost during hemodialysis treatment. As a rule CFU-GM were not inhibited by patients' sera either predialysis or postdialysis (Fig. 3).

When tested in a bioassay two out of 6 sera decreased Ep stimulated ^{59}Fe incorporation in polycythemic mice.

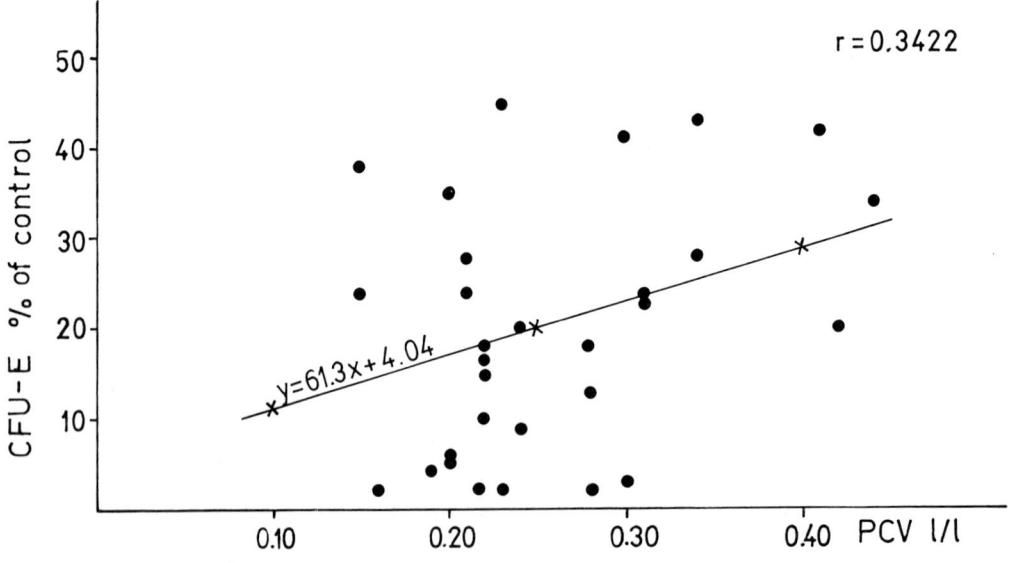

Fig. 1. Relationship between inhibition of erythroid colony formation and hematocrit in patients with chronic renal failure receiving maintenance hemodialysis treatment. Patients' sera (10%) was added to the cultures.

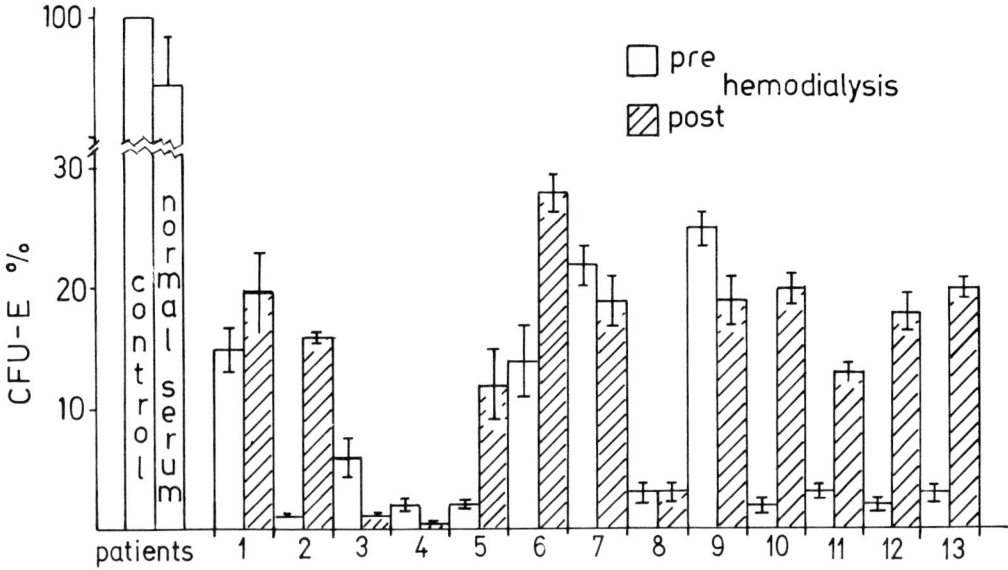

Fig. 2. Percent inhibition of CFU-E colony formation by patients' pre and post hemodialysis sera (10%). Control number of CFU-E from 10^5 cells with 0.25 U Ep expressed as 100%. Mean ± SE for 4 plates.

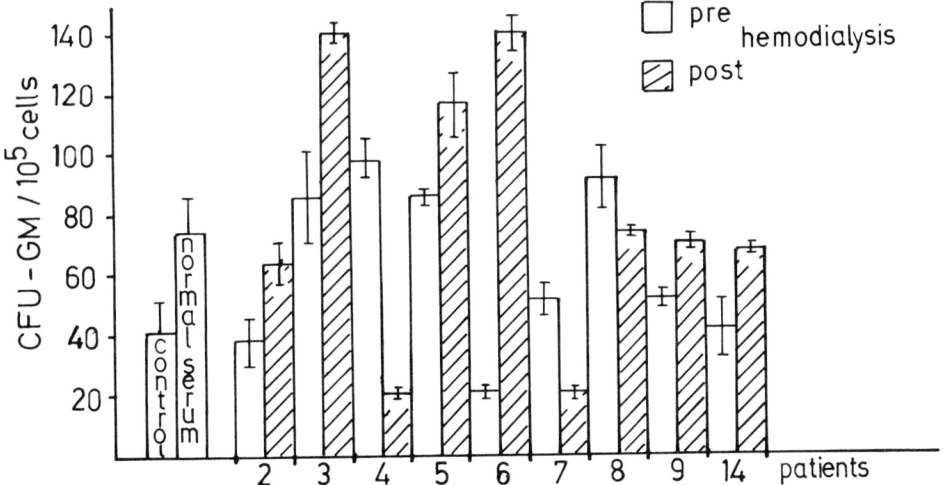

Fig. 3. Number of CFU-GM colonies in cultures incubated in the presence of 10% patients' sera pre and post hemodialysis. Control number of CFU-GM from 10^5 cells with 5% PHLCM. Mean ± SE for 4 plates.

DISCUSSION

The role of inhibitors of erythropoiesis in the anemia of chronic renal failure is not completely clarified although substances that inhibit growth of erythroid progenitors in vitro (Wallner, 1978) have been reported. McGonigle (1984a, 1985) found that serum inhibitors of erythropoiesis determine the degree of anemia in patients with renal failure. We have not found any correlation here between hematocrit values and inhibition of CFU-E derived colonies in vitro by sera from the patients studied. The same sera did not inhibit growth of CFU-GM from mouse bone marrow. This is in agreement with the observation of McGonigle (1984) on human marrow in vitro but differs from Delwiche (1986) who reported a lack of in vitro cell specificity of inhibitors in chronic renal failure. The results presented here indicate the heterogeneity of inhibitors. Most inhibitors cannot be detected in vivo in mice with normal renal function. Most are not dialysed out by hemodialysis treatment and do not affect CFU-GM but rather CFU-E growth in vitro. Although the level of erythropoiesis in patients with end stage renal disease depends predominantly on the availability of Ep and replacement therapy with recombinant Ep proved successful (Eschbach, 1987), the role of inhibitors of erythropoiesis in the anemia of chronic renal failure cannot be denied.

REFERENCES

Aye, M.T., Niho, Y., Till, J.E., McCulloch, E.A. (1974): Studies of leukemic cell populations in culture. Blood 44, 205-219.
Biljanović-Paunović, L., Milenković, P., Kapa, D., Pavlović-Kentera, V. (1985): Regeneration of erythroid progenitor cells in polycythemic mice treated with cyclophosphamide. Exp. Hematol. 13,67-73.
Delwiche, F., Segal, G.M., Eschbach, J.W., Adamson, J.W. (1986): Hematopoietic inhibitors in chronic renal failure: Lack of in vitro specificity. Kidney Int. 29, 641-648.
Eschbach, J.W., Egrie, J.C., Downing, M.R., Browne, J.K., Adamson, J.W.(1987): Correction of the anemia of end-stage renal disease with recombinant human erythropoietin. N. Engl. J. Med. 316,73-78.
Jovčić, G., Stojanović, N., Biljanović-Paunović, L., Hajduković, S. (1986): Comparative study of the effect of serum containing granulocyte-stimulating activity and colony-stimulating activity on hematopoiesis in triple-set diffusion chambers. Exp. Hematol. 14, 298-303.
Lindemann, R. (1976): Erythropoiesis-inhibiting factor(s) (EIF): Methodic studies. Blood 47, 155-163.
McGonigle, R.J.S., Wallin, J.D., Shadduck, R.K., Fisher, J.W. (1984): Erythropoietin deficiency and inhibition of erythropoiesis in renal insufficiency. Kidney Int. 25, 437-444.
McGonigle,R.J.S., Husserl,F., Wallin,J.D., Fisher,J.W. (1984a):Hemodialysis and continous ambulatory peritoneal dialysis effects on erythropoiesis in renal failure. Kidney Int. 25, 430-436.
McGonigle,R.J.S., Boineau,F.G., Beckman,B., Chene-Frempong,K., Lewy, J.E., Shadduck,R.K., Fisher,J.W. (1985): Erythropoietin and inhibitors of in vitro erythropoiesis in the development of anemia in children with renal disease. J. Lab. Clin. Med. 105, 449-458.
Wallner,S.F., Vautrin,R.M., Kurnick,J.E., Ward,H.P. (1978): The effect of serum from patients with chronic renal failure on erythroid colony growth in vitro. J. Lab. Clin. Med. 92, 370-375.

Suppression of hematopoiesis in acute leukemias

Inhibition de l'hématopoïèse dans les leucémies aiguës

Biomolecules associated with suppression of myelopoiesis in normal conditions and during myeloid leukemia and other related disorders

Hal E. Broxmeyer [1,2], Douglas E. Williams [1], Li Lu [1], Saroj Vadhan [3], Scott Cooper [1], David C. Bicknell [1], Peter Ralph [4], Jordan Gutterman [3] and Guido Tricot [1]

From the Departments of [1] Medicine, [2] Microbiology and Immunology, Indiana University School of Medicine, Indianapolis, IN 46223, the [3] Department of Clinical Immunology, M.D. Anderson Hospital and Tumor Institute, Houston, TX, 77030 and the [4] Department of Cell Biology, Cetus Corporation, Emeryville, CA, 94608, USA.

ABSTRACT

This paper reviews the myelopoietic suppressive roles in vitro for several purified and well-characterized molecules, acidic isoferritin (AIF), lactoferrin (LF), tumor necrosis factors (TNF)- alpha and beta, interferon (IFN)-gamma and prostaglandin E (PGE). Data is presented demonstrating that: A) granulocyte-macrophage (CFU-GM) progenitors are more sensitive to inhibition with TNF-alpha when they are stimulated with granulocyte (G)- colony stimulating factor (CSF) than with granulocyte-macrophage (GM)-CSF, macrophage CSF (CSF-1) or medium conditioned by 5637 cells which contains both GM- and G-CSF, B) AIF, TNF-alpha, IFN-gamma and PGE can suppress colony formation by cells very highly enriched for progenitors and depleted of accessory cells, C) IFN-gamma can synergize with AIF and PGE to suppress colony formation of mouse and human CFU-GM, and D) cells of patients with leukemia, myelodysplasia, solid tumors and other hematological disorders can vary from normal in their responses to the suppressive effects of AIF, TNF-alpha, IFN-gamma and PGE.

KEY WORDS

Lactoferrin, Acidic Isoferritin, Tumor Necrosis Factors, Interferon-Gamma, Prostaglandin E, Myeloid Leukemia, Hematopoietic Colony Stimulating Factors.

INTRODUCTION

Molecules that stimulate or suppress the growth and differentiation of granulocyte-macrophage (CFU-GM), erythroid (BFU-E) and multipotential (CFU-GEMM) progenitor cells have been isolated, purified and characterized. In many cases the genes for these bioregulatory molecules have been isolated, cloned and the proteins expressed and then purified (reviewed in Broxmeyer, 1986, Metcalf, 1986). Some investigators have suggested that the regulation of blood cell production can be entirely considered as a function of stimulating molecules where feedback suppression of a proliferative event is due to either the down regulation of a receptor for the stimulating molecule and/or the disappearance of a stimulating molecule. Such mechanisms might be involved in dampening of proliferative responses. However, there is a large literature which also demonstrates a role for molecules with direct or indirect suppressive activities (Broxmeyer, 1986). Moreover, these suppressive molecules can be induced or primed for release from accessory cells by colony stimulating factors (CSF) such as the macrophage CSF (CSF-1) (Pelus et al., 1979, Moore et al., 1984, Broxmeyer et al., 1985b, Warren and Ralph, 1986). These studies in vitro clearly suggest that regulation of blood cell production entails an interaction between stimulating and suppressing molecules. Evidence for this is that molecules can have more than one function and one molecule can have both suppressor and stimulatory capabilities depending on the concentration of the molecule and the cell type with which they are in contact (reviewed in Broxmeyer, 1986). This present paper describes the suppressive effects of purified acidic isoferritin (AIF), lactoferrin (LF), tumor necrosis factors (TNF) alpha and beta (also referred to as lymphotoxin), interferon-gamma (IFN-gamma) and prostaglandin E (PGE).

AIF is a metal binding protein with 24 subunits containing a greater proportion of heavy ($\sim$ 21,000, Mr) to light ($\sim$ 19,000, Mr) subunits and suppresses colony formation in vitro of CFU-GM, BFU-E and CFU-GEMM from normal donors, but not from patients with leukemia (Broxmeyer, 1982, Broxmeyer et al., 1981, 1982, 1986b). LF is a metal- binding glycoprotein that decreases the release from monocytes-macrophages of CSFs or monokines (probably interleukin-1) that can trigger the release of CSFs from T-lymphocytes, endothelial cells and fibroblasts (Broxmeyer et al., 1978, 1986a, Bagby et al., 1981, 1986, Zucali et al., 1987). LF can decrease release of growth factors from cells of patients with leukemia, but in some cases the LF is less effective against patient than normal cells (Broxmeyer et al., 1984). Increased responsiveness to LF can be induced by IFN-gamma in established cell lines (Broxmeyer et al., 1986c). TNFs cause the necrosis and regression of some tumors in mice (Old, 1985) and can decrease colony formation by CFU-GM, BFU-E and CFU-GEMM (Degliantoni et al., 1985, Broxmeyer et al., 1986d, Murphy et al., 1986). There is evidence that primary progenitor cells of patients with leukemia are more sensitive to inhibiton by TNF than cells from normal donors (Broxmeyer et al., 1986d, Peetre et al., 1986, Munker and Koeffler, 1987). IFN-gamma is a glycoprotein produced by cells in responses to various stimuli and can suppress colony formation by hematopoietic progenitors (Broxmeyer et al. 1983, 1985a, Raefsky et al. 1985). IFN-gamma can also suppress progenitors from patients with leukemia. PGE suppresses colony formation by CFU-GM but at more physiological concentrations this effect is more selective for monocyte-macrophage progenitors (Pelus et al., 1979). Progenitors from patients with leukemia may be less sensitive than normal progenitors to suppression by PGE (Pelus et al., 1980, Aglietta et al. 1980). Synergism between some of these molecules has also been noted (some of the above references and reviewed in Broxmeyer, 1986). This present report presents information characterizing further the suppressive effects of TNF-alpha, IFN- gamma, PGE and AIF on cells of normal donors and patients with leukemia.

MATERIALS AND METHODS

Bone marrow cells were obtained with informed consent from healthy donors and patients with leukemia at the Indiana University School of Medicine, Indianapolis, IN, and the M.D. Anderson Hospital and Tumor Institute, Houston, TX. Mice of the BDF_1 strain were obtained from Cumberland View Farms, Clinton, TN.

Assays for hematopoietic progenitor cells have been described elsewhere (Broxmeyer et al., 1982, 1986d). Cells were cultured at low (5%) Oxygen tension. Purified recombinant human AIF was obtained from Dr. P. Arosia, Milan, Italy (Broxmeyer et al. 1986b) and natural AIF was purified (Broxmeyer et al. 1981). Purified recombinant human TNF-alpha preparations were obtained from Dr. H.M. Sheppard, Genentech, Inc., South San Francisco, CA, and from Dr. L. Lin, Cetus Corp., Emeryville, CA. TNF was titred for unitage against L929 cells in the absence of agents such as Actinomycin D. Both preparations were equal in activity against hematopoietic progenitor cells (Broxmeyer, unpublished observations). Purified recombinant mouse and human IFN-gamma were also obtained from Dr. H.M. Sheppard. PGE_1 was purchased from Sigma Chem. Co., St. Louis, MO. 5637- and pokeweed mitogen mouse spleen cell (PWMS) conditioned medium (CM) were prepared as described elsewhere (Broxmeyer et al. 1982). Erythropoietin was purchased from Toyoba, NY. Purified recombinant human GM-CSF was obtained from Dr. S. Gillis, Immunex Corp., Seattle, WA, and purified recombinant G-CSF and human CSF-1 were obtained from Cetus Corp. Mouse marrow CFU-GM and human marrow CFU-GM and BFU-E were enriched as described (Williams et al. 1987, Lu et al. 1987).

RESULTS

Normal human bone marrow low density cells (< 1.077 gm/cm^3) stimulated by 5637 conditioned medium or by purified preparations of either recombinant human GM-CSF, G-CSF or CSF-1 were compared for their responsiveness to the suppressive effects of purified recombinant human TNF-alpha. As shown in Table 1, for day 7 and day 14 CFU-GM, cells stimulated by G-CSF were most sensitive, and cells stimulated by GM-CSF were least sensitive to the suppressive effects of TNF-alpha.

Table 1. Comparative Effects of Purified Recombinant Human Tumor Necrosis Factor-Alpha (TNF) on CFU-GM in Normal Human Bone Marrow Stimulated by 5637 Conditioned Medium or by Purified Preparations of Either Recombinant Human GM-CSF, G-CSF or CSF-1.

	Colonies + Clusters (% Change from Controls)			
	5637 CM	GM-CSF	G-CSF	CSF-1
A) Day 7 CFU-GM				
Control Medium	92 ± 4	88 ± 5	95 ± 6	85 ± 7
TNF 100 units	33 ± 3(-64*)	47 ± 3(-47*)	6 ± 0.9(-94*)	28 ± 2(-67*)
TNF 10 units	72 ± 1(-22*)	79 ± 12(-10)	12 ± 1(-87*)	50 ± 6(-41*)
TNF 1 unit	95 ± 3(+3)	86 ± 5(-2)	39 ± 4(-59*)	87 ± 2(+2)
TNF 0.1 unit	99 ± 4(+8)		52 ± 2(-45*)	
B) Day 14 CFU-GM				
Control Medium	57 ± 4	65 ± 6	49 ± 2	50 ± 0.5
TNF 100 units	28 ± 1(-51*)	36 ± 2(-45*)	3 ± 0.3(-94*)	11 ± 1(-78*)
TNF 10 units	52 ± 2(-9)	59 ± 1(-9)	14 ± 1(-71*)	31 ± 2(-38*)
TNF 1 unit			23 ± 4(-53*)	42 ± 7(-16)
TNF 0.1 unit				35 ± 2(-29*)

* Significant percent change from control medium, p at least < 0.05.

There is no strong evidence yet as to whether the suppressive effects noted for TNF-alpha, IFN-gamma, PGE_1 or AIF can be mediated directly at the level of the progenitor cells themselves. To assess this possibility directly, human (Table 2) and mouse (Table 3) bone marrow cells were highly enriched for progenitor cells and were assessed for sensitivity to the effects of the molecules. Human non-adherent low density T-lymphocyte-depleted ($NALT^-$) cells were sorted for varying cell surface density expression of My10 and HLA-DR. The highest cloning efficiency, in methylcellulose in the presence of 1 unit erythropoietin, 0.2 mM hemin and 10% v/v 5637 CM, was obtained for CFU-GM and BFU-E in the My10 population of cells expressing a low level of HLA-DR antigens (Lu et al., 1987), and these cells with a cloning efficiency of up to 47% were at least as sensitive to the suppressive effects of TNF-alpha, human IFN-gamma and PGE_1 as were the unsorted population of $NALT^-$ cells which demonstrated a cloning efficiency of 0.4% (Table 2).

Table 2. Effect of Recombinant Human Tumor Necrosis Factor -Alpha (TNF), Recombinant Human Interferon Gamma (IFN) and Prostaglandin E_1 (PGE_1) on Colony Formation by CFU-GM and BFU-E in Non-adherent Low Density T-Lymphocyte Depleted ($NALT^-$) and Sorted Bone Marrow Cell Fractions Stimulated with Erythropoietin, Hemin and 5637 Conditioned Medium[a]

	Percent Change from Control Medium				
	CFU-GM			BFU-E	
Molecules	$NALT^-$	$My10^{+to+++} DR^+$		$NALT^-$	$My10^{+to+++} DR^+$
TNF (100 units)	$-79 \pm 5*$	$-90 \pm 3*$		$-80 \pm 2*$	$-91 \pm 3*$
IFN (100 units)	$-60 \pm 4*$	$-78 \pm 5*$		$-54 \pm 4*$	$-75 \pm 11*$
PGE_1 (10^{-7}M)	$-30 \pm 1*$	$-43 \pm 4*$		$+11 \pm 4$	$+8 \pm 3$

[a] Cloning efficiencies for CFU-GM + BFU-E were 0.4% for $NALT^-$ cells and up to 47% for $My10^{+to+++} DR^+$ sorted cells. $My10^{+to+++}$ refers to all My10 positive cells.

* P at least < 0.05.

Bone marrow cells isolated, 3 days after mice were given a sublethal dosage of cyclophosphamide (200 mg/kg), by Ficol-Hypaque and centrifugal elutriation (Fraction, 28 ml/hr) (Williams et al., 1987) demonstrate cloning efficiencies of CFU-GM up to 99%. These cells were stimulated by 10% v/v PWMSCM and assessed for responsiveness to recombinant human AIF, recombinant human TNF-alpha, recombinant murine IFN-gamma and PGE_1. The purified CFU-GM, with cloning efficiencies of up to 94% (in these experiments), were at least as sensitive to the suppressive effects as were the unseparated populations of cells (Table 3).

Table 3. Suppressive Influence of Biomolecules on Colony Formation by CFU-GM in Purified and Unseparated Mouse Bone Marrow Preparations Stimulated by Pokeweed Mitogen Mouse Spleen Cell Conditioned Medium.[a]

Molecules[b]	Percent Significant Inhibition of Colony Formation[c]	
	Purified CFU-GM	Unseparated Marrow
AIF (10^{-9}M)	34 ± 4	23 ± 2
TNF (1000 units)	43	30
IFN-gamma (100 units)	38 ± 10	29 ± 3
PGE_1 (10^{-7}M)	40 ± 9	32 ± 3

[a] Cloning efficiencies of up to 94% for purified CFU-GM with 100 cells per plate.

[b] AIF, purified recombinant human H-subunit (acidic) ferritin
TNF, purified recombinant human tumor necrosis factor -alpha
IFN, purified recombinant murine interferon -gamma
PGE_1, prostaglandin E_1

[c] p at least < 0.05

When IFN-gamma was assessed in combination with either purified natural AIF or recombinant human or mouse IFN-gamma, synergism was noted in suppression of colony formation by day 7 human CFU-GM and day 5 mouse CFU-GM (Table 4).

Table 4. The Synergistic Suppressive Effects of Interferon (IFN) Gamma with Purified Natural Acidic Isoferritin (AIF) or Prostaglandin E_1 (PGE_1) on Colony Formation by Human or Mouse Bone Marrow CFU-GM[a]

	Day 7 Human CFU-GM Colonies		Day 5 Mouse CFU-GM Colonies	
Control = McCoys Medium	112 ± 5		67 ± 4	
IFN (0.01 unit)	111 ± 4	(-1)	64 ± 5	(-4)
AIF (10^{-12}M)	84 ± 6	(-25*)	41 ± 2	(-39*)
PGE_1 (10^{-11}M)	114 ± 3	(+2)	60 ± 3	(-10)
IFN (0.01 unit) + AIF (10^{-12}M)	57 ± 3	(-49*)	30 ± 3	(-55*)
IFN (0.01 unit) + PGE_1 (10^{-11}M)	61 ± 1	(-46*)	33 ± 2	(-51*)

[a] 10^5 Human bone marrow cells were stimulated by 10% v/v 5637 CM and 7.5 x 10^4 unseparated mouse bone marrow cells were stimulated by 10% v/v PWMSCM. Recombinant murine IFN-gamma was used for mouse cell assay and recombinant human IFN-gamma was used for human cell assay.

* Significant percent change from control, p at least < 0.05.

Low density bone marrow cells (< 1.077 gm/cm^3) from normal donors and patients with leukemia, myelodysplasia, solid tumors and other disorders were stimulated by 10% v/v 5637 CM and compared for their responsiveness to purified recombinant-human AIF, -human TNF-alpha, -human IFN-gamma and PGE_1 (Table 5). Day 7 marrow CFU-GM colonies plus clusters were suppressed by these molecules in a manner similar to that previously reported by us. The cells from the 11 patients with leukemia and myelodysplasia were less sensitive than normal to the suppressive effects of AIF, IFN-gamma and PGE_1 but were more sensitive to the suppressive effects of TNF-alpha. The cells from the 7 patients with solid tumors and other disorders were slightly less sensitive than normal cells to IFN-gamma and PGE_1 but were more sensitive to TNF-alpha and equal in sensitivity with normal cells to AIF. Cells from one myelodysplastic patient with Refractory Anemia with excess blasts (RAEB) were hyporesponsive to the suppressive effects of all these molecules.

Table 5. Influence of Acidic Isoferritin, Tumor Necrosis Factor-Alpha, Interferon-Gamma or Prostaglandin E_1 on Colony and Cluster Formation by Day 7 Marrow CFU-GM from Normal Donors and Patients with Hematopoietic Disorders.[a]

	Percent Change from Control			
Biomolecules	Normal (N=4-6)	Leukemia and Myelodysplasia (N=8-11)	Solid Tumors and other Disorders (N=7)	Patient with RAEB
AIF (10^{-10}M)	-35 ± 3	$-4 \pm 2*$	-31 ± 4	0*
TNF (100 units)	-70 ± 7	-85 ± 2	-80 ± 3	$-27*$
TNF (10 units)	-49 ± 7	$-73 \pm 3*$	$-68 \pm 5*$	0*
TNF (1 unit)	-26 ± 4	$-56 \pm 4*$	$-56 \pm 4*$	
TNF (0.1 unit)	-6 ± 4	$-25 \pm 4*$	$-32 \pm 6*$	
IFN (100 units)	-43 ± 5	$-27 \pm 8*$	$-30 \pm 5*$	$-4*$
IFN (10 units)	-26 ± 6	$-4 \pm 4*$	$-11 \pm 4*$	
IFN (1 unit)	-11 ± 6		$+2 \pm 4$	
PGE_1 10^{-7}M	-38 ± 6	$-13 \pm 8*$	-19 ± 10	0*
PGE_1 10^{-8}M	-30 ± 3	$-7 \pm 5*$	$-8 \pm 6*$	
PGE_1 10^{-9}M	-9 ± 4			

[a] Purified preparations of recombinant human acidic (H-subunit) ferritin (AIF), recombinant human tumor necrosis factor-alpha (TNF), recombinant human interferon-gamma (IFN) and prostaglandin E_1 (PGE_1) were used.

Leukemia/myelodysplasia group included patients with AML, APL, RAEB, RA and accelerated CML. Solid tumor/other group included patients with adenocarcinoma of colon with liver metastasis, breast cancer with liver metastasis, malignant melanoma, squamous cell carcinoma of lung, Hodgkins disease, liposarcoma, and myelofibrosis.

* Significant difference from normal donors, p at least < 0.05.

DISCUSSION

There is an expanding literature demonstrating a role for suppressor molecules in the regulation of myelopoiesis. This present communication only mentions some of the reports concerning these molecules and only a small part of the literature of the molecules described here have been cited. There is evidence from studies in animals that these molecules may be relevant physiologically (reviewed in Broxmeyer, 1986) and some of the molecules may be useful in treatment of certain selected diseases. In fact, clinical trials have already begun with TNF-alpha and IFN-gamma.

One obvious point from the studies reported by us and others, and from the data presented here, is that several molecules can act together with additive or synergistic effects and a true understanding of the actions of these molecules probably requires that they be evaluated in combination rather than only as single molecules. The hematopoietic progenitors from normal donors were most sensitive to the suppressive effects of TNF-alpha when cells were stimulated with G-CSF, in comparison with cells stimulated with GM-CSF or CSF-1. Cells stimulated by 5637 CM, which contains both GM- and G-CSF, were also less sensitive to the action of TNF-alpha than were cells stimulated with G-CSF. IFN-gamma synergizes with TNF-alpha (Degliantoni et al., 1985, Broxmeyer et al., 1986d) and with IFN-alpha (Broxmeyer et al., 1985a, Raefsky et al., 1985a) to suppress colony formation and the results here have shown that it also synergizes with AIF and PGE_1 to suppress colony formation. It cannot be determined from these studies whether the synergistic effects are mediated directly on the progenitors or indirectly via an action on accessory cells, but from our data using very highly enriched populations of progenitors (Tables 2 and 3) it is very likely that the action of single molecules such as TNF-alpha, IFN-gamma, PGE_1 and AIF can act directly at the level of the progenitors. These studies do not rule out the possibility that the suppressive actions of single molecules can also be mediated via accessory cells.

The efficacy of action of molecules during clinical trials may vary depending on the type of disease, stage of disease, prior treatment of the patients and may even vary within specific subsets of patients, facts previously established with chemotherapeutic drugs. In such cases, assessment of the responsiveness of hematopoietic progenitor cells from patients in vitro to the candidate molecules may help to define subpopulations of patients that stand a greater chance of demonstrating clinical responsiveness to the factors. It is clear from our data elsewhere and from that shown in Table 5 that responsiveness to one molecule does not predict responsiveness to another molecule. Some groups of patients are more responsive to some of the molecules than others, but large variations in responsiveness to the same molecule can be seen in patients with a similar disease. Thus, screening tests for factor responsiveness should probably be done with the cells from the individuals, before and during the clinical trials for maximum information and most effective planning of the use of the molecules.

ACKNOWLEDGEMENTS

We wish to thank Shirley Duke and Stephanie Moore for typing the manuscript. These studies were supported by Public Health Service Grants R01 CA 36740, R01 CA 36464 and T32 AM 07519 to H.E.B.

REFERENCES

Aglietta, M., Piacibello, W., Gavosto, F. (1980): Decreased sensitivity of chronic myeloid leukemia cells to inhibition of growth by prostaglandin E. Cancer Res. 40: 2509-2511.

Bagby, G.C., Regas, V.D., Bennett, R.M., Vandenbark, A.A., Garewal, H.S. (1981): Interaction of lactoferrin, monocytes and T lymphocyte subsets in the regulation of steady state granulopoiesis in vitro. J. Clin. Invest. 68: 56-63.

Bagby, G.C., Dinarello, C.A., Wallace P., Wagner, C., Hefeneider, S., McCall, E. (1986): Interleukin-1 stimulates GM-CSA release by vascular endothelial cells. J. Clin. Invest., 78: 1316-1323.

Broxmeyer, H.E. (1982): Relationship of cell cycle expression of Ia-like antigenic determinants on normal and leukemia human granulocyte-macrophage progenitor cells to regulation in vitro by acidic isoferritins. J. Clin. Invest. 69: 632-642.

Broxmeyer, H.E. (1986): Biomolecule-cell interactions and the regulation of myelopoiesis. Int. J. Cell Cloning 4: 378-405.

Broxmeyer, H.E., Smithyman, A., Eger, R.R., Meyers, P.A., deSousa, M. (1978): Identification of lactoferrin as the granulocyte-derived inhibitor of colony-stimulating activity (CSA) production. J. Exp. Med. 148: 1052-1067.

Broxmeyer, H.E., Bognacki, J., Dorner, M.H., deSousa M. (1981): The identification of leukemia-associated inhibitory activity (LIA) as acidic isoferritins: a regulatory role for acidic isoferrtins in the production of granulocytes and macrophages. J. Exp. Med. 153: 1426-1444.

Broxmeyer, H.E., Bognacki, J., Ralph, P., Dorner, M.H., Lu, L., Castro-Malaspina, H. (1982): Monocyte-macrophage derived acidic isoferritins: normal feedback regulators of granulocyte-macrophage progenitor cells. Blood 60: 595-607.

Broxmeyer, H.E., Lu, L., Platzer, E., Feit, C., Juliano, L., Rubin, B.Y. (1983): Comparative analysis of the influences of human gamma, alpha and beta interferons on human multipotential (CFU-GEMM), erythroid (BFU-E) and granulocyte-macrophage (CFU-GM) progenitor cells. J. Immunol. 131: 1300-1305.

Broxmeyer, H.E., Gentile, P., Cooper, S., Lu, L., Juliano, L., Piacibello, W., Meyers, P.A., Cavanna, F. (1984): Functional activities of acidic isoferritins and lactoferrin in vitro and in vivo. Blood Cells 10: 397-426.

Broxmeyer, H.E., Cooper, S., Rubin, B.Y., Taylor, M.W. (1985a): The synergistic influences of human interferon-γ and interferon-α on suppression of hematopoietic progenitor cells is additive with the enhanced sensitivity of these cells to inhibition by interferons at low oxygen tension in vitro. J. Immunol. 135: 2502-2506.

Broxmeyer, H.E., Juliano, L., Waheed, A., Shadduck, R.K. (1985b): Release from mouse macrophages of acidic isoferritins that suppress hematopoietic progenitor cells is induced by purified L cell colony stimulating factor and suppressed by human lactoferrin. J. Immunol. 135: 3224-3231.

Broxmeyer, H.E., Bicknell, D.C., Gillis, S., Harris, E.L., Pelus L.M., Sledge, G.W. Jr. (1986a): Lactoferrin: affinity purification from human milk and polymorphonuclear neutrophils using monoclonal antibody (II2C) to human lactoferrin, development of an immunoradiometric assay using II2C, and myelopoietic regulation and receptor-binding characteristics. Blood Cells 11: 429-446.

Broxmeyer, H.E., Lu, L., Bicknell, D.C., Williams, D.E., Cooper, S., Levi, S., Salfeld, J., Arosio, P. (1986b): The influence of recombinant human heavy-subunit and light-subunit ferritins on colony formation in vitro by granulocyte-macrophage and erythroid progenitors. Blood 68: 1257-1263.

Broxmeyer, H.E., Piacibello, W., Juliano, L., Platzer, E., Berman, E., Rubin, B.Y. (1986c): Gamma interferon induces colony-forming cells of the human monoblast cell line U937 to respond to inhibition by lactoferrin, transferrin and acidic isoferritins. Exp. Hematol 14: 35-43.

Broxmeyer, H.E., Williams, D.E., Lu, L., Cooper, S., Anderson, S.L., Beyer, G.S., Hoffman, R., Rubin, B.Y. (1986d): The suppressive influences of human tumor necrosis factors on bone marrow hematopoietic progenitor cells from normal donors and patients with leukemia: synergism of tumor necrosis factor and interferon- . J. Immunol. 136: 4487-4495.

Degliantoni, G., Murphy, M., Kobayaski, M., Francis, M.K., Perussia, B.,

Trinchieri, G. (1985): Natural killer (NK) cell-derived hematopoietic colony inhibiting activity and NK cytotoxic factor. Relationship with tumor necrosis factor and synergism with immune interferon. J. Exp. Med. 162: 1512-1530.

Lu, L., Walker, D., Broxmeyer, H.E., Hoffman, R., Hu, W., Walker, E. (1987): Characterization of adult human marrow hematopoietic progenitors highly enriched by two-color sorting with My10 and major histocompatibility class II monoclonal antibodies. Submitted for Publication.

Metcalf, D. (1985): The granulocyte-macrophage colony stimulating factors. Science 229: 16-22.

Moore, R.N., Pitruzzello, F.J., Larsen, H.S., Rouse, B.T. (1984): Feedback regulation of colony stimulating factor (CSF-1)-induced macrophage proliferation by endogenous E prostaglandins and interferon α/β. J. Immunol. 133: 541-543.

Munker, R., Koeffler, P. (1987): In vitro action of tumor necrosis factor on myeloid leukemia cells. Blood. 69: 1102-1108.

Murphy, M., Louden, R., Kobayashi, M., Trinchieri, G. (1986): γ-interferon and lymphotoxin, released by activated T cells, synergize to inhibit granulocyte/monocyte colony formation. J. Exp. Med. 164: 263-279.

Old, L.J. (1985): Tumor necrosis factor (TNF). Science 230: 630-632.

Peetre, C., Gullberg, U., Nilsson, E., Olsson, I. (1986): Effects of recombinant tumor necrosis factor on proliferation and differentiation of leukemic and normal hemopoietic cells in vitro. Relationship to cell surface receptor. J. Clin. Invest. 78: 1694-1700.

Pelus, L.M., Broxmeyer, H.E., Kurland, J.I., Moore, M.A.S. (1979): Regulation of macrophage and granulocyte proliferation: specificities of prostaglandin E and lactoferrin. J. Exp. Med. 150: 277-292.

Pelus, L.M., Broxmeyer, H.E., Clarkson, B.D., Moore, M.A.S. (1980): Abnormal responsiveness of colony forming cells from patients with chronic myeloid leukemia to inhibition by prostaglandin E. Cancer Res. 40: 2512-2515.

Raefsky, E.L., Platanias, L.C., Zoumbas, N.C., Young, N.S. (1985): Studies of interferon as a regulator of hematopoietic cell proliferation. J. Immunol. 135: 2507-2512.

Warren, M.K., Ralph, P. (1986): Macrophage growth factor CSF-1 stimulates human monocyte production of interferon, tumor necrosis factor and myeloid CSF. J. Immunol. 137: 2281-2285.

Williams, D.E., Straneva, J.E., Shen, R-N, Broxmeyer, H.E. (1987): Purification of murine bone-marrow-derived granulocyte-macrophage progenitor cells. Exp. Hematol. 15: 243-250.

Zucali, J.R., Broxmeyer, H.E., Dinarello, C.A., Gross, M.A., Weiner, R.S. (1987): Regulation of early human hematopoietic (BFU-E and CFU-GEMM) progenitor cells in vitro by interleukin 1-induced fibroblast conditioned medium. Blood 69: 33-37.

Résumé

Dans cet article, les auteurs présentent une revue générale sur les effets inhibiteurs in vitro de plusieurs substances qui sont maintenant isolées et purifiées : iso-ferritine acide (AIF), lactoferrine (LF), tumor necrosis factor (TNF) alpha et beta, interféron gamma (IFN) et prostaglandines E (PGE). On peut actuellement constater que :
A) les progéniteurs granulo-macrophagiques (CFU-GM) sont plus sensibles à l'effet inhibiteur du TNF alpha en présence de G-CSF, qu'en présence de GM-CSF, de M-CSF (CSF-1) ou en présence d'un milieu conditionné par des cellules de la lignée 5637 qui contient à la fois du GM et du G-CSF.
B) l'isoferritine acide, le TNF alpha, l'interféron gamma et la prostaglandine E peuvent empêcher le développement de colonies à partir de populations cellulaires très enrichies en progéniteurs et appauvries en cellules accessoires.
C) l'interféron gamma a une action inhibitrice synergique avec l'isoferritine acide et la prostaglandine E sur la formation de colonies granulo-macrophagiques chez la souris et chez l'homme.
D) on constate de grandes variations dans la sensibilité des progéniteurs granulo-macrophagiques de sujets atteints de leucémie, myélodysplasie, cancers et lymphomes aux effets suppresseurs de l'isoferritine acide, du TNF alpha, de l'interféron gamma et de la prostaglandine E.

Erythropoiesis in human acute non lymphoblastic leukemias: place of stimulating and inhibitory factors

Albert Najman, Ladan Kobari, Nicholas Dainiak *, Claude Baillou, Jean Philippe Laporte, Luc Douay and Norbert Claude Gorin

Laboratoire d'Hématologie, Faculté de Médecine St-Antoine, Paris, France
* Departments of Medicine and Biomedical Research, St-Elizabeth's Hospital of Boston, Tufts University School of Medecine, Boston, USA

ABSTRACT

Suppression of normal erythropoiesis during acute non lymphoblastic anemia (ANLL) is still poorly understood. By studying the effect of plasmas from patients with ANLL, we have found a high amount of several hemopoietic factors stimulating erythroid progenitor growth and murine pluripotent stem cells (CFU-S) proliferation. Conversely blast conditioned medium (BCM) prepared with leukemic blast cells has a strike inhibitory effect on peripheral blood BFU-E and only a slight effect on bone marrow BFU-E. The inhibitory effect is mediated through adherent cells. No correlation between this inhibitory effect and the expression of HLA-DR determinants or cell cycling was found. These data are in favour of a blocking of normal stem cells in ANLL, by a factor depending of leukemic cells which prevents normal stimulating factor to act, at a level which has to be determined.

KEY WORDS

BFU-E, CFU-S, Growth factors, Blast conditioned medium, Human acute leukemias.

INTRODUCTION

Anemia is a common finding in acute non lymphoblastic leukemia (ANLL) which is due to a quite total disappearance of erythroblasts in the bone marrow at the time of diagnosis. Erythroid colonies formation is also disturbed as shown by in vitro studies (Handler and Handler 1976). When complete remission is achieved, conversely, a normal production of red cells is restored and a normal development of erythroid progenitors in vitro can be observed. The underlying mechanisms of the disturbance of erythropoiesis in ANLL are not yet clarified. Several hypothesis are still discussed. Evidence supporting an inhibitory effect of the leukemic blast cells on normal differentiation, as demonstrated by the effect of conditioned medium or direct cell-cell interaction on cultures of hemopoietic progenitors, has been advanced (Broxmeyer et al. 1978, Morris et al. 1975, Quesenberry et al. 1978). A same effect was demonstrated in vivo by studying the development of erythroid colonies from normal bone marrow seeded in plasma clot diffusion chambers implanted in rats with the Shay chloroleukemia (Steinberg and Handler 1979). This inhibitory effect decreased after the lysis of leukemic cells (Bernstein et al. 1987) or with leukemic cell differentiation (Steinberg et al. 1985). The nature of this inhibitory activity remains unclear : acidic isoferritins identified to leukemic inhibitory activity by some (Broxmeyer et al. 1981) and questioned by others (Sala et al. 1986) have been reported to inhibit the in vitro growth of BFU-E and also of multipotential progenitors (Lu and al. 1983), but the leukemic associated inhibitor recently purified by Olofsson (1986) has no effect on erythroid progenitors. It is also possible that leukemic infiltration of the bone marrow may impair the production or the release of growth factors or hemopoietic mitogens by stromal or accessory cells (Greenberg et al. 1978).

Therefore, we have decided to investigate the pattern of the regulation of erythropoiesis during human ANLL by evaluating in a first step the level of erythroid stimulating factors in these patients and, in a second step, the effect of leukemic blast cells on erythroid differentiation.

THE LEVEL OF ERYTHROPOIETIN AND HEMOPOIETIC STIMULATING FACTORS IN HUMAN ANLL.

The level of erythropoietin (Epo) was measured in the serum of 14 patients before any transfusion and chemotherapy by the fetal mouse liver assay (Najman et al. unpublished data). It was elevated in each case, with a mean value of 379 mU/ml (extremes : 56 to 872 mU/ml). The level of erythropoietin in normal controls was 14 $\pm$ 4 mU/ml. It was possible in one patient to follow the level of erythropoietin after transfusions. Variations were demonstrated with a progressive decrease of Epo level from 552 mU/ml to 38.5 mU/ml whereas hemoglobin level rose from 6 gr/dl to 12 gr/dl (Figure 1).

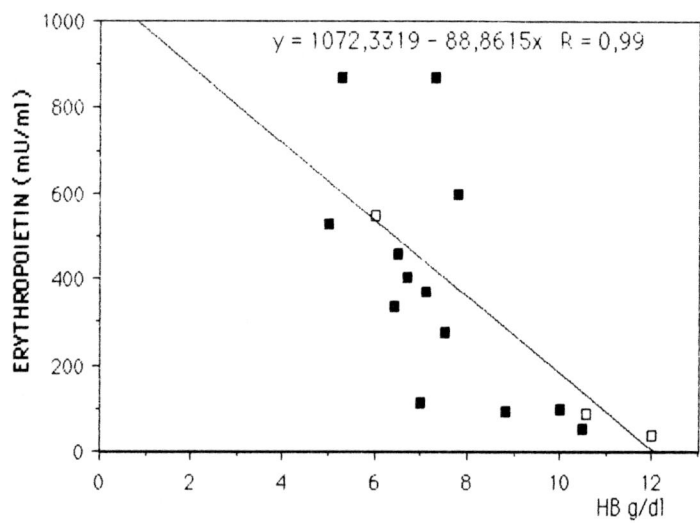

Figure 1 - Serum erythropoietin level in patients with acute non lymphoblastic leukemia in relation to hemoglobin level. (□). Variation of Epo after tranfusion in the same patient.

Increased amounts of hemopoietic stimulating factors were also found in the plasmas of these patients, prepared before any transfusion and treatment (Kobari et al. 1987). The effect of plasma was assayed on human early erythroid progenitors from bone marrow or peripheral blood : it increased in each case the number of erythroid colonies as compared to normal AB plasma (figure 2).

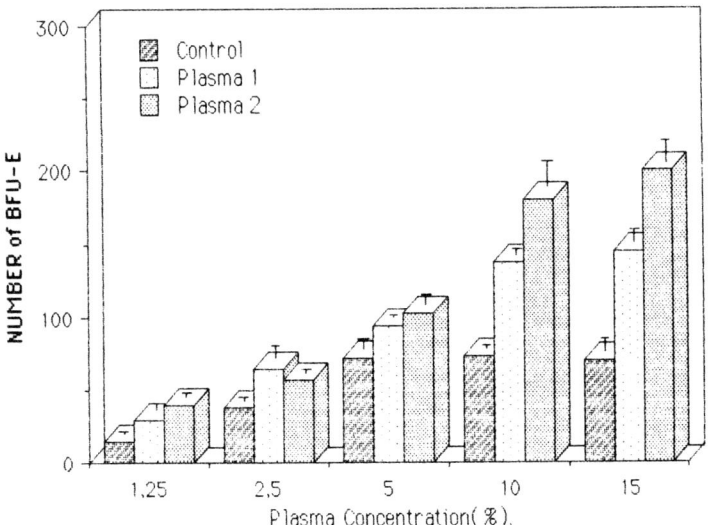

Figure 2 - Effect of leukemic plasmas on human peripheral blood BFU-E.

This effect disappeared after absorption of the plasmas with IgG anti-lymphocyte membranes that are known to neutralize burst promoting activity (Dainiak et al. 1985), while no modification was observed after preabsorption of the leukemic plasmas with anti-erythropoietin IgG even if Epo activity was totally abrogated (figure 3).

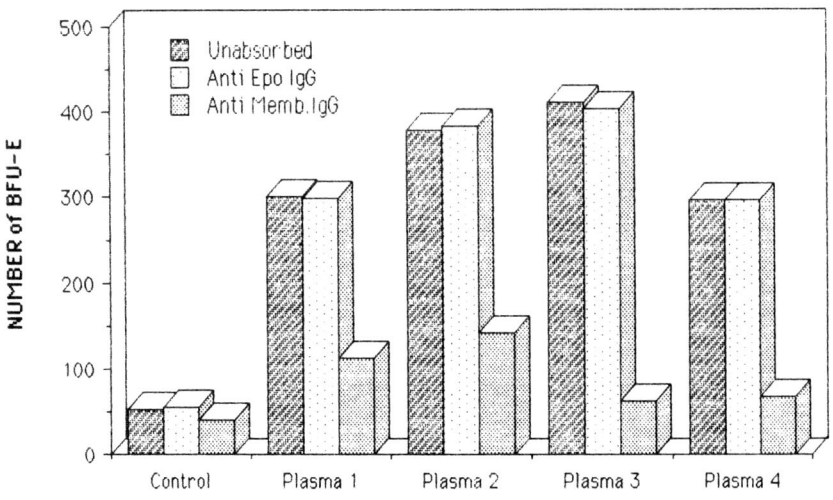

Figure 3 - Absorption of erythropoietin and growth factors from normal and leukemic plasmas. Effect on normal bone marrow BFU-E growth.

Leukemic plasmas were found also to increase the percentage of murine pluripotent stem cells (CFU-S) in DNA synthesis, effect which was not abrogated by pre-absorption of the plasmas with anti-lymphocyte membranes IgG (Kobari et al. 1987).

THE EFFECT OF LEUKEMIC BLASTS CELLS ON ERYTHROID DIFFERENTIATION.

To study the effect of leukemic blast cells on erythroid differentiation, we have prepared a conditioned medium from blast cells from patients with ANLL. Blood and bone marrow were collected from patients before any treatment or transfusion. Normal blood or bone marrow were obtained from healthy volunteers. Cells with a density of 1077 g/ml separated by Ficoll Hypaque density sedimentation were used throughout these studies.

The conditioned media were prepared using the technique of Olofsson and Olsson (1980). Cells were incubated in IMDM with 5 % fetal calf serum at 37°C. The optimal cell concentrations for the production of an inhibitory activity was $5-10^6$ cells/ml and the optimal time of incubation was 5 hours. After centrifugation, the supernatants were harvested, filtered through 0.8 um Nalgene filter and stored frozen at -40°C till use.

Blast conditionned medium (BCM) assays on human erythroid progenitors were done according to Iscove and Sieber technique (1975). Briefly peripheral blood cultures ($5-10^5$ cells/ml) and bone marrow cultures (10^5 cells/ml) were seeded in methyl cellulose (Fluka 4000) at a final concentration of 0.9 %, supplemented with IMDM, deionized BSA 10 mg/ml (Sigma), alpha-thioglycerol 7.5×10^{-5} M (Sigma), 15 % fetal calf serum and 1 unit/ml plasma pork erythropoietin (CNTS Paris). In some experiments adherent cells were removed using the technique of Messner et al. (1973). The blast conditioned medium was added at different concentrations from 0.075 to 10 % (v/v). Quadriplicate cultures were incubated at 37°C, 5 % CO_2 in humidified air and scored at 14 days with an inverted microscope : orange red cell agregates containing more than 50 cells or isolated groups of 3 colonies were counted as a burst.

The effect of the blast conditioned medium on the percentage of cells in DNA synthesis was studied after preincubation of the cells with BCM during 60 minutes and then washing, by the suicide technique with Ara-C (Cytarabine Upjohn) (Preisler 1980) or with tritiated thymidine (Eaves and Eaves 1978).

Means $\pm$ standard errors (SE) were calculated for burst numbers scored in each of four cultures. Data were compared by the Student t test.

<u>The blast conditioned medium was first tested on the early erythroid progenitors of the patients in complete remission.</u>

When BCM was added at various concentrations to the culture of bone marrow cells, a slight decrease in the number of BFU-E derived colonies was noted, with 79 % of the colonies as compared to the control; this result was observed with a concentration of 0.3 % of BCM only.

The same assay on the peripheral blood BFU-E has shown, on the contrary, a strike decrease to 60 % of the colonies as compared to the control, at a concentration of 0.6 %, with a linear dose-response correlation at very low concentrations (between 0.075 to 0.6 %).

Removal of adherent cells have altered clearly the effect of BCM. When BCM was tested on non adherent cells from bone marrow or from peripheral blood, no decrease in the numbers of erythroid colonies was observed (Figure 4).

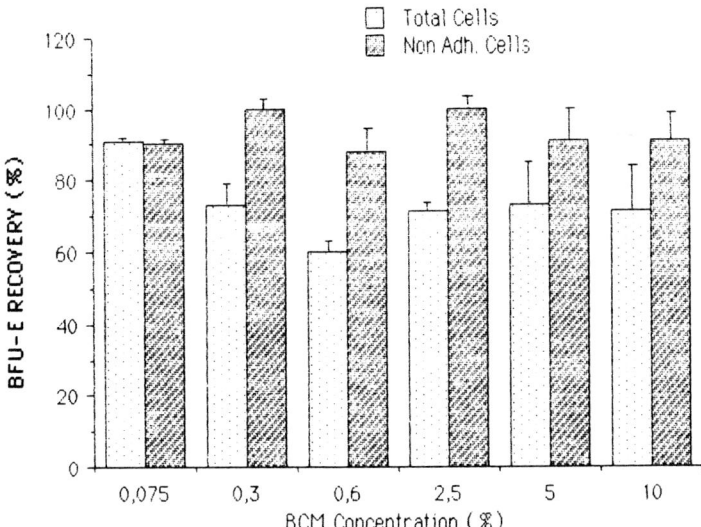

Figure 4 - Effect of blast conditioned medium (BCM) on peripheral blood BFU-E from patients with ANLL in complete remission.

BCM was then tested on mononuclear cells from normal controls.

The effect of BCM was always evident on peripheral blood BFU-E, whereas the number of BFU-E derived colonies did not change in the bone marrow cultures.

To explain such difference of sensitivity between the bone marrow and the peripheral blood erythroid progenitors to BCM, we have explored the proliferative state of the BFU-E and the expression of HLA-DR antigen at their surface.

The percentage of BFU-E in DNA synthesis as studied by Ara-C suicide technique was of 52 ± 7.7 % in peripheral blood and of 30 ± 7.1 % in bone marrow. When the peripheral blood cells were first incubated with BCM during one hour and then with Ara-C during 20 minutes, the number of BFU-E derived colonies after 14 days of culture was quite similar to the control, but the population in DNA synthesis decreased from 52 ± 7.7 % to 12.7 ± 2.5 % ($p < 0.001$). Conversely, the pre-incubation of bone marrow cells with BCM did not modify the number and the percentage of BFU-E in DNA synthesis (control: 30.5 ± 7.1 % and after BCM incubation 33.9 ± 8 %). Similar results were obtained when suicide experiments were performed with tritiated thymidine instead of Ara C (Figure 5).

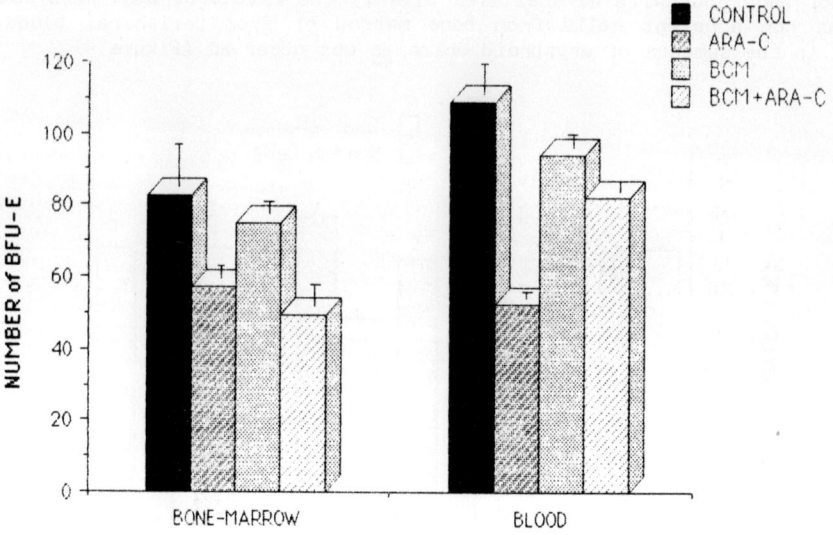

Figure 5 - Effect of blast conditioned medium (BCM) on the population of BFU-E in DNA synthesis in peripheral blood and bone marrow.

The expression of HLA-DR antigen at the surface of BFU-E was studied by complement dependent cytotoxicity, by using a murine monoclonal antibody anti-human HLA-DR (Dr M. Fellous, Institut Pasteur, Paris). Eighty per cent of peripheral blood BFU-E and 92 % of bone marrow BFU-E expressed HLA-DR antigen. When suicide technique with Ara-C was performed after treatment of the cells by anti HLA-DR and complement, the decrease in the number of BFU-E derived colonies was low in peripheral blood and in bone marrow as well.

DISCUSSION

Normal hematopoiesis is characterized by a simultaneous and well balanced production of cells from several different lineages under the control of positive and negative factors acting at different stages of differentiation. This equilibrium is disturbed in ANLL where the accumulation of leukemic cells blocked in their differentiation is associated with a reduction of other cell lineages. Therefore it seems important to investigate the possible modifications in the balance between hemopoietic factors in ANLL which can in part explain the abnormal pattern of hematopoiesis in the disease.

In the first part of our study, we have demonstrated that high amounts of hemopoietic growth factors can be found in the plasmas of leukemic patients. Serum erythropoiesis level was elevated in relation to the degree of anemia ($r = 0.99$) and variations of Epo level after transfusions indicate that the regulation of its production seems to be normal.

The use of monospecific antibody capable to neutralize burst stimulating activity (Dainiak et al. 1985) has allowed to distinguish among the stimulating activity of these plasmas on early erythroid progenitors the presence of a factor immunologically related to burst promoting activity. By the same procedure, this factor was separated from the one which triggered into cycle murine pluripotent stem cells (CFU-S).

As for erythropoietin whose clinical relevance is now well demonstrated (Iscove 1977, Winearls et al. 1986, Eschbach et al. 1987), it could be assumed that elevated circulating growth factors acting on stem cells and on the first steps of differentiation during cytopenias are also the result of a feedback regulatory mechanism. In a previous study (Najman et al. 1987 submitted) on the variations of serum erythropoietin level during anemias, we have found a clear correlation between the absence of erythroblasts in the bone marrow and the highest levels of Epo. Indeed all these patients were anemic and their bone marrow devoid of erythroblasts. Although each of these different stimulating activities was obtained with total plasma and not with a purified molecule, the results achieved on BFU-E growth from either patients in complete remission or normal controls, and on CFU-S proliferation are in favour of functionally unaltered factors. Steinberg et al. (1979) likewise have observed in chloroleukemic rats that the high circulating level of erythropoietin in these animals (Derelanko et al. 1978) was really effective on normal erythroid progenitors. Therefore, the absence of erythroid differentiation in ANLL could not be explained by a lack of stimulating factors. It is more likely that they are prevented to act on their target cells. This could be the result of a disturbance in the microenvironnement. An alteration of surface specific receptors or intra cellular modifications are other possible mechanisms.

We have found indeed that a medium conditioned by blastic cells from patients with ANLL was able to inhibit the in vitro growth of BFU-E from the same patients in complete remission or from normal controls. The main property of BCM was to protect the cells from the action of Ara-C as shown by the decrease in the number of cells killed during incubation with the drug. Protection could result from the blocking of an Ara-C receptor or the blocking of DNA synthesis in these cells. The effect seems to be reversible since the number of BFU-E derived colonies after incubation with BCM was not different from the controls. Conversely, one can assume that in vivo a permanent contact with such an inhibitory factor is able to block for a long time normal stem cells and to prevent differentiation of these cells.

Inhibition was more evident on peripheral blood BFU-E than on bone marrow BFU-E. This difference could not be explained by a different proliferative state or by a difference in HLA-DR antigens expression at the surface of erythroid progenitors as it was proposed by Lu et al. (1983), since we have found quite the same populations in peripheral blood or in the bone marrow. BFU-E are however an heterogeneous population (Gregory and Eaves 1977, Eaves and Eaves 1978, Shekhter-Levin et al. 1985) and some other properties can explain our results.

Accessory cells obviously play an important role in the action of BCM since removal of adherent cells from peripheral blood abrogated the inhibitory effect. The role of peripheral blood monocytes on BFU-E growth was a debated question (Rinehart et al. 1978, Zuckerman 1981) till the recent work of Feldman et al. (1986) who demonstrated an in vivo release of different factors that exert opposing effects on BFU-E. If the effect of BCM is mediated through monocytes, one can understand the absence of inhibition of BCM on bone marrow BFU-E since bone marrow monocytes have primarly a stimulating effect on BFU-E growth (Gordon et al. 1980). These results have to be considered as preliminary and more information is needed on the exact place of accessory cells in the inhibitory effect of BCM (Broxmeyer et al. 1979).

Another usual finding was an incomplete inhibition of BFU-E growth even with peripheral blood cells. Similar results were already reported on granulo-macrophagic progenitors (Broxmeyer et al. 1978, Olofsson and Olsson 1980). Moreover, at high concentrations of BCM, this inhibitory effect disappeared suggesting the presence of other opposite activities in the conditioned medium. Recently, Young et al. (1987) have reported the presence of biologically active CSF in supernatant conditioned by leukemic cells from a subset of patients with AML whose cells expressed GM-CSF transcripts.

Comparison of our results with other inhibitory activities has to be done very carefully. Since BCM has been only assayed on early erythroid progenitors, it is difficult to compare it to other inhibitory factors found in leukemias such as LIA (Olofsson and Olsson 1980) or LAI (Broxmeyer et al. 1979) which mostly affect granulopoiesis. Crude LAI has however some effect on a subset of erythroid progenitors restricted to the expression of HLA-DR determinants (Lu et al. 1983). Since BCM has not yet been purified, it is also premature to relate its activity to other inhibitors acting on erythroid progenitors, as the negative regulatory protein described by Axelrad et al. (1983), TNF-alpha (Broxmeyer et al. 1986), or on other stem cells (Laerum et al. 1984, Wdzieczak-Bakala et al. 1983).

In conclusion, the disturbance of erythropoiesis in ANLL seems really to be due to a factor released by the leukemic blast cells. The exact mechanism has to be more investigated. The high amount of hematopoietic stimulating factors found in the plasma of patients with ANLL, although functionnally active, cannot operate during the disease on their normal target cells. A clear explanation of this phenomenon would have important clinical consequences. The prevention of the inhibition of stem cells normal differentiation will be an important step in the treatment of acute leukemias. A new equilibrium between normal and leukemic cells will be achieved which is certainly one of the main conditions for a successful therapy of these diseases.

ACKNOWLEDGEMENTS

This work was supported by a grant 6184 from the Association pour la Recherche sur le Cancer (Villejuif). We are grateful to Dr Marc FELLOUS for providing monoclonal antibodies. We thank Dr Martine Guigon for her helpful critical review.

REFERENCES

Axelrad A., Croizat H., Eskinazi D., Stewart S., Vaithilingam and Van der Gaag H. (1983). Genetic regulation of DNA synthesis in early erythropoietic progenitor cells: mechanism and consequences for polycythemia. In haemopoietic stem cells, Alfred Benzon Symposium 18, ed Killmann S.V. Aa, Cronkite E.P., Muller Berat C.N. p 234-251. Numbsgaard Copenhagen.

Bernstein I.D., Singer J.W., Andrews R.G., Keating A., Powel J.S., Bjornson B.H., Cuttner J., Najfeld V., Reaman G., Raskind W., Sutton D.M.C. and Fialkow P.J. (1987). Treatment of acute myeloid leukemia cells in vitro with a monoclonal antibody recognizing a myeloid differentiation antigen allows maximal progenitor cells to be expressed. J. Clin. Invest. 79 : 1153-1159.

Broxmeyer H.E., Jacobsen N., Kurland J., Mendelson N. and Moore M.A.S. (1978). In vitro suppression of normal granulocytic stem cells by inhibitory activity derived from human leukemia cells. J. Nat. Cancer Inst. 60: 497-511.

Broxmeyer H.E., Ralph P., Margolis V.B., Nakoinz I., Meyers P., Kappor N. and Moore M.A.S. (1979). Characteristics of bone marrow and blood cells in human leukemia that produce leukemia inhibitory activity (LIA). Leukemia Res. 3: 193-203.

Broxmeyer H.E., Bognacki J., Dorner M.H. and Sousa M. (1981). Identification of leukemia associated inhibitory activity as acidic isoferritins. J. Exp. Med. 153: 1426-1444.

Broxmeyer H.E., Williams D.E., Lu L., Cooper S., Anderson S.L., Beyer G.S., Hoffman R., Rubin B.Y. (1986). The suppressive influences of human tumor necrosis factors on bone marrow hematopoietic progenitors cells from normal donors and patients with leukemia: synergism of tumor necrosis factors and interferon . J. Immunol. 12: 4487-4495.

Dainiak N., Feldman L. and Cohen C.M. (1985). Neutralization of erythroid burst promoting activity in vitro with anti-membrane antibodies. Blood 65: 877-885.

Derelanko M.J., Lobue J., Gordon A.S., Camiscoli J.F., Gizzi C. and Fredrickson T.N. (1978). Measurement of erythropoietin in chloroleukemic rats. Exp. Hematol (Copenh.) 6: 91-95.

Eaves C.J., and Eaves H.C. (1978). Erythropoietin (Ep) dose response curves for three classes of erythroid progenitors in normal human marrow and in patients with polycythemia vera 4. Blood 52 : 1196-1210.

Eschbach J.W., Egrie J.C., Downing M.R., Browne J.K. and Adamson J.W. (1987). Correction of the anemia of end-stage renal disease with recombinant human erythropoietin: result of a phase I and II clinical trial. N. Engl. J. Med. 316: 73-78.

Feldman L., Cohen C.M., and Dainiak N. (1986). In vivo release of physically separable factors from monocytes that exert opposing effects on erythropoiesis. Blood 67 : 1454-1459.

Gordon L.I, Miller W.J., Branda R.F., Zanjani E.D and Jacob H.S (1980). Regulation of erythroid colony formation by bone marrow macrophages. Blood, 55: 1047-1050.

Gregory C.J., and Eaves A.C. (1977). Human marrow cells capable of erythropoietic differentiation : definition of three erythroid colony responses. Blood 49 : 855-846.

Greenberg P.L., Mara B. and Heller P. (1978). Marrow adherent cell colony stimulating activity production in acute myeloid leukemia. Blood 52: 362-369.

Handler E.S. and Handler E.E. (1976). In vitro erythroid colony formation in acute myelogenous leukemia in the rats. J. Natl. Cancer Inst 56: 851-853.

Iscove N.N. and Sieber F. (1975). Erythroid progenitors in mouse bone marrow detected by macroscopic colony formation in culture. Exp. Hematol. 3: 32-40.

Iscove N.N. (1977). The role of erythropoietin in regulation of population size and cell cycling of early and late erythroid precursors in mouse bone marrow. Cell Tissue Kinet. 10: 323-334.

Kobari L., Dainiak N., Najman A., Kreczko S., Gorin N.C., Duhamel G. and Frindel E. (1987). Stimulatory effects of plasma from patients with ANLL on early erythroid progenitors and pluripotent stem cells. Exp. Hematol. (in press).

Laerum O.D., and Paukovits W.R. (1984). Inhibitory effects of a synthetic pentapeptide on hemopoietic stem cells in vitro and in vivo. Exp. Hematol. 12 : 7-17.

Lu L., Broxmeyer H.E., Meyers P.A., Moore M.A.S. and Thaler H.T. (1983). Association of cell cycle expression of Ia-like antigenic determinants on normal human multipotential (CFU-GEMM) and erythroid (BFU-E) progenitor cells with regulation in vitro by acidic isoferritins. Blood 61: 250-256.

Messner H.A., Till J.E. and Mc Culloch E.A. (1973). Interacting cell population affecting granulopoietic colony formation by normal and leukemic human marrow cells. Blood 42 : 701-710.

Morris J.C., Mc Neill T.A. and Bridges J.M. (1975). Inhibition of normal human in vitro colony forming cells by cells from leukemic patients. Brit. J. Cancer 31: 641-648.

Olofsson T. and Olsson I. (1980). Suppression of normal granulopoiesis in vitro by a Leukemic-Associated inhibitor (LAI) of acute and chronic leukemia. Blood 55: 975-982.

Olofsson T. (1986). Purification of leukemia associated inhibitor (LAI) by use of monoclonal antibodies against HL60 cells. Exp. Hematol. 14: 555.

Preisler H.D. (1980). Prediction of response to chemotherapy in acute myelocytic leukemia. Blood 56 : 361-365.

Quesenberry P.J., Rappeport J.M., Fountebuoni A., Sullivan R., Zuckerman K. and Ryan M. (1978). Inhibition of normal murine hematopoiesis by leukemic cells. New Engl. J. Med. 299: 71-75.

Rinehart J.J., Zanjani E.D., Nondedeu B., Gormus B. and Kaplan M.E. (1978). Cell-cell interaction in erythropoiesis : role of human monocytes. J. Clin. Invest. 69 : 979-983.

Sala G., Worwood M. and Jacobs A. (1986). The effect of isoferritins on granulopoiesis. Blood 67: 436-443.

Shekhter-Levin S., Amato D., Karrass L. and Axelrad A.A. (1985). Heterogeneity of buoyant density and proliferative state of circulating erythropoietic progenitor cell (BFU-E) in man. Exp. Hematol. 13 : 1138-1142.

Steinberg H.N. and Handler E.S. (1979). Leukemic host influence on normal erythrocytic and granulocytic colony formation in in vivo plasma clot diffusion chamber cultures. Cancer Res. 39 : 1575-1578.

Steinberg H.N., Tsiftsoglou A.S. and Robinson S.H. (1985). Loss of suppression of normal bone marrow colony formation by leukemic cell lines after differentiation is induced by chemical agents. Blood 65: 100-106.

Wdzieczak-Bakala J., Guigon M., Lenfant M. and Frindel E. (1983). Further purification of a CFU-S inhibitor: in vivo effects after Cytosine Arabinoside treatment. Biomedicine and Pharmacoth. 37: 467-471.

Winearls G.G., Pippard M.J., Downing M.R., Oliver D.O., Reid C. and Mary Cotes P. (1986). Effect of human erythropoietin derived from recombinant DNA on the anaemia of patients maintained by chronic haemodialysis. The Lancet ii 1175-1178.

Young D.C., Wagner K. and Griffin J.D. (1987). Constitutive expression of the granulocyte-macrophage colony stimulating factor gene in acute myeloblastic leukemia. J. Clin. Invest. 79 : 100-106.

Zuckerman K.S. (1981). Human erythroid burst promoting units. Growth in vitro is dependent on monocytes but not T lymphocytes. J. Clin. Invest. 67: 702-709.

Résumé

La disparition de l'hématopoièse normale au cours des leucémies aigues non lymphoblastiques est un phénomène encore inexpliqué. L'étude de plasmas de malades atteints de LANL nous a permis de constater la présence de taux élevés de plusieurs facteurs hématopoiétiques stimulant le développement in vitro des progéniteurs érythroïdes précoces (BFU-E) et la prolifération des cellules souches pluripotentes de la souris (CFU-S). Nous avons par ailleurs constaté qu'un milieu conditionné par des cellules leucémiques blastiques (BCM) avait un effet inhibiteur très net sur les BFU-E du sang et peu marqué sur les BFU-E de la moelle. Cet effet inhibiteur nécessite la présence de cellules adhérentes. Nous n'avons pas trouvé de corrélation entre cet effet inhibiteur, l'état de prolifération des cellules et l'expression de l'antigène HLA-DR à leur surface. Cette double constatation nous fait penser qu'au cours des LANL les facteurs de stimulation sont normalement élevés mais qu'ils ne peuvent agir sur leur cellules cibles qui sont bloquées à un niveau qu'il faut déterminer dans leur différenciation par un facteur dépendant des cellules leucémiques.

Suppression of normal hemopoiesis in leukemia : *in vivo* and *in vitro* studies

Howard N. Steinberg

Charles A. Dana Research Institute, Harvard-Thorndike Laboratory, Department of Medicine, Harvard Medical School and Beth Israel Hospital, Boston, Massachusetts, 02215, USA

ABSTRACT

The pattern of normal hemopoiesis was significantly altered in plasma clot diffusion chamber cultures implanted into rats with acute myelogenous leukemia. Since the normal bone marrow cells are compartmentalized within the chamber, the suppression of granulocyte and erythroid colony formation appears to be mediated by the release of diffusible inhibitory substances from the leukemic cells. Inhibition of normal stem cells growth was also observed in vitro in coculture studies with human leukemic cell lines. The ability of leukemic cell lines to suppress normal colony formation was lost after their exposure to differentiation-inducing agents. Ribavirin, a nucleoside analogue, induces the differentiation of HL-60 cells and alters their responsiveness to GM-CSF which results in enhanced differentiation.

KEY WORDS

Acute Leukemia, Diffusion Chamber, Differentiation, Inhibitory Regulator

INTRODUCTION

Normal hemopoiesis appears to be regulated by a complex network of stimulatory and suppressive factors that act in concert to maintain peripheral blood cell numbers at constant levels (Broxmeyer, 1986; Spivak, 1986). These regulatory molecules may exert their influence directly on the pluripotent and lineage specific committed progenitor cells or may modulate the activity of accessory cells in the microenvironment that have an important role in stem cell proliferation and differentiation. The balance of stimulatory and inhibitory influences on progenitor or accessory cells may be temporarily shifted to accommodate an increase in the demand for hemopoiesis in response to infection or hypoxia (Steinberg, 1976) or, as may occur in nonhematological or hematologic malignancy (Takahashi, 1985), may be more chronically altered resulting in defective hemopoiesis (Broxmeyer, 1983; Moore, 1979). In AML, normal myeloid and erythroid precursor cells are replaced by rapidly dividing undifferentiated cells derived from the clonal expansion of a transformed hemopoietic precursor cell (Fialkow, 1980). The decline in normal cellular elements is associated with disturbances in in vitro colony formation (Handler, 1974; Moore, 1973). Both the reduction in granulocyte-macrophage colonies and the formation of abortive clusters have been reported. Whether this suppression of normal progenitor cells in situ is the result of a direct leukemic cell to cell interaction or is mediated by an inhibitory molecule secreted from the leukemic cells has been the subject investigation in our laboratory, as reviewed in this chapter.

Although the cellular mechanisms that coordinate proliferation and differentiation in normal cells may be uncoupled in leukemic cells (Sachs, 1980), studies suggest that the block in differentiation may be reversible (Tsifsiglou, 1985) under certain experimental conditions. Differentiation of established human and murine leukemia cell lines has been effected in vitro by a variety of chemical agents (Collins, 1978), antineoplastic drugs (Ishikura, 1985), naturally occurring molecules (Weinberg, 1986) and growth factors like G-CSF (Nicola, 1987) that regulate the differentiation of normal myeloid stem cells. However, it remains unclear whether leukemia cells induced to undergo differentiation to "non-malignant" cells lose their ability to suppress normal hemopoiesis. Coculture studies examining the effect of human leukemic cell lines on the suppression of normal bone marrow progenitor cells before and after differentiation induction will be described. More recent studies have been concerned with the effects of Ribavirin, a nucleoside analogue and antiviral agent, and the role of GM-CSF in the induction of leukemic cell differentiation.

I. IN VIVO STUDIES: INFLUENCE OF LEUKEMIC ENVIRONMENT ON NORMAL HEMOPOIESIS:

The Shay chloroleukemia, an acute myelogenous leukemia (AML) in the Long-Evans rat, was used as a model system to examine the changes occurring in normal hemopoiesis during the onset and progression of the disease (Handler, 1970). Its similarities to human AML have made it a useful model system for these studies. To determine the influence of the leukemic environment on normal hemopoiesis, the development of normal erythroid (CFU-E, BFU-E) and granulocyte (CFU-G) colonies was assessed in plasma clot diffusion chamber cultures (PCDC) implanted into the peritoneal cavities of leukemic rats. Diffusion chamber cultures (DC) have been used to study cell growth within the physiological milieu of a host animal and to assess the effect of humoral factors on this growth (Cronkite and Carsten, 1980). As bone marrow cells are compartmentalized within the diffusion chamber, their proliferation and differentiation is influenced by diffusible factors from the host animal. Changes in the pattern of cell growth within the chamber reflect changes in the physiological milieu of the host animal (Steinberg, 1976). Thus, using the diffusion chamber culture system, the leukemic influence on normal hemopoiesis as mediated by cell to cell interaction or humoral products of leukemic cells can be distinguished.

The preparation, harvest, and fixation of PCDCs has been previously described in detail (Steinberg, 1985). Briefly, diffusion chamber cultures were prepared by adhering 0.22 uM GS-type Millipore (Millipore Corporation, Bedford, MA) or Nuclepore (Nuclepore Corporation, Pleasanton, CA) membrane filters to both sides of a lucite ring. Each chamber was seeded with 120 ul containing 5×10^5 normal bone marrow cells prepared from the femurs of male Long-Evans rats (200-240 g), followed by the addition of 20 ul of Bovine Citrated Plasma (BCP). Each chamber was immediately stoppered, heat sealed, shaken vigorously to insure the uniform distribution of cells within the chamber, and set aside for 3-5 minutes until clot formation was complete. Under anaesthesia, chambers were surgically implanted into the peritoneal cavities of the rat, each animal receiving 4 chambers. Clots from PCDC were harvested at varying intervals of time and were dehydrated with Whatman No. 1 filter paper, fixed with 5 percent gluteraldehyde, washed with water, and stained with benzidine and hematoxylin. PCDC colony formation in leukemic host rats was compared with that observed in normal host rats. In PCDC cultures implanted into normal rats, two classes of erythroid colonies (scored as benzidine-positive units) appeared, with kinetics and physical properties similar to those observed in plasma clot assays in vitro. Colonies appearing with an initial peak on day 2 of PCDC culture were similar to CFU-E derived formed in vitro and appeared as groups of 8-64 (low proliferative capacity) mature erythroid cells randomly dispersed in the clot matrix. Similar to the burst forming units described by Axelrad (1973), erythroid colonies appearing on day 7 of PCDC culture appeared as large entities of 100-300 immature cells divided into 6-8 smaller aggregates. The clonal efficiencies of CFU-E and BFU-E growth in PCDC of normal rat marrow were 38.1 ± 6.9 and 5.5 ± 1.9 per 1×10^5 cells respectively. Granulocyte colonies of greater than 100 cells appeared on day 4 of culture with an efficiency of 27.3 ± 3.3 per 1×10^5 bone marrow cells. After a slight decline on day 5, the number of colonies remained constant through day 9 and appeared to be associated with the expression of a second, more immature class of granulocyte progenitor cell (Steinberg, 1983). The physiologic validity of the PCDC assay was demonstrated by the changes in PCDC colony formation in animals made neutropenic by endotoxin treatment and anemic by phenylhydrazine-induced erythropoietic stress (Steinberg, 1976).

The results in Figures 1 and 2 show the development of granulocyte and erythroid roid colonies in PCDC cultures of normal rat marrow implanted into rats after the induction of AML by the I.V. injection of a suspension containing 10 to 15 $\times 10^6$ viable leukemia cells obtained from a transplantable chloroma. The progression of the disease is rapid with death occuring 12 to 14 days after induction. Thus, experiments were designed so that colony formation were evaluated over this 12 to 14 day period. Since the progression of the leukemia with time after induction correlates with a gradual increase in bone marrow myeloblasts, the day of the leukemia has been used to define the relative stage of the leukemia. In experiment I, "early stage" leukemia hosts received PCDCs several hours after tumor cell inoculation. Chambers from these animals were harvested on days 1 to 7 of the leukemia (early and middle phase). "Middle" and "late" stage leukemic recipients were implanted with PCDCs 4 and 6 days after disease induction (experiments 2 and 3 respectively). PCDC colony formation was evaluated from day 6 to day 12 of the leukemia (middle to late phase).

In experiments I and II significant suppression ($p< 0.003$; $p< 0.005$) of granulocyte colony formation was observed over the entire course of the disease (Fig. 1). Colony number decreased to a mean of 36 percent of control values with a range of 10 to 60 percent. Colony size was also dramatically reduced to well under 100 cells. However, the maturation of cells within the colony appeared to be unaffected, with the appearance of normal metamyelocyte-neutrophil colonies. In experiment III, 5 of 7 rats showed a mild stimulation ($p< 0.005$) in the number of granulocyte colonies on day 10 (late stages) of the disease. This stimulation may be related to a mild increase in the serum levels of CSA in response to the leukemia-induced neutropenia.

In experiment I, mean erythroid colony formation in the PCDC was mildly depressed ($p< 0.03$) during the early stages of the disease. However, significant inhibition (> 50 percent) was observed on day 4 of the leukemia. Hematocrits remained relatively normal (mean, 47 percent) during this early phase of AML with some rats beginning to show a mild depression to 32 percent. In experiment II, a significant stimulation of CFU-E colony formation (100 to 150 percent of control) was observed on day 7 ($p< 0.05$) and day 8 ($p< 0.0005$) after the induction of the leukemia (middle stage of progression). Colony formation in some individual hosts was as high as 350 percent of control values. In experiment III, this level of erythroid stimulation (350 to 500 percent of control) was also evident during the middle stage of AML progression. During this phase of the disease, the onset of anemia was evident, with hematocrits as low as 26 percent observed (range = 26 to 43 percent). During the terminal stages of the disease (day 9 to 12 in experiment II) growth both CFU-E and BFU-E was markedly suppressed, to 65 percent of control values (p< 0.0003). The number of individual subunit colonies comprising the erythroid burst colonies as well as the total number of cells within each subcolony was noticeably reduced.

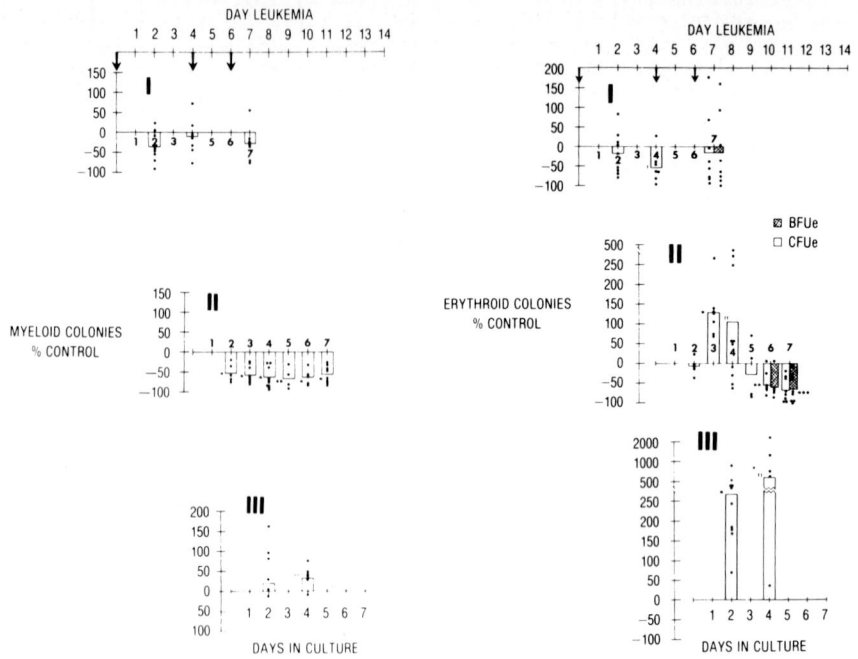

Figs 1 & 2. Granulocyte and erythroid colony formation in PCDCs implanted into leukemic hosts on day 0 (experiment I), day 4 (experiment II) and day 6 (experiment III) of the leukemia. Each point represents mean colony number as a percent of control values from 2-4 chambers harvested from a single leukemic host. The bars represent the mean values percentages of control values for all chambers harvested on a particular day.

The changes evident in hemopoiesis in PCDC cultures at various stages in the progression of AML may reflect the physiological alterations in hemopoiesis occurring within the bone marrow of the leukemic host animal. The suppression of erythroid and granulocyte colony formation in PCDC is similar to the disturbances in in vitro colony formation observed in humans and animals with AML (Broxmeyer,1983; Moore, 1979). However, since the decrease in hemopoiesis occurs within a compartmentalized chamber inserted in the leukemic host animal, the inhibition observed in these experiments appears to be mediated by a diffusible leukemia-associated inhibitory (LAI) factor or factors and does not require cell-cell interaction. This conclusion is further supported by the observation that inhibition of PCDC colony formation is first detected during the early phase of the disease, at a time when less than 5 percent leukemic myeloblasts are observed in the host bone marrow. This implies that the shutdown in hemopoiesis observed during the early phase of AML in humans and in animal models is not the result of a "crowding out" effect by the increasing numbers of leukemic myeloblasts in the bone marrow.

The early suppression of colony formation in PCDC is transiently reversed at a time in the course of the disease when the progressive increase in the percentage of leukemic myeloblasts in the bone marrow may be expected to exacerbate the suppression of hemopoiesis and lead to the onset of neutropenia and anemia (hematocrit, 26 percent). The stimulation of PCDC colony formation is probably related to a positive response to circulating levels of CSA and erythropoietin that become increased as the animals become more cytopenic. The fact that an increase in PCDC colony formation is observed following its initial suppression suggests that the effect of the LAI on normal progenitor cells is reversible and does not result in the destruction or permanant inactivation of these cells. Thus, under the influence of LAI, erythroid progenitors in PCDC's remain suppressed and do not respond to erythropoietin until the endogenous levels are high enough to override the effect of the inhibitor. In terms of a mechanism for the induction of anemia in AML, the presence of erythropoietin-responsive cells in the PCDC may support the view that, during the progression of the disease, medullary erythroid progenitor cells, in situ, do not decline in absolute numbers but become refractory to the high circulating levels of erythropoietin (Derelanko, 1978). The ability of erythropoietin to stimulate the differentiation of potentially active erythroid progenitor cells is impaired by the presence of LAI. Thus, disturbances in the balance between stimulatory and inhibitory regulatory factors and the response of the progenitor cells to the dominating factor appears to characterize the changes occurring in normal hemopoiesis with the onset of AML. The reemergence of normal progenitor cells during remission supports this idea.

To further characterize the LAI mediated suppression of normal hemopoiesis, a suspension of 1×10^5 Shay chloroleukemia cells were isolated in diffusion chambers (LDC) and implanted into normal hosts (5 DC's/rat). Rats implanted with empty diffusion chambers (EDC) served as controls. 6 days after implantation, the bone marrow of these host animals was harvested and assayed for the number of granulocyte and erythroid colonies in plasma clot cultures in vitro or in PCDC cultures. These PCDC were implanted into normal hosts in the absence and presence of 5 LDC implanted 3 days prior to the assay of the donor bone marrow. As shown in Table 1, when compared with the bone marrow of rats implanted with EDC, a reduction in the number of erythroid clusters (27 percent), erythroid colonies (30 percent) and granulocyte colonies was observed in vitro from the bone marrow of rats bearing LDC. This decline in colony formation was not observed when LDC donor bone marrow was assayed in PCDC cultures that were implanted into normal rats. It appears that removal of the leukemic influence when LDC donor marrow was cultured in normal hosts allowed the development of normal progenitors cells. To test this possibility, bone marrow from LDC bearing rats was assayed in PCDC cultures implanted into normal hosts that had received 5 LDC 3 days prior to marrow culture. Thus, LDC donor marrow continued to be exposed to the influence of isolated leukemic cells. As shown in Table 1, the continued presence of leukemic cells markedly inhibited both erythroid and granulocyte colony formation in PCDC cultures. The results further support the notion of suppression of normal hemopoiesis by an inhibitory regulator elaborated by the expanding leukemia cell population. The advantages of this suppressor mediated mechanism is clear. By suppressing normal progenitor cells in advance of its invasive growth, the leukemic cells successfully set up conditions that favor their own proliferation. However, the effect on normal hemopoiesis is abrogated when the leukemic influence is removed.

Table 1. In vitro plasma clot and PCDC cultures of bone marrow harvested 6 days after LDC or EDC implantation. Bone marrow from untreated rats also served as controls. Colony number in EDC and untreated control rats were similar. Data are expressed as the ratio of LDC/EDC. PCDC cultures containing donor bone marrow were implanted into either normal hosts or normal hosts implanted with 5 LDC 3 days prior to assay.

	In Vitro Plasma Clot	PCDC in Normal Host		PCDC in Normal Host + LDC	
	Day 2	Day 2	Day 4	Day 2	Day 4
Erythroid clusters	0.73 a	0.84 e	0.92 f	0.34 a	0.21 c
Erythroid colonies	0.71 a	0.74 e	0.84 f	0.13 b	0.15 c
Granulocyte clusters	0.81 b	1.19 e	1.09 f	0.94 f	0.34 c
Granulocyte colonies	—	1.49 d	0.87 f	0.61 d	0.23 c

a. $p < 0.001$; $p < 0.002$
b. $p < 0.005$
c. $p < 0.01$; $p < 0.02$
d. $p < 0.05$
e. $p < 0.2$; $p < 0.3$
f. $p < 0.5$; $p < 0.6$; $p < 0.7$

II. IN VITRO STUDIES: EFFECT OF LEUKEMIC CELL DIFFERENTIATION ON NORMAL HEMOPOIESIS:

Based on studies of murine and human leukemic cell lines (Tsiftsoglou, 1985), the induction of differentiation of leukemic cells has been proposed as an alternative approach to cytotoxic therapy which has a detrimental effect on normal progenitor cells (Sachs, 1980). Although the induction of leukemic cell lines to undergo differentiation gives rise to cells with the appearance and properties of "normal" blood cells, it is not clear whether these differentiated cells are "non-malignant" with respect to their inhibitory effect on normal progenitor cells. Thus, the ability of human leukemic cell lines to suppress normal human bone marrow progenitor cells was examined before and after the induction of differentiation.

Suspensions of normal human bone marrow cells were prepared from the ribs of patients undergoing thoracic surgery. Standard methods, including ficoll-hypaque separation of the mononuclear cells, were used to prepare the bone marrow cells for culture. Leukemic cell lines HL-60 and K562 were grown in RPMI-1640 supplemented with 10 percent heat inactivated fetal bovine serum (Hyclone, Logan, Utah) and glutamine. Induction of K562 cell differentiation was initiated by treatment with either hemin or sodium butyrate. K562 cells were treated for 2 to 5 days with 2 mM sodium butyrate and for 5 days with 30 uM to 70 uM hemin before coculture with normal bone marrow cells. After exposure to inducer, K562 cells were harvested, washed twice, and resuspended in fresh Iscove's Modified Delbecco's Medium (IMDM) with 2% heat-inactivated human AB serum. Similarly, HL-60 cells were treated for 2 to 5 days with 2 uM cis-retinoic acid before coculture with normal bone marrow cells. The growth of HL-60 and K562 before and after exposure to chemical inducers was assessed in suspension and plasma clot cultures by cell counts or colony counts respectively. Differentiation was assessed by changes in functional assays such as nitroblue tetrazolium dye (NBT) reduction and hemoglobin synthesis for HL-60 and K562 respectively.

CFU-E and BFU-E growth in the cocultured normal marrow was assayed in plasma clot cultures in 1% BSA, 30% human AB serum, 1×10^{-4} M mercaptoethanol, and erythropoietin (Eaves, Vancouver, British Colombia) 0.25 U/ml and 2 U/ml, respectively 2.5×10^4 to 5.0×10^4 normal bone marrow cells were plated in 0.3 ml plasma clot in 2 ml Linboro microtiter wells (Flow Labs, McLean, Virginia) in the presence and absence of K562 or HL-60 cells. Chemically-induced or untreated K562 or HL-60 cells (1×10^4) were cocultured with the normal bone marrow cells either within the same plasma clot or in underlayers in 0.5 percent agar. Plasma clots containing CFU-E and BFU-E were harvested on days 7 and 12 respectively, dehydrated, fixed in 5% gluteraldehyde, and stained with benzidine-hematoxylin. CFU-E and BFU-E colonies were scored using previously described criteria (Axelrad, 1973).

Direct coculture of increasing numbers of untreated K562 or HL-60 cells with normal bone marrow cells in the same plasma clot resulted in a progressive decrease in the number of CFU-E derived colonies (Fig. 3). A 40 to 80 percent inhibition of CFU-E growth was generally observed when 1×10^4 leukemic cells were cocultured with the normal bone marrow cells. Similar results were obtained when leukemic cells were cocultured with normal bone marrow cells in a 0.5 percent agar underlayer. BFU-E were also suppressed in a dose dependent manner when increasing numbers of K562 cells were cocultured with normal bone marrow cells (Table 2). However, while a gradual decline in BFU-E colony number was also observed with K562 cells induced with sodium butyrate for 5 days prior to coculture, the degree of suppression at each cell concentration was significantly less than that observed with untreated K562 cells ($P < 0.005$). Only 58.2 percent of the BFU-E were inhibited with 10^6 induced cells as compared with their total suppression when cocultured with 10^6 uninduced cells. A gradual loss in the ability of K562 and HL-60 cells to suppress BFU-E growth was observed with the time of exposure to sodium butyrate and retinoic acid respectively (Table 3). This correlated well with the degree of differentiation of the leukemic cells after exposure to retinoic acid for 2 and 4 days. 35 and 80 percent of the HL-60 cells were NBT+ at these times of retinoic acid exposure. Only 5 percent of untreated HL-60 cells were NBT+.

The effect of 5 days of exposure of K562 cells to varying concentrations of hemin prior to coculture with normal marrow cells on the development of CFU-E is shown in Fig. 4. Uninduced K562 cells caused a 40 percent suppression of CFU-E growth. Suppression was no longer evident in cocultures containing K562 cells pretreated with different concentrations of hemin. Preincubation of K562 cells for 5 days with 30 to 50 uM hemin had little or no effect on K562 cell growth in suspension cultures or colony formation in plasma clot assays (data not shown). However, the increase in hemoglobin synthesis in these pretreated cells to 40 ug/10^7 cells as compared with no detectable hemoglobin in untreated cells suggested the induction of K562 cell differentiation.

In these in vitro studies, the suppression of human CFU-E and BFU-E colony formation by cocultured leukemic cell lines is mediated by an inhibitory regulator elaborated by these cells. Two observations support this conclusion. First, leukemic cells cocultured in a 0.5 percent underlayer inhibited CFU-E and BFU-E to the same extent as leukemic cells cocultured directly with normal bone marrow cells within the same plasma clot culture. Second, conditioned medium (CM) from cultures of K562 and HL-60 cells suppressed normal colony formation (data not shown). The induction of leukemic cell differentiation

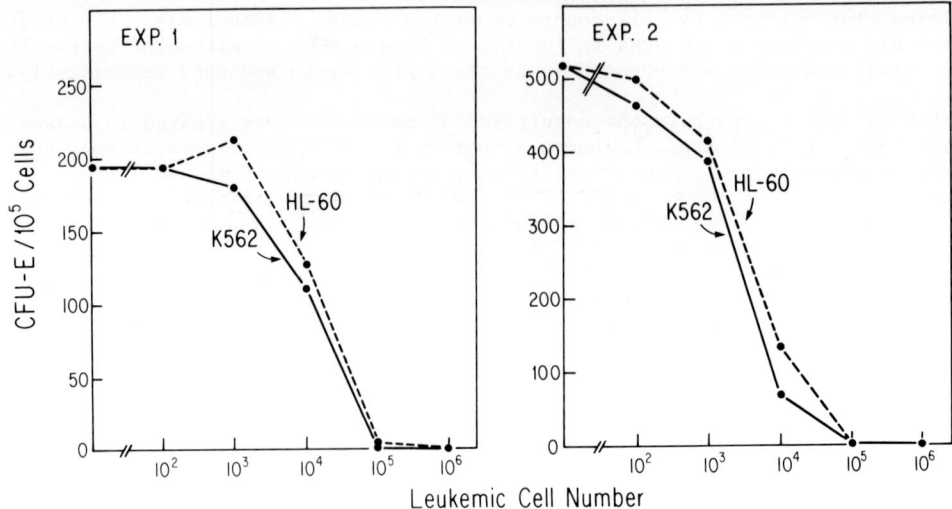

Fig 3. Effect of K562 and HL-60 leukemic cells on growth of normal bone marrow CFU-E. Each point represents the mean of nine plasma clot cultures.

Table 2. Suppression of normal BFU-E growth by K562 cells before and after five days of pretreatment with 2 mM sodium butyrate.

K562 Cell Concentration	Untreated K562 Cells	Pretreated K562 Cells
0	39.2 ± 1.2 (0)*	39.2 ± 1.2 (0)
10^4	22.0 ± 1.1 (43.6)	35.5 ± 1.6 (9.5)
10^5	15.3 ± 0.82 (61.0)	27.2 ± 2.1 (30.6)
10^6	0 (100)	16.4 ± 1.6 (58.2)

*Absolute number of BFU-E colonies per 10^5 bone marrow cells. Numbers in parentheses represent percentage of suppression for BFU-E growth compared with values in the absence of K562 cells.

Table 3. The effect of time of leukemic cell line exposure to chemical inducers on their ability to suppress normal BFU-E growth. HL-60 and K562 cells were pretreated with 2 uM retinoic acid and 2 mM sodium butyrate respectively prior to coculture of 1 x 10^4 cells. Results are expressed as % control and represent the mean ± SE of 3 experiments.

Day of Pretreatment	HL-60 Cells	K562 Cells
0	*57.5	*43
2	*81.0	*72
3	-	84
4	101	90

*p<0.05

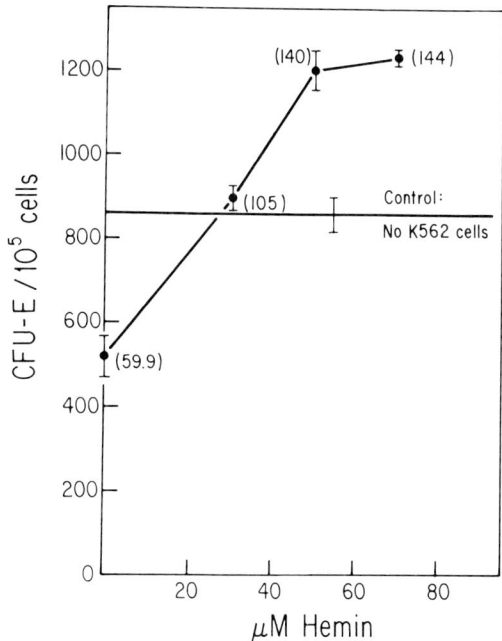

Fig. 4. Effect of pretreatment of K562 cells with varying concentrations of hemin on their ability to suppress normal CFU-E growth. Each point represents the mean of nine plasma clot cultures. Numbers in parentheses represent percentage of control. †Significance, $P < .001$; *significance, $P < .005$.

results in the loss in their ability to suppress normal hemopoiesis. The degree to which leukemic cell suppression of normal CFU-E and BFU-E growth is reversed correlates with the duration of previous exposure to the inducing agent and is no longer evident after 5 days of prior treatment with the chemical inducers. Thus, an increasing fraction of leukemic cells appear to lose their inhibitory capacity with time of treatment. This loss of suppression can be correlated with a decrease in the elaboration of the inhibitory molecule as the cells undergo differentiation (Table 4). This is confirmed by a gradual decline in the amount of inhibition detected in the conditioned medium of K562 cells induced with 30 uM hemin from 60% in uninduced cells to 28% and 0% two and three days after treatment. 30 uM hemin alone had no effect on colony formation. Reversal of cell suppression is due to the more differentiated state of the leukemic cells and their apparently "less malignant" nature after chemical induction and not to changes in the numbers of cocultured leukemic cells. This conclusion is supported by the fact that K562 cell growth in suspension culture and clonal assay is unaltered after exposure to 30 uM hemin but the ability of these cells to suppress normal hemopoiesis is significantly deminished. In addition, no detectable inhibitory activity is observed in the conditioned medium of K562 cells 3 days after exposure to 30 uM hemin. Dorner (1983) showed a significant decrease in the production of acidic isoferritin (a putative negative regulator) from HL-60 cells at a concentration of DMSO that induced differentiation and caused a decline in proliferative capacity. Therefore, it is likely that the loss in the ability of butyrate and retinoic acid-pretreated K562 and HL-60 cells to suppress normal hemopoiesis is related to a decrease in the production of the inhibitory factor. The differentiation approach to chemotherapy may, therefore, prove to be an effective means of treating leukemia for two reasons: elimination of the malignant cell population and elimination of its negative influence on hemopoiesis, thus sparing the normal progenitor cells required to repopulate the bone marrow.

III. INDUCTION OF DIFFERENTIATION OF LEUKEMIC CELLS BY THE NUCLEOSIDE ANALOGUE RIBAVIRIN AND GM-CSF:

The success of leukemic cell differentiation as a therapeutic approach will depend on the identification of agents that have a preferential effect on the leukemic cell population and little or no detrimental activity against normal progenitor cells. Ribavirin is a nucleoside analogue that was originally developed as an antineoplastic drug but, because of its activity against a broad spectrum of viruses, is presently used in the treatment of viral illnesses (Smith, 1980). Ribavirin (RBV) treatment is associated with an anemia that resolves after the administration of the drug is terminated. No effect on white blood cells or platelets is observed. The in vitro effects of REV on normal human progenitor cells reflect these clinical findings and suggest that RBV exerts a selective but reversible effect on the CFU-E (Steinberg, 1986). Because of RBV's limited toxicity for normal progenitor cells, RBV's potential as an inducer of leukemic cell differentiation was investigated. Its effectiveness as an inducer when combined with GM-CSF was also examined.

Ribavirin's effect on HL-60 cell growth and differentiation was assessed in clonal assay in the presence of 20% fetal bovine serum, 1% BSA and 1×10^{-4}M mercaptoethanol. GM-CSF derived from giant cell tumor-conditioned medium (GCT) was added at a 10% final concentration to duplicate cultures. Nitroblue tetrazolium positive (NBT+) colonies were scored after 7 to 10 days of culture. As shown in Table 4, RBV effected a dose dependent reduction in the proliferative ability of HL-60 cells and induced their differentiation to functionally normal cells. This differentiation inducing effect was enhanced by the addition of GCT derived GM-CSF so that the combination of agents resulted in a greater number of NBT+ colonies at any concentration of RBV. This effect was manifested by a shift to the left in RBV's dose-response induction of HL-60 differentiation. Thus, greater than 80% of the colonies were NBT+ at 1 ug/ml RBV in the presence of GM-CSF as compared with only 32% in its absence. GM-CSF did not influence HL-60 cell differentiation in the absence of RBV.

Table 4. Effect of Ribavirin on the growth and differentiation of HL-60 cells in plasma clot culture in the presence and absence of GCT derived GM-CSF.

RBV Concentration ug/ml	Colonies/300 Cells -CSA	+CSA	% NBT + Colonies -CSA	+CSA
0	136 $\pm$ 3	125 $\pm$ 6	0	5 $\pm$ 0.8
0.1	83 $\pm$ 3	100 $\pm$ 4	0	14 $\pm$ 1
1.0	62 $\pm$ 6	47 $\pm$ 4	32 $\pm$ 8	82 $\pm$ 3
2.0	23 $\pm$ 3	16 $\pm$ 3	77 $\pm$ 3	100
3.0	7 $\pm$ 0.7	9 $\pm$ 0.7	100	100

Ribavirin's influence on the proliferation and differentiation of HL-60 cells may be mediated by an effect on guanine nucleotide biosynthesis and its intracellular pool size (Smith, 1984). In some cell systems, Ribavirin blocks the activity of the enzyme inosine monophosphate dehydrogenase (IMPD), which mediates guanylate biosynthesis from IMP. Inhibitors of IMPD have been shown to decrease HL-60 cell proliferation and the induce their terminal differentiation (Lucas, 1983). This maturation was associated with the down regulation of guanine nucleotide biosynthesis suggesting that changes in purine pool sizes, the production of guanine nucleotide, and the activity of IMPD may be important to the regulation of terminal differentiation of myeloid cells. IMPD inhibitors may influence terminal maturation by altering the responsiveness of HL-60 cells to GM-CSF (Wright, 1986) which, normally, has a positive effect on their proliferation but no influence on their differentiation in culture (Ruscetti, 1981). RBV appears to increase the responsiveness of HL-60 cells to the differentiating effects of GM-CSF resulting in the acceleration in the terminal maturation of these cells. Thus, both agents act in concert to influence the differentiation of HL-60 cells. While these preliminary studies suggest a potential clinical role for GM-CSF, its effectiveness as an inducing agent may only be realized in combination with other agents that initially exert an influence on the proliferation and induction of leukemic cells. The modulation of guanine nucleotides by IMPD inhibitors suggests that its effect on HL-60 differentiation may involve a GTP-regulated signal transduction pathway for hemopoietic growth factors (Wright, 1986). As such, RBV's effect on HL-60 cell responsiveness to GM-CSF may be mediated by an effect on this pathway. The limited effect of RBV on normal stem cells and its ability to act in concert with GM-CSF to induce leukemic cell differentiation suggests a role for ribavirin or compounds similar to it as a therapeutic agent for AML.

REFERENCES.

1. Broxmeyer, H.E. (1986): Biomolecule cell interactions and the regulation of myelopoiesis. Int. J. Cell Cloning 4, 379-405.
2. Spivak, J.L. (1986): The mechanism of action of erythropoietin. Int. J. Cell Cloning 4, 139-166.
3. Steinberg, H.N., Handler, E.S., Handler, F.E. (1976): Assessment of erythrocytic and granulocytic colony formation in an in vivo plasma clot diffusion chamber system. Blood 47, 1041-1051.
4. Takahashi, M., Fujiwara, M., Kishi, K., et. al. (1985): CSF producing gall bladder cancer: Case report and characteristics of the CSF produced by tumor cells. Int. J. Cell Cloning 3, 294-302.
5. Broxmeyer, H.E. (1983): Colony assays of hemopoietic progenitor cells and correlations to clinical situations. CRC Crit. Rev. Oncol. Hematol. 1, 227-257.
6. Moore, M.A.S. (1979): Humoral regulation of granulopoiesis. Clinics in Hematology 8, 287-309.
7. Fialkow, P.J. (1980): Clonal and stem cell origin of blood cell neoplasms. In: Contemporary Hematology/oncology. eds. J. LoBue, A.S. Gordon, R. Silber, F.M. Muggia, pp 1-46. New York, Plenum Press.
8. Handler, E.S., Handler, E.E., Bacon E.R. (1974): In vitro colony formation in bone marrow and spleen cell suspensions from rats with an acute myelogenous leukemia. J. Lab. Clin. Med. 84, 249-257.
9. Moore, M.A.S., Williams, N., Metcalf, D. (1973): In vitro colony formation by normal and human hemopoietic cells: Interaction between colony forming and colony stimulating cells. J. Natl. Cancer Inst. 50, 591-602.
10. Sachs, L. (1980): Constitutive uncoupling of pathways of gene expression that control myeloid growth and differentiation in myeloid leukemia: A model for the origin and progression of malignancy. Proc. Natl. Acad. Sci. USA. 77, 6152-6156.

11. Tsiftsoglou, A.S. (1985): Differentiation of leukemic cell lines: A review focusing on murine erythroleukemia and human HL-60 cells. Int. J. Cell Cloning 3, 349-366.
12. Collins, S.J., Ruscetti, F.W., Gallagher, R.E, Gallo, R.C. (1978): Terminal differentiation of human promyelocytic leukemia cells induced by dimethylsulfoxide (DMSO) and other polar compounds. Proc. Natl. Acad. Sci. USA. 75, 2458-2462.
13. Ishikura, H., Okazaki, T., Mochizuki, T., et al (1985): Effect of antimetabolites and thymidine blockage on the induction of differentiation of HL-60 cells by retinoic acid or 1,25 dihydroxyvitamine D_3. Exp. Hematol. 13, 981-988.
14. Weinberg, J.B., Misukonis, M.A., Hobbs, M.M., Borowitz, M.J. (1986): Cooperative effects of gamma interferon and 1,25 dihydroxyvitamin D_3 in inducing differentiation of human promyelocytic leukemia (HL-60) cells. Exp. Hematol. 14, 138-142.
15. Nicola, N.A. (1985): Granulocyte colony-stimulating factor and differentiation induction in myeloid leukemic cells. Int. J. Cell Cloning 5, 1-15.
16. Handler E.E., Handler, E.S. (1970): Experimental leukemia. In: Regulation of Hematopoiesis. ed. A.S. Gordon, pp 1273-129. New York, Appelton-Crofts.
17. Cronkite, E.P., Carsten, A.L. (1980): Diffusion Chamber Culture. Berlin, Heidelberg. New York, Springer-Verlag.
18. Steinberg, H.N., Robinson, S.H. (1985): Laboratory methods in experimental hematology: I. The diffusion chamber technique in experimental hematology. Exp. Hematol. 13, 16-22.
19. Axelrad, A.A., McLeod, D.L., Shreeve, M.M., Heath, D.S. (1973): Properties of cells that produce erythrocytic colonies in plasma clot cultures. In: Hemopoiesis in Culture. Second International Workshop. ed. W.A. Robinson. pp 226-234. National Institutes of Health. Washington, D.C.
20. Steinberg,H.N. Page, P.L., Robinson, S.H. (1983): Two classes of murine granulocyte progenitor cells expressed in plasma clot diffusion chamber cultures. Blood 59, 838-846.
21. Derelanko, M.J., LoBue, J., Gordon, A.S. et al (1978): Measurement of erythropoietin in chloroleukemic rats. Exp. Hematol. 6, 91-95.
22. Tsiftsoglou, A.S., Wong, W. (1985): Molecular and cellular mechanism of leukemic hemopoietic cell differentiation: An analysis of the friend system. Anticancer Res. 5, 81-100.
23. Sachs, L. (1984): The reversibility of neoplastic transformation: Regulation of clonal growth in differentiation in hemopoiesis and the normalization of myeloid leukemia cells. In: Advances in Viral Oncology 4. ed. G. Klein, pp 307-329. New york, Raven Press.
24. Dorner,M.H., Broxmeyer, H.E., Silverstone, A., Andreeff, M. (1983): Biosynthesis of ferritin subunits from different cell lines of HL-60 human promyelocytic leukemia cells and the release of acid isoferritin-inhibitory activity against normal granulocyte-macrophage progenitor cells. Brit. J. Hematol., 55, 47-55.
25. Smith, R.A., Kirkpatrick, W. (eds.) (1980): Ribavirin: a broad spectrum antiviral agent. New York. Academic Press.
26. Steinberg, H.N., Crumpacker, C.S., Furst, A., Zeldis, J.B. (1985): The effect of ribavirin on erythroid progenitors. Blood 66, 122a (abstr.).
27. Lucas, D.L., Webster, H.K., Wright, D.G. (1983): Purine metabolism in myeloid precursor cells during maturation. J. Clin. Invest. 72, 1889-1900.
28. Wright, D.G., Jett, D.C. (1986): Evidence that intracellular supplies of guanine nucleotides (G-NTD) regulate the growth and maturation of human myeloid progenitors via a phospholipase C (PLC) mediated signal transduction pathway for GM-CSF. Blood 68, 185a (abstr).
29. Ruscetti, F.W., Collins, S.J., Woods, A.M., Gallo, R.C. (1981): Clonal analysis of the response of human myeloid leukemic cell lines to colony-stimulating activity. Blood 58, 285-292.

Résumé

Le développement de l'hématopoièse étudiée en caillot de plasma dans des chambres de diffusion implantées chez des rats développant une leucémie aigue myéloblastique induite est très anormale. On constate en effet une diminution du nombre de colonies érythroïdes et granulo-macrophagiques, qui étant donné l'isolement des cellules souches dans la chambre, ne peut être dûe qu'à la diffusion de substances inhibitrices provenant des cellules leucémiques. La même inhibition a été aussi observée in vitro quand des cocultures ont été faites avec des lignées leucémiques humaines. La capacité des lignées leucémiques d'empêcher le développement de colonies normales disparaît quand ces cellules sont soumises à l'action d'agents de différenciation. La Ribavirine, analogue nucléosidique, induit la différenciation de la lignée HL60 et modifie sa réponse au GMCSF qui entraine une différenciation accrue.

Leukemia associated inhibitor (LAI) : biological characterization and purification of the active subunit

Tor B. J. Olofsson

Division of Hematology, Department of Medicine, University of Lund, S-221 85 Lund, Sweden

ABSTRACT

A culture system was designed for the production of large volumes of HL-60 cell conditioned medium for purification of a leukemia associated inhibitor, LAI, which is constitutively produced by HL-60 cells. Monoclonal antibodies against HL-60 cells were produced and tested for their ability to bind LAI. One of the reactive antibodies, MoAb-58, was coupled to CNBr-activated Sepharose and used in affinity chromatography for purification of LAI from HL-60 cell conditioned medium. The biologically active subunit of LAI was isolated by preparative SDS-PAGE and had a molecular weight of 125K. MoAb-58 was also used to isolate a subpopulation of normal LAI-producing cells from peripheral blood by a panning technique. Such cells inhibited the colony/cluster formation by normal human bone marrow cells. LAI reversibly reduced the fraction of human CFU-GM in S-phase, whereas the S-phase fraction of BFU-E was unaffected. A monoclonal antibody against acidic ferritin, 2A4, did not bind or inhibit the activity of LAI.

KEY WORDS

Leukemia, granulopoiesis, CFU-GM, BFU-E, growth regulation.

INTRODUCTION

A characteristic feature of acute leukemia is the suppression of normal hematopoiesis leading to anemia, trombocytopenia and neutropenia with complications such as bleedings and severe infections, which is a greater clinical problem than the presence of large numbers of leukemic cells. However, the mechanisms for this suppression are not fully understood. Broxmeyer and co-workers have shown (1981, 1982a) that acidic isoferritins from leukemia cells inhibit the growth of granulocyte-macrophage and erythroid progenitor cells in vitro. It has been suggested that this inhibitory mechanism could be responsible for the suppression of normal hematopoiesis in leukemia. Acidic ferritins are also produced by a subpopulation of adherent mononuclear cells in normal peripheral blood which has been suggested as a normal feedback regulatory mechanism for granulocyte-macrophage progenitor cells (Broxmeyer, 1982b, 1984, 1986).
Another inhibitor originally found in conditioned media from leukemia cells and therefore called leukemia associated inhibitor, LAI, (Olofsson, 1980 a,b) is biochemically different from acidic ferritins and also expresses another type of inhibitory activity. Whereas acidic ferritins irreversibly inhibits the growth of

CFU-GM day 7 by an action restricted to HLA-DR-positive cells in S-phase (Broxmeyer, 1982a), LAI reversibly reduces the fraction of normal CFU-GM in S-phase and make these cells less sensitive to S-phase specific agents such as tritiated thymidine, cytosine arabinoside or hydroxyurea. Both inhibitors show only minor inhibition of leukemia stem cells, which could explain the growth advantage of leukemia cells over normal cells in leukemia. In addition to leukemia cells, a subpopulation of normal non-adherent, non-phagocytic, non-T, non-B, non-NK, Fc-receptor positive cells in blood and bone marrow produce and release LAI (Olofsson, 1985), which has been suggested as a regulatory mechanism for the proliferation rate of normal CFU-GM. The physiological role of these inhibitors, if any, has not yet been established.

The human promyelocytic cell line HL-60 constitutively produces LAI (Olofsson, 1980c), which has made it possible to produce conditioned media in amounts sufficient for purification of LAI by use of a monoclonal antibody against HL-60 cells that binds LAI. In this report the initial attempts to isolate the biologically active subunit of LAI is described as well as the isolation of normal LAI-producing cells from peripheral blood.

MATERIAL AND METHODS

Bone marrow and blood cells
Bone marrow was obtained from the posterior iliac crest of normal volunteers, collected in heparinized McCoy's medium and separated in Isopaque-Ficoll (Lymphoprep, Nycomed AS, Oslo, Norway). The light density cells were washed twice in McCoy's medium with 1% fetal bovine serum (FBS). Heparinized venous blood was diluted with an equal volume of saline before separation in Lymphoprep; the light desity cells were used for isolation of LAI-producing cells by panning technique (see below).

CFU-GM colony assay
Bone marrow cells (<1.077 g/ml) were cultured in 0.3% agar in McCoy's medium with 15% FBS and 10% human placenta conditioned medium (Burgess, 1977) on top of a 0.5% agar underlayer without cells. In some experiments bone marrow cells were cultured on top of adherent cells isolated by panning technique and in these experiments the 0.5% agar underlayer was left out. Cell concentration was 1.5×10^5/ml if not otherwise indicated. Four replicates were made and incubated at 37°C in 5% CO_2 in fully humidified atmosphere. Colonies (>40 cells) and clusters (10-40 cells) were counted on day 7 and day 10 in an inverted microscope.

BFU-E assay
Light density cells from peripheral blood were cultured in 0.8% methyl-cellulose (Methocel, A4M, Dow Chemical Co) in Iscove's modified Dulbecco medium (IDMEM) with 30% FBS, 5% Mo-cell conditioned medium (a gift from Dr. David Golde, UCLA, Los Angeles), 1% BSA, iron-saturated transferrin 0.36 mg/ml, 2-mercaptoethanol 5×10^{-5}M, and erythropoietin 1.2U/ml (purified from human urine; a gift from Dr. Miloslav Beran, M.D. Anderson Hospital, Houston, Texas). Cell concentration was 3×10^5/ml. Four replicates were made and incubated as described for bone marrow cells. Typical hemoglobinized single and multicentric erythroid colonies were counted on day 14-16 of culture.

LAI-assay
This has been described in detail previously (Olofsson, 1980a); briefly, normal bone marrow cells isolated from Lymphoprep gradients, or peripheral blood cells in the case of BFU-E, were incubated with and without conditioned medium (HL-60 cells, normal LAI-producing cells etc) or different fractions of purified LAI (from affinity columns, extracts of polyacrylamid gels etc) at the concentration of 1.5×10^6 cells/ml with an equal volume of the material to be tested. After incubation at 37° for 60 min 2 ug/ml of cytosine arabinoside was added

for another 45 min and the cells washed three times with McCoy's medium 1% FBS. The cells were then cultured in agar at 3×10^5 cells/ml. Four replicates were made and colonies were counted on day 10. The difference in colony numbers between the sample incubated with test material but without cytosine arabinoside and the sample incubated with both test material and cytosine arabinoside was taken as the fraction of CFU-GM (or BFU-E) in S-phase. A control S-phase determination was made on cells incubated with medium alone. In the presence of LAI the fraction of CFU-GM in S-phase is reduced.

Isolation of cells by panning technique
This technique has been described previously (Olofsson, 1985). Peripheral blood cells isolated in Lymphoprep gradients were further separated by removing adherent cells by adherence to plastic petri dishes for 60-90 min and removal of phagocytic cells by incubation with carbonyl iron for 20 min. The non-adherent, non-phagocytic cells were then incubated with the appropiate monoclonal antibody seeded onto plastic petri dishes coated with affinity purified goat-anti-mouse-IgG 50 ug/ml in 0.1M $NaHCO_3$ pH 9.2 (Cappel Worthington Biochemicals). The cells were left to adhere at 4^oC for 60 min. The negative cell population, i.e. the unlabelled cells, were pipetted off and collected in a tube. Both adherent cells (labelled cells) and non-adherent cells (unlabelled cells) were used to condition medium by adding equal volumes of McCoy's medium 1% FBS to the cells and incubated at 37^oC for 3 hours. The cell free conditioned medium was taken to tests for LAI-activity. In some experiments the adherent cells (labelled cells) were overlayered with normal bone marrow cells in 0.3% agar, 15% FBS and 10% human placenta conditioned medium in McCoy's medium and cultured for 7-10 days to study the inhibitory effects of the adherent cells on colony growth. To one set of the dishes hydrocortisone, 1 uM, was added.

HL-60 cells
The human promyelocytic cell line HL-60 (Collins, 1977) was cultured in plastic T-flasks in RPMI with 10% FBS and subcultured every 5-7 day. HL-60 cells were used to produce standard sources of conditioned medium containing LAI by incubating fresh cells at 10^6/ml in RPMI with 1% FBS for 2-3 hours after which the cell free conditioned medium was collected by centrifugation. HL-60 cells were also used to produce LAI for purification purposes (see below).

Conditioned medium
Cell free conditioned media from HL-60 cells or normal peripheral blood cells isolated by panning technique were washed on XM300 Diaflo filters (Amicon Co) with McCoy's medium 1% FBS and reconstituted to the original volume before sterilization by Millipore filtration and testing of LAI-activity.

Large scale production of HL-60 cell conditioned medium
For production of large quantities of HL-60 cell conditioned medium to be used as starting material for purification of LAI, HL-60 cells were expanded in RPMI medium with 10% FBS in a 10 liter glass bottle (MCS-104 Microcarrier stirrer, Techne, Cambridge, UK) equipped with a stirring rod which is intermittently rotated by a magnetic stirrer. Conditioned medium was harvested through a glass pipe dipping into the culture flask connected to a plastic tubing and a peristaltic pump. The cell suspension was passed through a fiber plasmapheresis filter (Gambro, Lund, Sweden), which allows the collection of the cell free conditioned medium from an outlet on the side of the filter. The concentrated cell suspension was recirculated to the culture flask. After each harvest the concentrated cell suspension remaining in the bottle was diluted to the original volume with fresh medium and 1.5% FBS. In this way 7-8 liters of conditioned medium can be harvested every day for more than one month without interruption if the cell concentration is kept at $0.5 - 1.0 \times 10^6$/ml by removing cells intermittently. The culture system is shown in Fig. 1.

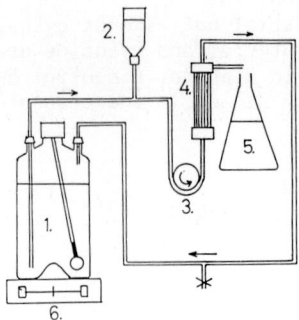

1. culture vessel
2. NaCl
3. pump
4. fiber plasmapheresis filter
5. filtrate collector
6. magnetic stirrer

Figure 1. Principal outline of the culture system for production of large volumes of HL-60 cell conditioned medium.

Production of monoclonal antibodies
Female Balb/cBom mice were immunized with 10^7 fresh HL-60 cells in saline by subcutaneous injection and received booster injections intraperitoneally of 10^7 cells after 2, 4 and 8 weeks. Four days after the last injection spleen cells were fused with Sp2/0 myeloma cells and cultured in Dulbecco's Modified Eagle's Medium (DMEM) with 20% FBS and HAT (Flow Laboratories) in 96-well microtitration plates on feeder layers of murine peritoneal macrophages (3000 per well). Hybridomas were tested for antibody production by immunofluorescence assay of the supernatants using fresh HL-60 cells as target cells. Hybridomas producing antibodies against HL-60 cell surface antigens were cloned in HAT-medium. When a stable clone was obtained the cell supernatant was tested for anti-LAI activity in the following way. One ml of HL-60 conditioned medium (known to contain LAI) was incubated with 0.5 ml hybridoma supernatant at 4°C over night. To each sample was added 0.5 ml goat-anti-mouse-Ig-Sepharose (2 mg affinity purified antibody was coupled to 2 g CNBr-activated Sepharose) corresponding to 0.3 ml packed gel beads; the tubes were capped and rotated at room temperature for 2-3 hours and then centrifuged. The supernatant was collected, filter sterilized and tested for remaining LAI-activity. Larger quantities of monoclonal antibodies were produced by collecting ascites fluid from mice injected with hybridoma cells. Antibodies were purified by ammonium sulphate precipitation and affinity chromatography on protein A-Sepharose when applicable.

Purification of LAI by affinity chromatography
The conditioned medium from HL-60 cells was first concentrated using a Pellicon Cassette System (Millipore Corp) with a PTHK Cassette filter with a cut off at approximately 100,000. The medium was concentrated about tenfold before being applied to affinity chromatography. One of the monoclonal antibodies with anti-LAI activity, hereafter called MoAb-58, was purified from ascites fluid and coupled to CNBr-activated Sepharose (Pharmacia Fine Chemicals, Uppsala, Sweden), 5 mg antibody per 1 g of gel. The gel was packed in small columns of 5 ml each and equilibrated with 5 mM HEPES in 0.15 NaCl, pH 8.2. The concentrated conditioned medium was adjusted to pH 8.2 and applied to the columns at a rate of 15-20 ml/

hour. Ten to 15 liters of concentrated conditioned medium were passed through each column before the gel was extensively washed with HEPES-NaCl buffer and the absorbed material eluted by 4 x 5 ml of 0.1M glycine in 0.5M NaCl, pH 3.0, into borosilicate glass tubes. Tween 20 was added to a final concentration of 0.05% and pH was adjusted immediately to pH8 with 1M NaOH. Aliquots from the eluted fractions that were to be tested for LAI-activity were taken off before addition of Tween 20 and diluted with McCoy´s medium 1% FBS and washed on XM300 filters.

SDS-PAGE

SDS-PAGE was performed using a 5-20% gradient gel (Laemmli, 1970). Aliquots (5-150 ul) of the fractions eluted from affinity chromatography on MoAb-58 Sepharose were reduced with 2-mercaptoethanol and electrophoresed for analytical or preparative purposes. In the former case gels were silver stained. When preparative SDS-PAGE was performed the gel was immediately cut by a razor blade into 1-2 mm slices. The separate gel slices were transferred to borosilicate glass tubes, minced with a stainless steel spatula and covered with 2-3 ml of 5 mM HEPES in 0.15M NaCl, pH 7.4. The tubes were tightly capped and left rotating in the cold over night to elute the protein. Aliquots of the eluate was taken to test of LAI-activity after mixing with McCoy´s medium 1% FBS and XM300 filtration, and to SDS-PAGE to assess the purity of the isolated protein.

Monoclonal anti-ferritin antibody

A monoclonal antibody, 2A4, directed against acidic ferritin (Luzzago, 1986) (a generous gift from Dr. Paolo Arosio, University of Milano, Italy) was tested for anti-LAI acitivty. To one ml of HL-60 conditioned medium was added 10 ul antibody (2.5 mg/ml) and incubated at $4^{o}C$ over night. Goat-anti-mouse-Ig-Sepharose (150-200 ul gel beads) was added and the sample treated as described for testing of anti-LAI activity of monoclonal antibodies against HL-60 cells.

RESULTS

LAI in HL-60 cell conditioned medium

We have previously observed that when HL-60 cells are suspended in fresh medium and left to sediment slowly in the culture flask LAI is produced over the first 2-5 hours but then the production ceases; if, however, the cells are kept in suspension by continuous stirring they continue to produce LAI for at least 50 hours (Olofsson, 1980c). Table 1 shows the results of one of several tests for LAI-activity in conditioned medium, plasmapheresis filtrate and the outcome of ultrafiltration.

Table 1. LAI-activity in HL-60 cell conditioned medium harvested by fiber plasmapheresis filtration.

Sample	CFU-GM in S-phase (%)
Conditioned medium (1)	23.0 ($p<0.05$)
Plasmapheresis filtrate	23.7 ($p<0.05$)
Ultrafiltration concentrate (2)	17.8 ($p<0.01$)
Ultrafiltration filtrate	30.4 NS
Control (McCoy´s medium)	32.6

(1), cell free supernatant directly from the bottle; (2), ultrafiltrated on Pellicon Cassette System, diluted 10x before tested for LAI.

As expected the conditioned medium obtained by centrifugation and the plasmapheresis filtrate contained the same LAI-activity and all activity was recovered in the ultrafiltration concentrate.

LAI and BFU-E

Two preparations of highly purified LAI from affinity chromatography on MoAb-58 Sepharose and material eluted from SDS-polyacrylamid gels were tested independently for LAI-activity against BFU-E. LAI did not inhibit erythroid colony formation nor did it reduce the fraction of BFU-E in S-phase (exp.1: 21.3% control, 21.3% with LAI; exp.2: 13.3% control, 13.6 with LAI; the number of BFU-E/dish was 235 and 240 in exp.1, and 90 and 88 in exp.2 respectively in the cultures not pretreated with cytosine arabinoside).

Monoclonal antibodies against HL-60 cells

Out of a total of 576 wells 152 showed growth of hybridoma cells of which 37 produced antibodies that stained surface antigens on HL-60 cells in the immunofluorescence assay. Twenty-three of the 37 hybridomas produced stable clones. The fraction of positively stained cells varied from 5-95 per cent and the staining pattern varied from a uniform lining of the entire cell to a sparse distribution of small fluorescent dots. Five of the antibodies showed anti-LAI activity of varying intensity; they also reacted differently when tested in immunofluorescence using normal peripheral blood cells as target as shown in Table 2.

Table 2. Monoclonal antibodies against HL-60 cells with anti-LAI activity

MoAb	Ig subclass	Per cent stained MNC (1)
58	IgG_1	15-25
105	IgG_2	70-80
112	IgM	20-25
118	IgM	1-2
143	IgG_1	1-2

The anti-LAI activity was ascertained in two independent tests as described in Material and Methods. (1), immunofluorescence staining of peripheral blood mononuclear cells, the figures are the approximate range obtained from several independent tests.

Affinity chromatography on monoclonal antibody-coupled Sepharose

The monoclonal antibodies shown in Table 2 were all produced in quantities sufficient for coupling to CNBr-activated Sepharose and tested for their capacity to bind LAI in affinity chromatography of concentrated HL-60 cell conditioned medium. They all absorbed LAI but with different efficiency and MoAb-58 was selected for further experiments aiming at the purification of LAI. Table 3 shows a typical experiment with MoAb-58 Sepharose affinity chromatography.

In the experiment shown in Table 3, three liters of concentrated conditioned medium corresponding to 30 liters fresh CM was processed. In other experiments more than 10 liters of concentrated CM has been applied to a 5 ml column with similar efficiency.

Table 3. Affinity chromatography test of MoAb-58 Sepharose

Sample	CFU-GM in S-phase (%)
Conc HL-60 CM (before) (1)	16.4 (p<0.01)
Conc HL-60 CM (after) (2)	38.0 NS
Glycine eluted material (3)	20.9 (p<0.01)
Control (McCoy´s medium)	31.6

(1), conditioned medium applied to the column; (2), CM passing through the column; (3), 0.1M glycine in 0.5M NaCl, pH 3.0, pooled fractions; all material was diluted 10-20 x in McCoy´s medium 1% FBS and XM300 filtered before testing of LAI on normal bone marrow cells.

SDS-PAGE of affinity purified LAI
Fig. 2 shows a typical silver stained gel after SDS-PAGE of the four glycine eluted fractions from a MoAb-58 Sepharose column. Despite extensive washing of the column before elution with glycine buffer fractions 1-3 contain several impurities of which a considerable amount is elution of the material originally coupled to the gel as evidenced by the elution pattern of gels without any sample applied (gel not shown). The band corresponding to a molecular weight of 125K contains the LAI-activity as evidenced by the preparataive SDS-PAGE shown in Fig. 3. Several experiments identical with that shown in Fig. 3 gave the same result. In addition, testing of gel slice eluates from the rest of the entire gel did not show any LAI-activity.

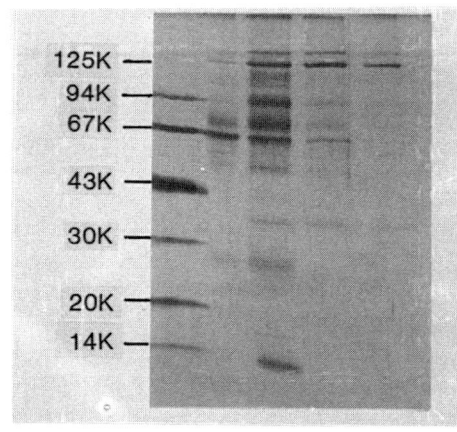

Figure 2

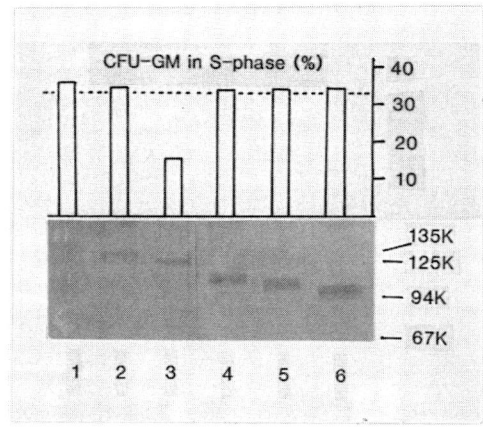

Figure 3

Isolation of LAI-producing cells by panning technique
The different monoclonal antibodies shown in Table 2 were all tested for their efficiency in a panning technique with the aim to isolate a subpopulation of peripheral blood cells with the capacity to produce LAI. The purity of the non-adherent cell population was ascertained by immunofluorescence microscopy. MoAb-58 regularly resulted in less than 1 per cent stained cells in the non-adherent fraction, whereas the other antibodies did not result in a good enough

separation between stained and unstained cells to be applicable in further experiments. Table 4 shows the results of two typical experiments where cells isolated by panning utilizing MoAb-58 resulted in one fraction of non-adherent MoAb-58 negative cells and another fraction of adherent MoAb-58 positive cells. The conditioned media of the two cell populations were tested for LAI-activity and as shown in the table only the adherent MoAb-58 positive cells produced LAI.

Table 4. LAI-activity in CM from cells isolated by panning technique

	CFU-GM in S-phase (%)
Exp. 1 control (McCoy's medium)	37.3
CM MoAb-58 positive cells	20.1 ($p<0.01$)
CM MoAb-58 negative cells	33.9 NS
Exp. 2 control (McCoy's medium)	30.8
CM MoAb-58 positive cells	17.1 ($p<0.01$)
CN MoAb-58 negative cells	33.1 NS

NS, not significant; p-values are from Student's t-tes.

Inhibition of colony growth by MoAb-58 positive cells

MoAb-58 positive cells isolated by panning technique were overlayered with normal bone marrow cells in agar to study the effects of MoAb-58 positive cells on colony growth. To one set of dishes hydrocortisone succinate was added to inhibit the production of LAI. As shown in Table 5 (two representative experiments) MoAb-58 positive cells inhibited colony/cluster growth on day 7 and this inhibition was abrogated by the addition of hydrocortisone.

Table 5. Inhibition of colony/cluster growth by MoAb-58 positive cells

	No. of colonies + clusters d.7	
Exp. 1 control	276 (11)	
MoAb-58 positive cells	218 (17)	($p<0.01$)
MoAb-58 positive cells + HC	263 (8)	NS
Exp. 2 control	171 (4)	
MoAb-58 positive cells	96 (3)	($p<0.01$)
MoAb-58 positive cells + HC	169 (6)	NS

The figures shown are mean values of four replicates with the SD shown in brackets. HC, hydrocortisone succinate, 1 uM. NS, not significant; the p-values are from Student's t-test. Clusters of more than 10 cells and colonies of more than 40 cells were counted.

Anti-ferritin vs LAI

The monoclonal antibody 2A4, specific for acidic ferritin, was tested for its capacity to inhibit LAI-activity. The results of 5 experiments are shown in Table 6. This antibody did not abrogate the inhibitory activity of LAI in HL-60 cell conditioned media.

Table 6. Anti-acidic ferritin vs LAI in HL-60 cell CM

	CFU-GM in S-phase (%)
Controls (McCoy's medium)	31.3 ± 3.0
HL-60 CM	21.0 ± 4.3 (p<0.01)
HL-60 CM + 2A4	18.8 ± 5.4 (p<0.01)

Mean $\pm$ SD of 5 experiments. Both McCoy's medium, HL-60 CM and HL-60 CM + 2A4 were treated with goat-anti-mouse-Ig-Sepharose before testing of LAI-activity. The p-values are from Student's t-test.

DISCUSSION

Earlier attempts to purify LAI from HL-60 conditioned media have not been successful using conventional methods such as gel chromatography on Sepharose 6B, affinity chromatography on ConA-Sepharose and SDS-PAGE (Olofsson, 1980b). This can mainly be ascribed to the time consuming procedures and that each purification step can only handle smaller aliquots of the material at a time, which gives rise to considerable losses of activity. Another problem is the minute quantities of LAI present in HL-60 cell conditioned media, which required first, the production of large amounts of conditioned medium, and second, a new strategy for the purification procedure.

The production of large quantities of HL-60 cell conditioned medium was solved by a special culture system where the important feature is the fiber plasmapheresis filter which allows repeated harvesting of conditioned medium and recirculation of the cells under sterile conditions. We felt that a monoclonal antibody with the capacity to bind LAI would be useful in affinity chromatography in our attempts to purify LAI and therefore, we produced monoclonal antibodies against HL-60 cells with the hope of finding an antibody that would bind LAI. Intact HL-60 cells were used for immunization because the cells probably expose LAI on their surface (Olofsson, 1980b). Surprisingly, several monoclonals were produced with the capacity to bind LAI. The binding characteristics of these antibodies to mononuclear blood cells varied considerably and the fact that they all react with LAI can possibly be explained by the fact that LAI is a large glycoprotein and that the different monoclonals react with different sugar moieties in the molecule, which may be shared with several other surface glycoprotein or glycolipid components of the HL-60 cell surface. This restricts the specificity of the antibodies and for several reasons one of them, MoAb-58, was selected for further attempts to purify LAI by affinity chromatography.

MoAb-58 coupled to Sepharose proved to very useful for these purposes. Large quantities of HL-60 cell conditioned medium could be chromatographed on relatively small columns with acceptable recovery of LAI. By preparative SDS-PAGE it was demonstrated that the inhibitory activity of LAI is confined to a single band of approximately 125K. Based on the relative strength of the molecular weight markers and LAI on silver stained gels it was calculated that the range of activity of the isolated subunit is $10^{-9} - 10^{-12}$ M and that the HL-60 cell conditioned medium contain approximately 100 ng per liter. Since it is not possible to measure LAI with a greater accuracy in the different preparations during purification no estimations of the recovery of LAI can be given. The native LAI molecule is greater than 500K on Sepharose 6B gel chromatography (Olofsson, 1980b). It is, however, not possible at this point to determine if the native molecule is composed of several 125K subunits or if the entire molecule contains other molecular weight components as well. The 135 K component visible in Fig. 2 and 3 is a FBS component that can be removed by rabbit-anti-FBS-Sepharose (data not shown).

MoAb-58 also proved valuable for isolation of a subpopulation of non-adherent, non-phagocytic mononuclear blood cells that produce LAI into conditioned medium. Furthermore, when these cells were overlayered with normal bone marrow cells in agar, they reproducibly inhibited colony/cluster growth on day 7 of culture. This is an interesting observation since earlier attempts to show inhibition of colony growth by addition of even highly purified LAI to agar cultures have failed to show any inhibition, which has been explained by the fact that the action of LAI is reversible and therefore may be overcome during a 7-14 day culture. Furthermore, recent studies have shown that normal bone marrow cells absorb LAI effectively during a 60 min incubation and that this absorption presumably is excerted by a receptor-like mechanism that can be abrogated by pretreatment of the marrow cells with trypsin or chymotrypsin (Olofsson, submitted for publication). However, if LAI-producing cells are present at sufficient concentration throughout the culture period as in the experiments reported here they obviously have the capacity to inhibit colony/cluster growth. The assumption that the inhibition is due to the production of LAI is substantiated by the observation that addition of hydrocortisone totally abrogated the inhibitory effect; corticosteroid hormones almost totally inhibit the production of LAI from both AML cells, normal cells and HL-60 cells (Olofsson, submitted for publication).

Conditioned medium from normal light density peripheral blood cells does not contain detectable amounts of LAI. However, if the adherent and phagocytic cells are removed the remaining cells reproducibly produce LAI. This LAI-production can readily be inhibited by PGE ($10^{-7}M$) as well as the LAI-production by HL-60 cells. If light density normal peripheral blood cells are treated with indomethacin ($10^{-6}M$) they produce LAI (Olofsson, submitted for publication). These observations all sustain the notion that LAI-production normally is regulated by adherent monocytic cells presumably through the action of PGE secreted by these cells.

The isolation of the active subunit of LAI as a 125K component and the demonstration that a highly specific monoclonal antibody against acidic ferritin (Luzzago, 1986) does not bind or inactivate LAI clearly shows that LAI and acidic ferritin are non-identical. It is therefore possible that LAI is a novel putative feedback regulator of granulocyte-macrophage production active in normal hematopoiesis as a regulator of the proliferative rate of granulocyte-macrophage progenitor cells, and in leukemia when overproduced by leukemia cells as an inhibitor of normal granulopoiesis that gives a growth advantage to the leukemia cells. Final purification of LAI and determination of the amino acid sequence will hopefully lead to identification of the molecule and open the way for more critical experiments for elucidation of the possible physiological role of LAI in hematopoiesis.

Acknowledgements
This study was supported by the Swedish Cancer Society, John and Augusta Persson Foundation, Magnus Bergvall Foundation and the Medical Faculty of Lund. I wish to thank Ingrid Gärtner, Elisabeth Perssson and Lill Ivarsson for skilful technical assistance, and May-Louise Andersson for preparation of the manuscript.

REFERENCES

Collins, S.J., Gallo, R.C., Gallagher, R.E. (1977): Continuous growth and differentiation of human myeloid leukemia cells. Nature 270, 347-349.

Broxmeyer, H.E., Bognacki, J., Dörner, M.H., de Sousa, M. (1981): Identification of leukemia-associated inhibitory activity as acidic ferritins. A regulatory role for acidic ferritins in the production of granulocytes and macrophages. J.Exp.Med. 153, 1426-1444.

Broxmeyer, H.E. (1982a): Relationship of cell-cycle expression of Ia-like antigenic determinants on normal and leukemia human granulocyte-macrophage progenitor cells to regulation in vitro by acidic ferritins. J.Clin.Invest. 69, 632-642.

Broxmeyer, H.E., Bognacki, J., Ralph, P., Dörner, M.H., Lu, L., Castro-Malaspina, H. (1982b): Monocyte-macrophage-derived acidic isoferritins: Normal feedback regulators of granulocytemacrophage progenitor cells in vitro. Blood 60, 595-607.

Broxmeyer, H.E., Juliano, L., Lu, L., Platzer, E., Dupont, B. (1984): HLA-DR human histocompatibility leukocyte antigens-restricted lymphocyte-monocyte interactions in the release from monocytes of acidic isoferritins that suppress hematopoietic progenitor cells. J.Clin.Invest. 73, 939-953.

Broxmeyer, H.E., Lu, L., Bicknell, D.C., Williams, D.E., Cooper, S., Levi, S., Salfeld, J., Arosio, P. (1986): The influence of purified recombinant human heavy-subunit and light-subunit ferritins on colony formation in vitro by granulocyte-macrophage and erythroid progenitor cells. Blood 68, 1257-1263.

Burgess, A.W., Wilson, E.M.A., Metcalf, D. (1977): Stimulation by human placental conditioned medium of hemopoietic colony formation by human marrow cells. Blood 49, 573-583.

Laemmli, U.K. (1970): Cleavage of structural proteins during assembly of the head of bacteriophage T4. Nature 227, 680-685.

Luzzago, A., Arosio, P., Iacobello, C., Ruggeri, G., Capucci, L., Brocchi, E., De Simone, F., Gamba, D., Gabri, E., Levi, S., Albertini, A. (1986): Immunochemical characterization of human liver and heart ferritins with monoclonal antibodies. Biochim.Biophys. Acta 872, 61-71.

Olofsson, T., Olsson, I. (1980a): Suppression of normal granulopoiesis in vitro by a leukemia-associated inhibitor (LAI) of acute and chronic leukemia. Blood 55, 975-982.

Olofsson, T., Olsson, I. (1980b): Biochemical characterization of a leukemia-associated inhibitor (LAI) suppressing normal granulopoiesis in vitro. Blood 55, 983-991.

Olofsson, T., Olsson, I. (1980c): Suppression of normal granulopoiesis in vitro by a leukemia associated inhibitor (LAI) derived from a human promyelocytic cell line (HL-60). Leukemia Res. 4, 437-447.

Olofsson, T., Nilsson, E., Olsson, I. (1984): Characterization of the cells in myeloid leukemia that produce leukemia associated inhibitor (LAI) and demonstration of LAI-producing cells in normal bone marrow. Leukemia Res. 8, 387-396.

Olofsson, T., Cedergren, B.M., Persson, E. (1985): Isolation of leukemia-associated inhibitor (LAI)-producing cells from normal peripheral blood. Scand.J. Haematol. 35, 511-517.

Résumé

Un système de culture a été établi pour la production en grande quantité de milieu conditionné par les cellules HL60 pour la purification de l'inhibiteur associé aux leucémies (LAI) qui est produit directement par ces cellules. Des anticorps monoclonaux contre les cellules HL60 ont été produits et testés pour leur capacité de fixer le LAI. Un des anticorps actifs, le MoAb-58, a été couplé avec de la Sepharose activée par du CNBr et utilisé en chromatographie d'affinité pour la purification du LAI à partir du milieu conditionné préparé à partir des cellules HL60 . La sous unité biologiquement active du LAI a été isolée sur électrophorèse en gel de polyacrylamide (SDS) et a un poids moléculaire de 125 K. L'anticorps MoAb-58 a été aussi utilisé pour isoler par panning une sous population de cellules produisant du LAI à partir du sang. Ces cellules inhibent la formation de colonies et de clusters par des cellules de moelle normale. Le LAI réduit de manière réversible la fraction des CFU-GM en phase S, mais n'a aucune action sur la fraction en phase S des BFU-E. Un anticorps monoclonal dirigé contre l'isoferritine acide, 2A4, ne se fixe pas sur le LAI et n'empêche pas son activité.

Discussion

Chairpersons/*Modérateurs*: A.A. Axelrad (Canada)
P. Boivin (France)

A. AXELRAD : Dr Olofsson, what are the biochemical characteristics of LAI?

T. OLOFSSON : Perhaps I did not point out clearly enough that the native molecule that we find in conditioned medium from AML cells or HL 60 cells is a large molecule of about 500,000. It is a glycoprotein that sticks to Con-A Sepharose, for instance. However, there is also a charge heterogeneity; if you run it on ion exchange chromatography, its comes off in several peaks, but the main observation is that the active subunit is a 125 K molecule. We don't know exactly if the native complete molecule is composed of several of these subunits or if it consists of other low molecular weight components as well, we don't know yet.

I. RICH : Just a comment. I was very interested in your first purification procedure, which I thought, is very similar to the one we have been using actually for purifying erythropoietin and its antibody, but using an FPLC system which is totally automated, same sort of affinity chromatography.

N. DAINIAK : I have a question for Dr Olofsson. What happens to cell viability when you trypsinize bone marrow cells?

T. OLOFSSON : They are still very viable. They are more than 95% viable and there is no effect on colony formation. We find the same number of colonies, it is a short incubation time, thirty minutes.

N. DAINIAK : And for a low concentration of trypsin?

T. OLOFSSON : Right, 0.1%.

M. AGLIETTA : Just a question to Dr Najman. Your inhibitor is active or not on myelopoietic cells or in other cell systems? How do you compare it with other known inhibitors of myelopoiesis?

A. NAJMAN : Till now, we have worked extensively on the BFU-E progenitors only and not on the other cells.

G. KONWALINKA : One more question, Dr Najman. Did you find any correlation between the burst promoting activity and erythropoietin concentration in sera or the inhibitory activity of your cultured blast cells and the clinical outcome of the patients tested?

A. NAJMAN : Not really.

D. ROODMAN : Dr Broxmeyer, I am not sure if I remember your data correctly. When you purified your progenitor cells and then looked at the effects of the different inhibitors on the highly purified subpopulations, was there a change in the sensitivity? As I remember, for example, TNF required much higher concentrations just to show inhibition, then you were doing it on your unfractionated population. If so, do you think there is an accessory cell helper effect on mediating the effects of these inhibitors or did I just misread your data?

H. BROXMEYER : Those studies were done with mouse cells. Human TNF alpha is much less active against mouse cells than it is against human cells. This is the reason why the inhibition was not as great as when we used human cells even though we had used high concentrations. There was no difference in sensitivity to the TNF of the purified (up to 94% cloning efficiency of mouse CFU-GM) versus the unseparated population of mouse cells. We didn't do full dose titrations of suppressor molecules against the highly-enriched (up to 47% cloning efficiency of CFU-GM and BFU-E) human progenitor cells because we didn't have many of those cells. We just picked a concentration of molecules that would give us a good inhibition. The mouse studies were done by Dr Douglas Williams and the human studies were done by Dr Li Lu. I believe that the data demonstrate that the purified population of progenitor cells is at least as sensitive as the progenitors in the unseparated population of cells.

J. FLETCHER : Dr Broxmeyer, could you speculate as to the differences between active lactoferrin from normal PMN and the unactive lactoferrin from PMN of patients with CML? Have you got any idea what the difference in these two molecules might be?

H. BROXMEYER : The monoclonal anti-human lactoferrin revealing purified lactoferrin from PMN of normal donors and patients with CML migrated exactly the same on SDS-PAGE, even though some of the sugars were removed. When you saturate the CML PMN lactoferrin exactly as you would the normal lactoferrin, the CML PMN lactoferrin does not become more active. These studies were done by David Bicknell. But we haven't done a definitive study yet to find out if the iron-binding capacity of those lactoferrin preparations is really different. We are evaluating this now. If it is not the iron-binding characteristics that are different, then we will probably have to do amino acid analysis of the molecules to see what the differences are between them.

M. AGLIETTA : Dr Broxmeyer, you showed that IL-3 is counteracting in vivo action of lactoferrin. My question is : can other growth factors counteract the lactoferrin effect in vivo and can you speculate why IL-3 is so effective in counteracting this inhibition?

H. BROXMEYER : The suppressive effect of lactoferrin in mice can be counterbalanced by purified recombinant murine GM-CSF and natural mouse and recombinant human CSF-1, as well as by purified recombinant murine IL-3 (Broxmeyer et al., J. Clin. Invest. 79 : 721, 1987 ; Blood Cells, in press, 1987). This is exactly what would be expected based on the in vitro data for lactoferrin which demonstrates suppression of the release of growth factors from accessory cells. We used lactoferrin to damper myelopoiesis in the mice. We assumed that it would be easier to see an effect of exogenously administered CSF if myelopoiesis was first decreased. We could also use the overbound phase after recovery from sublethal dosages of cyclophosphamide to damper myelopoiesis (Broxmeyer et al., Blood 69 : 913, 1987 ; Leukemia Res 11 : 201, 1987). We feel that using lactoferrin is a clearer system.

A. AXELRAD : Dr Broxmeyer, are you ready to accept the observations of Dr Olofsson that LIA is not acidic ferritin?

H. BROXMEYER : Until Dr Olofsson showed that he could not inactivate the LAI with a monoclonal antibody against acidic ferritin, I still wasn't convinced. But I think, assuming he used high enough concentrations of the antibody to neutralise any ferritin and I am sure he must have, that the two molecules must be different although it means that you have two extremely similar molecules in terms of action. This is certainly possible.

B. LORD : Dr Olofsson, as Dr Axelrad this morning, I was again interested in the kinetics of your system. 60 minutes is a very short time in which to drop cycling from 35% down to just as little as 10%. Are you also suggesting that the molecule you are talking about is one which suspends cells in DNA synthesis?

T. OLOFSSON : Yes, I think that is the effect. I cannot explain the effect if we don't assume that cells are arrested in S phase because it would not be enough if LAI just acted on cells moving from G1 into S. The effect would not be fully expressed within 60 minutes.

A. AXELRAD : I would like to make one comment. I was attracted by the statement
that you made that I wished I had made before, which is that it is very
difficult to trust data on percent kills when one is looking at a reduction,
there are so many non specific things that can do it. But you had one important
control, that was that the number of colonies did not go down, what went down
was the proportion of colonies that could be reduced by a cycle active agent.
And we use exactly the same control to make sure that we are not just looking at
non specific inhibition.

J.P. MARIE : Also inhibitory effects are observed during acute phase of
leukemia. Have you observed some inhibitory effect, residual inhibitory effect
of serum, or trace of LAI during remission as compared to residual disease?

T. OLOFSSON : There is no detectable LAI in serum from AML patients and many
would say that in that case LAI cannot be important, but you have to remember
that bone marrow cells do not live in serum, they actually never see serum.

H. BROXMEYER : We had looked at extracts of cells from patients with leukemia
for acidic isoferritin-inhibitory activity (formely called LIA, leukemia
associated inhibitory activity), not at serum. This cell derived activity was
much increased in cells of patients with non-remission acute leukemia but
decreased almost to normal levels when the patient went into remission. However,
in the non-remission acute leukemia, the progenitor cells that should have
responded to this inhibitory activity were not responsive, and even in
remission, many of the patients still had progenitor cells that would not
respond to the inhibitory activity. So, while the amount of inhibitors went down
during remission, the cells were still not very responsive. But that varied
between patients. These studies were published in the New England Journal of
Medecine (1979).

D. ROODMAN : Just a comment. I think that the data you just spoke of is
consistent with the data which showed that even in remission of acute leukemia,
there are leukemic cells as determined by DNA analysis. So you would expect
these remission cells would have some of the properties of the leukemic cells.
Even if they differentiate to granulocytes they still contain the DNA markers
for the leukemic clone.

H. BROXMEYER : I would agree with that. There is a lot of evidence that cells of
patients with leukemia in complete remission are still abnormal. For example,
when granulocytes come back in remission of acute leukemia, these cells still
don't have as much lactoferrin-inhibitory activity as do granulocytes from
normal donors.

Unidentified speaker : Dr Najman, you said that the inhibitory effect on the
erythropoiesis was problably mediated by adherent cells, what are your arguments
for this hypothesis?

A. NAJMAN : The inhibitory effect of the blast conditioned medium on peripheral
blood BFU-E disappeared in the absence of adherent cells. Conversely, in the
bone marrow, no modification was observed in the absence of adherent cells.

Brief reports

Communications brèves

Brief report

Macrophages secrete an inhibitory factor for mouse erythroleukemic cell differentiation : characterization and partial purification

Stephen D. Wolpe, Shigeru Sassa and Anthony Cerami

The Rockefeller University, New York, NY 10021, USA

Macrophages secrete a wide variety of biologically active compounds with disparate structures and activities (Takemura and Werb, 1984; Nathan, 1987). These products play an important role in normal homeostasis and in the regulation of the immune response to invasive stimuli. The secretion of such powerful mediators is not without risk, however; recent studies in this laboratory have documented the role of a macrophage hormone termed "cachectin" in the cachexia and shock that can accompany chronic illness or infection (Beutler and Cerami, 1987). This protein is identical to tumor necrosis factor, illustrating the multiplicity of actions a single mediator may have.

Chronic illnesses are also commonly observed to be accompanied by anemia of various degrees (c.f. Roadman et al, this volume). Results from this laboratory suggest that this too may be due in part to the production of an endogenous mediator by macrophages (Sassa et al, 1983). In this system, the differentiation of mouse erythroleukemic cells was shown to be inhibited by conditioned medium from stimulated macrophages. In the present report we demonstrate that this erythroid inhibiting activity (EIA) is different from other known monokines such as IL-1, cachectin/TNF, interferon, G-CSF, GM-CSF, TGFβ, PDGF or FGF. In addition, some of the physical characteristics of EIA as well as its partial purification are presented.

MATERIALS AND METHODS

Materials - Endotoxin (LPS W) from E. coli 0127:B8 was purchased from Difco Laboratories (Detroit, MI). A modified F12 medium was prepared in our laboratory (Sassa and Kappas, 1977). Fetal bovine serum was purchased from Hyclone Laboratories (Logan, UT). Me$_2$SO was a product of Eastman Organic Chemicals (Rochester, N.Y.). Recombinant human cachectin/TNF was obtained from Chiron Corp. TGFβ was purchased from R & D Systems, Inc. (Minneapolis, MN); FGF, ECGF and PDGF were purchased from Collaborative Research (Bedford, Mass.); murine α/β interferon was purchased from Lee Biomolecular (San Diego, CA). Recombinant murine IL-1 and recombinant human G-CSF were the generous gifts of Dr. P. Lomedico, Hoffman-La Roche and Dr. L. Souza, Amgen, respectively.

Cell Culture - Murine Friend virus-transformed erythroleukemia cells (clone DS-19) were cultivated in modified F12 medium supplemented with 10% heat inactivated fetal bovine serum. The macrophage cell line RAW 264.7 was obtained from the

American Type Culture Collection (Rockeville, Md.). They were grown to confluence in RPMI 1640 medium supplemented with 10% fetal bovine serum. The cell monolayers were washed five times in Hank's basic salt solution and the medium replaced with serum-free RPMI 1640 supplemented with 1 μg/ml LPS and incubated for 18 hours at 37°C before collecting the supernatant.

Determination of EIA Titers - MEL cells (5×10^4/ml) were incubated in a 5 ml suspension in a 25 cm^2 flask at 37°C in 5% CO_2 in humidified air for 18 hr. Me_2SO and test substances were added to cultures which were incubated for 96 hr. Cells were then harvested for cell counting and determination of heme content. Viable cell numbers were assessed by trypan blue staining and heme content was determined by a fluorometric assay of porphyrin derivatives after the removal of iron (Sassa et al, 1978). Cell numbers were counted using a Coulter Counter (Coulter Electronics, Hialeah, Fla.)

RESULTS AND DISCUSSION

Effects of Conditioned Medium from RAW 264.7 Cells - Figure 1 shows that conditioned medium from the RAW 264.7 cell line inhibits erythroid differentiation of MEL cells in a dose-dependant manner. Other mouse macrophage cell lines (P388D1 and WEHI-3) as well as mouse primary thioglycollate-elicited peritoneal macrophages also released EIA into the culture medium. In all these cases, EIA was present only when the cells were stimulated with endotoxin; no basal release of EIA was detected (data not shown). Inhibition of cell growth was not associated with an increase in the number of dead cells(<3% in all cases).

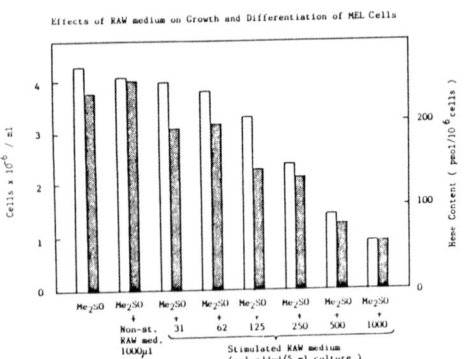

FIG. 1. DOSE RESPONSE OF MURINE ERYTHROLEUKEMIC (MEL) CELLS TO EIA. CELLS FROM THE MOUSE MACROPHAGE CELL LINE RAW 264.7 WERE GROWN IN SERUM-FREE MEDIUM IN THE PRESENCE OR ABSENCE OF LPS (1 UG/ML). SUPERNATANT TAKEN 16-18 HOURS AFTER STIMULATION CAUSE A DOSE-DEPENDENT INHIBITION OF BOTH CELL GROWTH AND DIFFERENTIATION AS MEASURED BY HEME CONTENT. SUPERNATANT FROM UNSTIMULATED RAW CELLS DOES NOT INHIBIT. THE VIABILITY OF THE MEL CELLS IS NOT REDUCED BY THE TREATMENT AS ASSESSED BY TRYPAN BLUE STAINING (NOT SHOWN).

In order to examine the physicochemical properties of EIA, endotoxin-stimulated RAW 264.7 medium was subjected to heat, freeze-thawing or proteases. Neither repeated freeze-thawing (5X) nor heating at 60°C for 10 minutes destroyed EIA. Heating at 100°C for 2 minutes resulted in 13% loss of EIA. In contrast, treatment with trypsin, chymotrypsin or proteinase K resulted in complete loss of EIA (data not shown).

Effect of Various Macrophage Products- In order to screen for EIA-like activity in known macrophage products, recombinant human cachectin/TNF, TGFβ, FGF, interferon, ECGF, G-CSF and rIL-1 were added to MEL cells treated with Me_2SO. rTNF (10-1000 ng/ml), TGFβ (0.3-3.0 ng/ml), FGF (0.1-100ng/ml), PDGF (0.005-0.5 U/ml) ECGF (3-300 ng/ml), G-CSF (2.5-2500 U/ml) and rIL-1 (0.05-5000 pg/ml) did not inhibit either cell growth or erythroid differentiation. In contrast, interferon α/β (1000-10000 U/ml) suppressed cell growth by 50% and stimulated heme synthesis by 25%. Combined additions of rTNF (10-1000 ng/ml) and interferon (1000-10000 U/ml) did not inhibit the interferon-mediated potentiation of erythroid differentiation (data not shown).

These data, combined with the physicochemical data above, suggest that EIA represents a novel monokine acting on the erythroid lineage. Because of the possible interference of other monokines (e.g., cachectin/TNF - c.f. Roodman et al, this volume) on erythropoietic cells our approach has been to attempt to purify this monokine before testing it in other systems.

Chromatographic Separation of EIA - Serum- and cell-free supernatants from RAW 264.7 cells (1-3 liters) was concentrated using a DC2 hollow fiber device with a 10,000 M.W. cutoff (Amicon). 20-fold concentrated supernatant was extensively diafiltrated (in the same device) with 20mM Tris, pH 8.0. The concentrated, diafiltrated supernatant was filtered through a 0.22μm filter an applied to a Mono Q (anion exchange) column attached to an FPLC (Pharmacia). Proteins were eluted with a gradient of 0 to 1 M NaCl in 20mM Tris. EIA was eluted in two peaks, i.e., a smaller peak at a NaCl concentration of 0.1 M and a major peak at 0.15 M (Fig. 2). The major peak eluted slightly earlier than the major protein peak which contained cachectin/TNF as assessed by either bioactivity or its position on silver-stained SDS-PAGE gels.

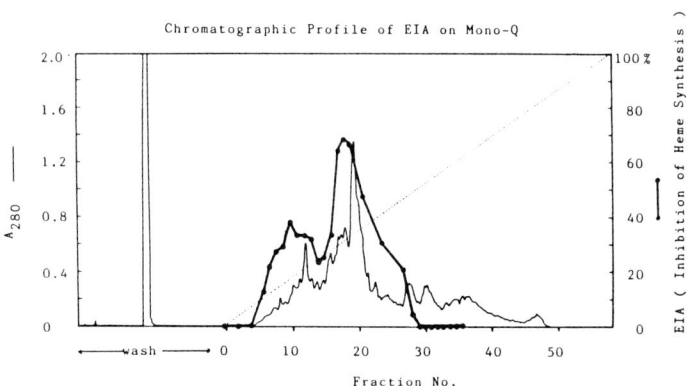

FIG. 2. ANION EXCHANGE CHROMATOGRAPHY OF EIA. RAW CELLS WERE GROWN IN SERUM-FREE MEDIUM PLUS 1 UG/ML LPS FOR 16-18 HOURS AND THE CELL-FREE SUPERNATANT WAS CONCENTRATED 15-20 FOLD USING A HOLLOW FIBER DEVICE. THE SUPERNATANT WAS EQUILIBRATED WITH 20 MM TRIS, PH 8.0 BY DIAFILTRATION AND FRACTIONATED ON A MONO-Q COLUMN ATTACHED TO AN FPLC. PROTEINS WERE ELUTED WITH A LINEAR GRADIENT OF 0 TO 1 M NACL IN THE SAME BUFFER. EIA ELUTED IN TWO PEAKS, I.E., A SMALLER PEAK AT 0.1 M NACL AND A MAJOR PEAK AT 0.15 M.

The major peak of EIA was concentrated by lyophilization and separated on a Superose 12 (gel filtration) column attached to an FPLC (Pharmacia). A single major peak was found corresponding to a molecular weight of approximately 40,000 daltons (Fig. 3).

Using a Dye-Matrex Gel kit (Amicon Corporation, MA), EIA was not found to bind to any of the dye-ligand columns (Blue A, Blue B, Red A, Orange A or Green A) when tested according to the manufacturer's directions. The Green A column did, however, bind a substantial portion of the proteins present in the crude supernatant (as assessed by SDS-PAGE gels) making it useful as a negative column. EIA was not bound to either concanavilin A or lentil-lectin columns but it was partially retained on heparin columns (data not shown).

Recent results have indicated that hydroxylapatite, cation exchange and native PAGE electrophoresis may be useful in the further purification of EIA (data not shown).

In summary, EIA is a novel monokine which inhibits the growth and differentiation of mouse erythroleukemic cells. We were fortunate in choosing this system as it appears to be relatively insensitive to the action of other, well-defined monokines. Studies with other monokines have revealed that most (e.g., IL-1, cachectin/TNF, interferons, etc.) exert a multitude of biologically important effects aside from the property which initially led to their identification. It is probable that once EIA is purified to homogeneity and tested in other systems (including other erythrogenic cell types) it will have other effects in addition to those on MEL cells. For this reason we are actively pursuing the further purification of this molecule.

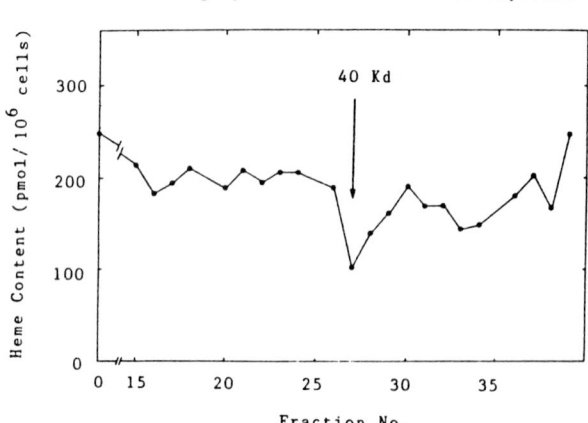

FIGURE 3. GEL FILTRATION OF EIA. THE PEAK FRACTIONS FROM MONO Q (FIG. 3) WERE CONCENTRATED AND RUN ON A SUPEROSE 12 (GEL FILTRATION) COLUMN EQUILIBRATED WITH PBS. A SINGLE MAJOR PEAK IS FOUND CORRESPONDING TO A MOLECULAR WEIGHT OF APPROX. 40,000 DALTONS.

REFERENCES

Beutler, B. and Cerami, A. 1987. Cachectin: More than a tumor necrosis factor. New. Eng. J. Med. 316:379-385.

Nathan, C.F. 1987. Secretory products of macrophages. J. Clin. Invest. 79:319-326.

Sassa, S., Granick, J.L., Eisen, H. and Ostertag, W. 1978. Regulation of heme biosynthesis in mouse Friend virus-transformed cells in culture. in In Vitro Aspects of Erythropoiesis, M.J. Murphy, Jr. (ed.), Springer-Verlag, New York pp. 135-142 & pp. 268-270.

Sassa, S. and Kappas, A. 1977. Induction of δ-aminolevulinate synthase and porphyrin in cultured liver cells maintained in chemically-defined medium. Permissive effects of hormones on the inductive process. J. Biol. Chem. 252:2428-2436.

Sassa, S., Kawakami, M. and Cerami, A. 1983. Inhibition of the growth and differentiation of erythroid precursor cells by an endotoxin-induced mediator from peritoneal macrophages. PNAS 80:1717-1720.

Takemura, R. and Werb, Z. 1984. Secretory products of macrophages and their physiological functions. Am. J. Phys. 246:C1-C9.

Blockage of erythroleukemia cell induced differentiation by autocrine growth factor(s)

Brigitte Fasciotto [1], Dusan Kanazir [1], Jon P. Durkin [2], James F. Whitfield [2] and Velibor Krsmanović [1]

[1] Unité de Virologie INSERM U51, UA 613 CNRS 1, Pl. du Pr. J. Renaut, 69371 Lyon Cedex 08, France
[2] Cellular Oncology Group, Division of Biological Sciences, NRC of Canada, Ottawa, Ontario, Canada

KEY WORDS

V-erbB oncogene, erythroleukemia cell secretion, autocrine factor activities, blockage of erythroleukemia cell differentiation, cell proliferation.

INTRODUCTION

In normal tissues, growth factors play key roles in the relationship between cell growth and differentiation. By interacting with specific cell-surface receptors, growth factors determine a number of intracellular regulatory events such as the initiation of cell cycle, and the various steps of cell differentiation process. This latter process in turn controls cell proliferation by progressively reducing cell growth potential. Studies of the relative growth factor autonomy of malignant cells, have shown that malignant transformation by various oncogenes can reduce or eliminate the cells' need for exogeneous growth factors by causing secretion of specific autocrine peptide factors which stimulate proliferation by binding to producer cell receptors (De Larco and Todaro, 1978; De Larco et al., 1981). In addition to, [a] mitogenic activity (responsible for cell proliferation), the autocrine factors have, [b] "transforming" activity enabling normal cells, such as embryo fibroblasts, to grow in soft agar, as well as [c] cell differentiation blocking activity.

GROWTH PROMOTING FACTORS SECRETED BY EYTHROLEUKEMIA CELLS

Oncogene *v-erbB* has essential function in the induction of erythroleukemia with avian erythroblastosis virus (AEV) as a vector (Frykberg et al., 1983; Yamamoto et al., 1983). By structural analysis, it has been shown that v-erbB oncogene product of AEV could be assimilated to a truncated EGF receptor, from which the extra cellular part corresponding to the EGF fixation site is missing. As shown for EGF receptor, the cytoplasmic domain exhibits both homology with *src* type gene and tyrosine kinase activity (Downward et al., 1984; Gilmore et al.,

1985). Thus, v-erbB gene product (that is truncated EGF receptor) does not seem to be regulated by EGF so that the function of this receptor as tyrosine protein kinase is constitutively activated. Transforming activity of v-erbB gene interferes with the differentiation of early target erythroid precursors such as BFU-E (Gazzolo et al., 1981) giving rise to erythroleukemia cells blocked at a stage of differentiation close to CFU-E (Samarut et Gazzolo, 1982; Beug et al., 1982), the nearest known stage to which differentiation leads from BFU-E (Samarut et al., 1979).

To study mitogenic activity of v-erbB oncogene, the LSCC HD3 erythroleukemia cells, transformed by v-erbB gene-containing ts34 mutant of AEV (Beug et al., 1982), have been examined for their capacities to grow under different culture conditions. For instance, proliferating LSCC HD3 cells transfered to a chemically defined medium without serum are able to grow for a limited period at 36°C, but not at 42°C (Kanazir et al., 1986). At this latter temperature the activity of v-erbB oncogene is suppressed (Frykberg et al., 1983; Yamamoto et al., 1983). However, cell growth at 42°C could take place either by addition of fetal calf serum or by addition of the serum-free medium derived from LSCC HD3 erythroleukemia cells grown at 36°C ("conditioned medium") (Kanazir et al., 1986). These results show that erythroleukemia cells, maintained under conditions in which the v-erbB oncogene is active, secrete growth factor(s).

Further experiments were performed with serum-free conditioned medium in order to test for its ability to promote growth of normal secondary chicken embryo fibroblasts (CEF) in soft agar (colony-stimulating activity). This conditioned medium, which is mitogenic for LSCC HD3 erythroleukemia cells arrested in G_0 at a restrictive temperature, was able to stimulate anchorage-independent growth of normal CEF (Fasciotto et al., 1987). By fractionating conditioned medium from LSCC HD3 erythroleukemia cell on Bio-Gel P-60 column chromatography, the mitogenic and soft agar colony-stimulating activities (tested with CEF cells) coeluted and were associated with components of 8, 12, 17, 20-27 and 44 kDa. Similar results were obtained using NIH 3T3 mouse cells instead of CEF in the assay. By contrast, the serum-free conditioned medium in which the LSCC HD3 cells were maintained at restrictive temperature (42°C), at which the ts v-erbB oncogene action is abrogated, was neither mitogenic nor able to induce growth in soft agar.

To our knowledge, this is the first example of the induction of anchorage-independent proliferation of uninfected, early passage CEF cells by transforming growth factors (Fasciotto et al., 1987). Since such cells do not have the EGF like receptors needed to bind type-α trasforming growth factors (Kryceve-Martinerie et al., 1982), it is unlikely that any of these mitogenic components belong to this family of transforming growth factors. This supports the observation according to which the human hematopoietic tumor cells do not express the TGF-α precursor gene, and suggests the same may be true for avian hematopoietic tumor cells (Derynck, 1986).

AUTOCRINE FACTOR(S) BLOCKING AVIAN AND MURINE ERYTHROLEUKEMIA CELL DIFFERENTIATION

An important characteristic of eythroleukemia cells is their inability to mature into hemoglobin synthesizing, proliferatively inactive, differentiated states. Exposure to increasing concentrations of unfractionated conditioned medium protein between 0 and 100 μg/ml increasingly blocked both the erythropoietin-induced differentiation of chicken erythroleukemia LSCC HD3 cells at the nonpermissive 42°C and the DMSO-induced differentiation of Friend murine erythroleukemia cells, into hemoglobin-synthesizing (i.e. benzidine-positive) erythroid cells. Interestingly, the differentiation-inhibiting activity for both LSCC HD3 erythroleukemia chicken cells and Friend erythroleukemia murine cells was associated not with any of the mitogenic components detected in Bio-Gel P-60 fractions, but with a non-mitogenic, approximately 40 kDa, component (Fasciotto et al., 1987). The ability of this autocrine factor to block both chicken and murine erythroid cell differentiation suggests that it is well conserved in evolution. It should be noted that a 43 kDa component inhibiting DMSO-induced Friend murine cell differentiation is secreted by murine embryonal carcinoma cells (Jakobovits et al., 1985), but relationship of these carcinoma cells with murine erythroid lineage is unknown ; also the relation of this mammalian inhibitor to the 40 kDa chicken inhibitor remains to be determined.

Experiments performed in collaboration with Dr. M.W. McBurney (Departments of Medicine and Biology, University of Ottawa) have shown that the process of murine embryonal carcinoma cell induced differentiation, by DMSO into muscle cells or by retinoic acid into neurons, is not affected by the presence of differentiation inhibiting factor secreted by LSCC HD3 erythroleukemia cells (unpublished results). In addition, we have used quail embryo myoblasts transformed with a Rous sarcoma virus *ts* mutant NY 68 (kindly supplied by Dr. M. Fiszman, Institut Pasteur, Paris) to examine induced differentiation of these cells at restrictive temperature. In the presence of erythroleukemia cell blocking factor, the process of differentiation, although somehow delayed, could succeed by giving rise to the formation of myotubes in the culture medium (unpublished data). These results show that LSCC HD3 erythroleukemia cell differentiation blocking factor acts specifically on cells of erythroid lineages.

CONCLUSION AND HYPOTHESES

Under control of viral oncogene, *ts*AEV-transformed chicken erythroleukemia LSCC HD3 cells secrete into their medium several autocrine factors that are both mitogenic and eliminate the cells' need to be anchored to a solid substrate in order to proliferate. These cells also secrete a non-mitogenic factor which indirectly promotes the proliferation of hematopoietic cells by preventing their differentiation into proliferatively inactive erythroid cells. Although we do not know how the 40 kDa factor blocked erythroid differentiation, the inappropriate expression of its gene in transformed cells might prevent the interaction of erythropoietin and/or other differentiation promoters with membrane receptors. This differentiation blocking factor secreted by avian erythroleukemia cells is well conserved in evolution and appears to be erythroid tissue specific since it is able to block differentiation of both chicken and murine erythroleukemia cells without affecting induced differentiation of several distinct tumor cell types.

As regards the biological significance of this autocrine factor capable of inhibiting erythroid cell differentiation, there could exist a possible relationship between its gene and those of some normal paracrine factor(s) which may serve as down regulating differentiation component(s) either in the maintaining of the equilibrium between self renewal and commitment of stem cells, or as modulating agents in the conversion of erythroid precursors from one to an other differentiation step. Further studies of the biological function of growth factors and differentiation inhibitors, and of their interaction with specific cell surface molecules, may enable us to understand the molecular mechanism endowing malignant cells with the capacity to proliferate persistently, i.e. to escape differentiation processes which direct them into non-proliferative states.

REFERENCES

Beug, H., Palmieri, S.,Freudenstein, C., Zentgraf, H., Graf, T.(1982): Hormone-dependent terminal differentiation in vitro of chicken erythroleukemia cells transformed by ts mutants of avian erythroblastosis virus. Cell 28, 907-919.

De Larco, J.E., Preston, Y.A., Todaro, G.J.(1981): Properties of a sarcoma-growth-factor-like peptide from cells transformed by a temperature-sensitive sarcoma virus. J. Cell Physiol. 109, 143-152.

De Larco, J.E., Todaro, G.J.(1978): Growth factors from murine sarcoma virus-transformed cells. Proc. Natl. Acad. Sci. USA 75, 4001-4005.

Derynck, R.(1986): Transforming growth factor-α: structure and biological activities. J. Cell. Biochem. 32, 293-304.

Downward, J., Yarden, Y., Mayes, E., Scrace, G., Totty, N., Stockwell, P.,Ullrich, A., Schlessinger, J., Waterfield, M.D.(1984): Close similarity of epidermal growth factor receptor and v-erbB oncogene protein sequences Nature 307, 521-527.

Fasciotto, B., Kanazir, D., Durkin, J.P., Whitfield, J.F., Krsmanovic, V. (1987): AEV-transformed chicken erythroid cells secrete autocrine factors which promote soft agar growth and block eythroleukemia cell differentiation. Biochem. Biophys. Res. Comm. 143, 775-781.

Frykberg, L., Palmieri, S., Beug, H., Graf, T., Hayman, M.J., Vennström, B. (1983): Transforming capacities of avian eyrthroblastosis virus mutants deleted in the erbA or erbB oncogenes. Cell 32, 227-238.

Gazzolo, L., Samarut, J., Bouabdelli, M., Blanchet, J.P.(1981): Early precursors in the erythroid lineage are the specific target cells of avian erythroblastosis virus in vitro. Cell 22, 683-691.

Gilmore, T., DeClue, J.E., Martin, G.S.(1985): Growth Factors and Transformation, In "Cancer Cells" (J. Feramisco, B. Ozanne and C. Stiles, ed.). 3, 25-32.

Jakobovits, A., Banda, M.J., Martin, G.R.(1985): Growth Factors and Transformation, In "Cancer Cells" (J. Feramisco, B. Ozanne and C. Stiles, ed.). 3, 393-399.

Kanazir, D., Fasciotto, B., Krsmanovic, V.(1986): Activité mitogène de facteurs sécrétés par des cellules transformées par le virus de l'érythroblastose aviaire. C. R. Acad. Sci. Paris 302, 63-66.

Kryceve-Martinerie, C., Lawrence, D.A., Crochet, J., Jullien, P., Vigier, P. (1982): Cells transformed by Rous sarcoma virus release transforming growth factors. J. Cell. Physiol. 113, 365-372.

Samarut, J., Blanchet, J.P., Nigon, V.(1979): Antigenic characterization of chick erythrocytes and erythropoietic precursors: identification of several definitive populations during embryogenesis. Dev. Biol. 72, 155-166.

Samarut, J., Gazzolo, L.(1982): Target cells infected by avian erythroblastosis virus differentiate and become transformed. Cell 28, 921-929.

Yamamoto, T., Hihara, T., Nishida, T., Kawai, S., Toyoshima, K.(1983): A new avian eythroblastosis virus, AEV-H,carries erbB gene responsible for the induction of both erythroblastosis and sarcomas. Cell 34, 225-232.

Hairy cell leukemia (HCL) : I) inhibitory effect of serum from HCL-patients on the normal bone marrow colony growth possibly removed by a prolonged α-interferon treatment

Francesco Lauria *, Gian-Paolo Bagnara **, Donatella Raspadori *, Marina Buzzi *, Anna Guarini *, Pier-Luigi Zinzani *, Lucia Catani *, Licia Gaggioli **, Marina Marini * and Maria-Antonietta Brunelli **

* Institute of Hematology L. and A. Seragnoli, University of Bologna, Bologna, Italy
** Institute of Histology, University of Bologna, Bologna, Italy

KEY WORDS:

CFU-GM, BFU-E, CFU-MK, COLONIES, SERUM, HCL, α-IFN

INTRODUCTION

It is well known that Hairy-cell leukaemia (HCL) is characteristically associated to myelofibrosis and mono or pancytopenia. In a preliminary report (Lauria et al. 1987), we demonstrated that in patients with HCL at the onset of the disease or before treatment, there was a reduced colony forming capacity of the bone marrow cells and that serum from HCL patients, even at very low concentration, produced a marked colony growth inhibition on all 3 normal cell lineages.

In this study, we evaluated more extensively the regulating effect of serum from HCL patients on the "in vitro" growth of normal granulocyte-macrophage (CFU-GM), erythroid (BFU-E) and megakaryocyte (CFU-MK) colonies before, during and after α-Interferon (α-IFN) treatment. The removal of the inhibitory activity was documented only in those patients in which a good haematological response was achieved.

PATIENTS AND METHODS

Seven patients suffering from HCL were treated daily with 3×10^6 units (3 MU) intramuscularly of α-IFN (Wellferon) till to the achievement of complete remission (CR) or stable partial remission (PR) (3-6 months). Then, the administration schedule was changed to 3 times a week for 3-6 months and finally the patients were randomized in two arms: no therapy or 3MU a week. CR was defined as absence of Hairy-cells (HC) in the bone marrow aspirate, bone biopsy and peripheral blood, recovery of Haemoglobin level to more than 12 g/dl, absolute granulocyte count to 1.5 x 10^9/liter or more, platelet count to more than 100 x 10^9/liter or more. PR was defined as a decrease in the HC infiltrate in the bone marrow and peripheral blood by more than 50 % of the pre-treatment values, an increase of 35 % of the normal bone marrow haemopoiesis, and a restoration of at least 2 of the peripheral blood values (as defined for CR) for at least 1 month. Minor response (MR) was defined as a decrease in the bone marrow infiltration and a restoration of at least 1 of the peripheral blood values as indicated above.

Sera samples, collected before starting α-IFN, after 3 months therapy and every 3 months, were inactivated at 56°C for 35 min. and stored at -80°C prior to seeding. Different serun concentrations (0.1, 1, 10 %) were added to each culture.

COLONY ASSAY

Normal bone marrow specimens were obtained from normal subjects undergoing to marrow harvesting for allogeneic bone marrow transplantation or from patients undergoing bone marrow examination for diagnostic purposes and showing no evidence of haematological involvement. In all cases, the patients had given their informed consent. Mononuclear cells were recovered from the interface of a Ficoll-Hypaque (Pharmacia, Uppsala, Sweden) gradient. Adherent cells were removed following overnight incubation at 37°C in plastic Petri dishes. The non-adherent cells were depleted of T-lymphocytes by rosetting with neuraminidase-treated sheep erythrocytes.

CFU-GM ASSAY

One ml of Iscove's Modified Dulbecco Medium (IMDM) containing 0.3 % agar and 2×10^5 cells was plated in 35 mm Petri dishes with a standardized source of CSA (GCT, GIBCO, Grand Island, USA) (100 ul/dish). Colonies were scored after 7 and 14 days incubation in the presence of 5 % CO_2 in humidified air (Pike and Robinson, 1970).

PLASMA CLOT CULTURES (BFU-E, CFU-MK)

Plasma clot cultures were performed according to McLeod et al. (1974) and Vainchenker et al. (1982). Briefly, 3×10^5 cells in 1 ml of IMDM containing 10 % bovine serum albumin (fraction V, Sigma St. Louis, USA) deionized by resin AG-501 x B (D)(bio-Rad Lab., USA), 10 % heat-inactivated pooled human AB serum, 10 % citrated bovine plasma (GIBCO, Grand Island, USA), 20 ug/ml (10 % final concentration), L-asparagine (Sigma, St Louis, USA), 10 % conditioned media from the T-lymphoblastic cell line Mo (Mo-CM), 3.4 ug Ca C12, were cultured per dish.

STATISTICAL ANALYSIS

The possibility of significant differences between samples was determined use of Student's t-test.

RESULTS

The effect of serum from 7 HCL patients, collected before and after α-IFN treatment, was evaluated on the colony forming capacity of normal bone marrow cells. Table 1 reports the clinical and haematological features, before and after 3 and 12 months of α-IFN therapy, correlated with the serum (at 0.1 % concentration) inhibitory effect on the normal CFU-GM, BFU-E and CFU-MK colonies.

Table 1 Clinical, haematological and percentage of serum inhibition before and after α-IFN treatment.

PATIENT (Age/Sex)		Hb g/dl	Neutr. $\times 10^9/1$	Plat. $\times 10^9/1$	Marrow Infiltr.	% Inhibition BFU-E 2×10^5	CFU-GM 2×10^5	CFU-MK 3×10^5	Respose to α-IFN
1	Pre	7.5	0.98	35	90%	60	40	47	CR
(51/M)	Post1	12.5	2.81	240	30%	69	14	46	
	Post2	15.1	4.73	289	5%	0	12	25	
2	Pre	13.1	1.12	237	60%	80	70	27	PR
(46/F)	Post1	11.1	0.94	97	40%	68	58	28	
	Post2	14.3	3.22	237	10%	20	20	21	
3	Pre	13.2	0.24	57	70%	82	62	22	CR
(36/M)	Post1	14.1	1.72	145	10%	69	14	46	
	Post2	15.4	2.63	175	5%	0	12	25	
4	Pre	10.1	0.21	37	90%	48	73	50	PR
(70/M)	Post1	12.5	0.52	102	30%	25	58	33	
	Post2	14.9	1.49	139	15%	3	36	43	
5	Pre	8.4	0.22	31	90%	28	15	37	PR
(61/M)	Post1	12.4	1.80	61	20%	/	/	/	
	Post2	14.7	2.75	100	10%	10	13	35	
6	Pre	8.8	0.92	125	80%	29	63	35	CR
(78/M)	Post1	10.9	1.91	179	20%	50	39	35	
	Post2	14.7	2.94	197	5%	0	20	37	
7	Pre	9.1	0.94	74	90%	35	50	41	MR
(62/M)	Post1	10.1	1.02	83	90%	48	51	38	
	Post2	10.8	1.22	94	80%	23	59	36	

CR = Complete Response; PR = Partial Response; MR = Minor response.

Independently of the serum concentration (data not shown), almost all sera showed, before treatment, an evident inhibitory activity rather unmodified after 3 months of therapy even in those patients in which an evident recovery of peripheral blood values was obtained. On the contrary, after 12 months of therapy, there was a significant reduction in the inhibition on all 3 cell lineages particularly evident on the erythroid compartment, in which the percentage of inhibition reached values as low as those observed in normal sera. The serum inhibitory effect was observed also on the CFU-MK colonies but, although an improvement statistically significant ($p < 004$) was documened after 12 months of therapy, the removal of the inhibitory activity was less evident. On the CFU-GM colonies, the percentage of inhibition decreased significantly in 2 patients after 3 months therapy and in all, after 12 months ($p < 0.0003$). Only 1 patient (case n° 7) showed a persistent inhibitory activity on all 3 haematopoietic lineages and this did correlate with the absence in this patient of any clinical and haematological response after treatment with α-IFN.

DISCUSSION

In a previous study (Geissler et al. 1986), a severe depression of the haemopoietic stem cell compartment has been observed in HCL patients. Moreover, serum from HCL, even at very low concentration, produced a marked colony growth inhibition of normal haemopoietic colonies (Lauria et al. 1987). Our study confirms and extends these preliminary results, suggesting that patients with HCL may contain a serological factor which produces a potent inhibitory effect on the normal CFU-GM, BFU-E and CFU-MK colonies at concentrations as low as 0.1 %. As the serum-derived inhibition shows only limited changes after 3 months of α-IFN therapy and that a significant reduction in the inhibitory activity is achieved only after 12 months of treatment, it is feasible to suggest that removal of inhibition is strictly correlated to a significant reduction in the number of neoplastic hairy cells.

In conclusion, from our data it appears that in HCL, the bone marrow stem cell deficiency may be related also to presence of a serological inhibitory factor probably produced by the hairy cells. A better characterization of this serological factor is under investigation.

ACKNOWLEDGMENTS

This work was partially supported by Associazione Italiana Ricerca sul Cancro, Milano.

REFERENCES

Geissler K., Hinterberger W., Bettelheim P., Nevmann E., Lechner K., Koller V., Knapp W. (1986). Myeloid progenitor cells in the peripheral blood of patients with hairy cell leukemia and other "leukemic" lymphoproliferative disorders. Leukemia Research 10: 677-681.

Lauria F., Guarini A., Bagnara G.P., Catani L., Gaggioli L., Gugliotta L., Raspadori D., Buzzi M., Zauli G., Tura S. (1987). Inhibitory effect of Hairy-cell Leukaemia (HCL) serum on the "in vitro" growth of haemopoietic precursors. Abstr. 4th Internat. Symp. Acute Leukemias 264: 7-12/2, Rome.

McLeod D.L., Shreeve M.M., Axelrad A.A. (1974). Improved plasma culture system for production of erythrocytic colonies in vitro: quantitative assay method for CFU-E. Blood 44: 517-534.

Pike B.L. and Robinson W.A. (1970). Human bone marrow colony growth in agar gel. J. Cell Physiol. 76: 77-84.

Vainchenker W., Deschamps J.F., Bastin J.M., Guichard J., Titeux M., Breton-Gorius J., McMichael A.J. (1982). Two monoclonal antiplatelet antibodies as markers of human megakaryocyte maturation: immunofluorescent staining and platelet peroxidase detection in megakaryocyte colonies and in vivo cells from normal and leukemic patients. Blood 59: 514-521.

Hairy cell leukemia (HCL): II) hairy cells produce factor(s) affecting the *in vitro* growth of normal hemopoietic stem cells

Licia Gaggioli *, Laura Bonsi *, Luisa Valvassori *, Lucia Catani °, Anna Guarini °, Pier-Luigi Zinzani °, Marina Buzzi °, Donatella Raspadori °, Marina Marini *, Francesco Lauria °, Gian Paolo Bagnara * and Carlo Rizzoli *

Institute of Histology, University of Bologna and ° Institute of Hematology L. and A. Seragnoli, University of Bologna, Bologna, Italy.

KEYWORDS
Hairy Cells, Colony Inhibiting Activity, Hemopoietic stem cells (CFU-GM, CFU-MK, BFU-E).

INTRODUCTION

Serum-borne inhibiting activity affecting normal hemopoietic progenitors has been described in patients affected by Hairy Cell Leukemia (HCL) (Lauria et al., 1987). The cellular source of these inhibiting factors is still unknown. The availability of peripheral blood cells from HCL patients having a high percentage of circulating Hairy Cells (HC) allowed us to study the production of Colony Inhibiting Activity by HC themselves

MATERIALS AND METHODS

Four selected patients with more than 80% circulating hairy cells were studied. Peripheral blood mononuclear cells, obtained from the interface of Ficoll Hipaque (Pharmacia, Uppsala, Sweden) density gradient, were incubated 2 hours in Iscove s Modified Dulbecco's Medium (IMDM) supplemented by 5% Fetal Bovine Serum (FBS) at 37°C, 5% CO_2 and 100% humidified air. The procedure was repeated twice. Mononuclear non aherent cells (MNAC) were collected and T-lymphocytes depleted by a double rosetting with neuroaminidase-treated sheep red blood cells. Collected cells showed a strong positivity ($>$ 95%) when tested with HC-2 and FMC-7, monoclonal antibodies which recognize HC; the same cells showed a very low positivity ($\leq$ 1%) to OK T11 monoclonal antibody (Pan T Lymphocytes) so T-lymphocytes contamination of cell suspension was rouled out.
Cells were counted and plated in synthetic medium at different concentrations (0.3 - 0.5 - 1 - 2 x 10^6 cells/ml) and incubated at 37°C and 5% CO_2 in humidified air for 2 days. Conditioned media (HC-CM) were harvested and tested at different concentrations (0.1%, 1%, 10%) on normal bone marrow cultures either unstimulated or stimulated by a standard source of Granulocyte Monocyte Colony Stimulating Activity (GM-CSA) or Burst Promoting Activity (BPA).
For Colony Forming Unit Granulocyte Monocyte (CFU-GM) assay, agar cultures were performed according to Moore et al. (1973). As source of standard CSA, Giant Cell Tumor Conditioned Medium (GCT-CM, Gibco, Grand Island, N.Y.) 100 µl/dish was used (Di Persio et al. 1980).

For Colony Forming Unit Megakaryocyte (CFU-MK) assay and Burst Forming Unit Erythroid (BFU-E) assay, plasma clot cultures were performed according to Mc Leod et al. (1974) and Vainchenker et al. (1979). As a source of BPA, conditioned medium from the T lymphoblastic cell line Mo (Mo-CM) (100 μl/dish) was used (Golde et al. 1980; Bagnara et al. 1987).

RESULTS AND DISCUSSION

The addition of synthetic HC-CM in unstimulated cultures of normal bone marrow did not induce hemopoietic colony formation (data not shown).
When HC-CM was added in bone marrow semisolid cultures stimulated by a standard source of GM-CSA and BPA, it induced a marked decrease of granulomonocyte, erythroid and megakaryocyte colony formation (Fig. 1).

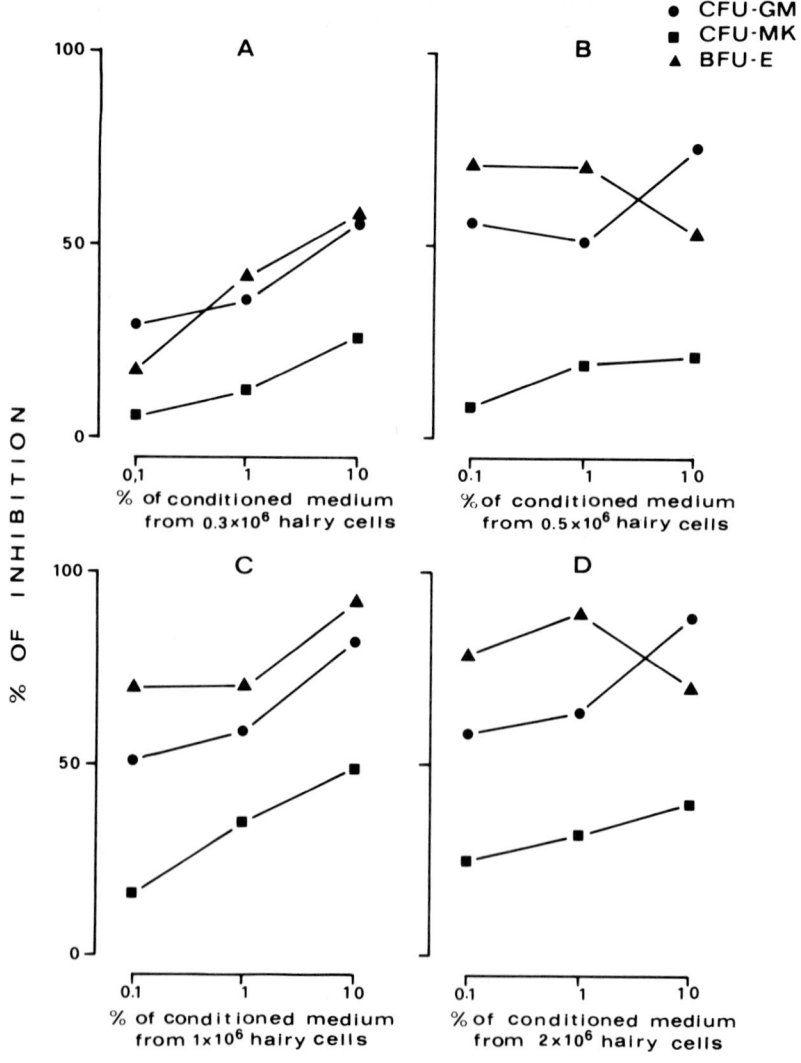

FIGURE 1

Medium conditioned by 0.3×10^6 HC displayed a lower inhibiting activity on CFU-GM and BFU-E when compared to that conditioned from 0.5×10^6 HC (Fig. 1 A-B). Higher HC concentrations did not proportionally increase the inhibiting activity of the CM (Fig. 1 C-D).
Erythroid progenitor cells appeared to be more sensitive then the other progenitor cells to the HC-CM inhibiting activity.
Only the higher concentration (10%) of conditioned media from 1×10^6 and 2×10^6 HC cultures significantly inhibited CFU-MK proliferation (Fig. 1 C-D).
In conclusion, our data strongly suggest the possibility that the serological inhibitory factor found in HCL patients, may be produced by the HC.

REFERENCES

Bagnara G.P., Guarini A., Gaggioli L., Zauli G., Catani L., Valvassori L., Zunica G., Gugliotta L., Marini M. (1987): Human T-lymphocyte-derived Megakaryocyte-Colony-Stimulating Activity. Exp. Hematol., in press.
Di Persio J.F., Brennan J.F., Lichtman M.A., Abond C.N., Kirkpatrik F.H. (1980): The fractionation, characterization, and subcellular localization or colony stimulating activities released by the human monocyte cell line GCT. Blood, 58, 717-727.
Golde D.W., Bersch N., Quan S.G., Lusis A.J. (1980): Production of erythroid-potentiating activity by a human T-lymphoblast cell line. Proc. Natl. Acad. Scie. USA, 77, 593-596.
Lauria F., Bagnara G.P., Raspadori D., Buzzi M., Guarini A., Zinzani P.L., Catani L., Gaggioli L., Marini M., Brunelli M.A. (1987): Hairy Cell Leukemia (HCL): I) Inhibitory effect of serum from HCL-patients on the normal bone marrow colony growth possible removed by a prolonged α-interferon treatment. This meeting.
Mc Leod D.L., Shreeve M.M., Axelrad A.A. (1974): Improved plasma culture system for production of Erythrocytic colonies in vitro: quantitative assay method for CFU-E. Blood, 44, 517-534.
Moore M.A.S., Williams N., Metcalf D. (1973): In vitro colony formation by normal and leukemic human hematopoietic cells. Characterization of colony forming cells. J. Natl. Cancer Inst., 50, 603-610.
Vainchenker W., Bouguet J., Guichard J., Breton Gorius J. (1979): Megakaryocyte colony formation from human bone marrow precursors. Blood, 54, 940-945.

AKNOWLEDGMENTS

This work was supported by Associazione Italiana per la Ricerca sul Cancro (A.I.R.C.) and C.N.R., grant n. 85.00694.04.

Inhibition of hematopoietic progenitor cells by a variant of the L1210 leukemia cell line

Norbert Frickhofen, Aruna Raghavachar, Ivan N. Rich, Wolfgang Heit and Hermann Heimpel

Department of Hematology/Oncology, University of Ulm, D-7900 Ulm, FRG

KEYWORDS

Inhibitory factors, hematopoietic progenitors, CFU-GM, BFU-E, CFU-E, L1210, cell line

INTRODUCTION

Cell lines are convenient sources of factors, regulating the growth of hematopoietic progenitor cells in vitro. Growth factors like colony stimulating factors (CSF) are produced by a variety of murine and human cell lines. Mass production of CSFs from cell lines has facilitated the characterization of these molecules and clones of cDNA for many of these factors have been isolated from cell lines (Metcalf, 1986). There are to our knowledge only two reports on cell line-derived inhibitory factors, both describing T cell lines (Trucco, 1984; Gullberg, 1986). We would like to describe a non T cell line, which was incidentally found to generate a potent inhibitory activity for murine and human hematopoietic progenitor cells in vitro.

MATERIALS AND METHODS

Cell lines: The mouse cell line X341 described in this study developed during subculture of bone marrow cells of a patient with malignant mastocytosis (Frickhofen, 1984). Established cell lines were kindly provided by Dr. F. Porzsolt, Ulm (L1210, K562) and Dr. A. Woelpl, Ulm (HL-60, CTV2, H9). All cell lines were set up in alpha medium without glutamin (Biochrom, Berlin) supplemented with 10% fetal calf serum (FCS, Gibco). L1210 and X341 were subcultured daily, the other cell lines twice a week. In some experiments cells were grown in RPMI 1640 or Iscove's modified Dulbecco's medium and horse serum (HS) instead of FCS (all from Gibco). For preparation of conditioned media cells were subcultured at 2×10^5/ml; culture supernatants were routinely collected after 20 hrs.

Colony assays for hematopoietic progenitor cells: Mouse CFU-GM, BFU-E and CFU-E were grown form C57BL bone marrow cells using heart conditioned medium as a source of CSF and step III erythropoietin (Connought, Cannada) as previously described (Rich, 1986). Human CFU-GM, BFU-E and CFU-E were grown from bone marrow cells of healthy bone marrow transplant donors using placenta-conditioned medium and step III erythropoietin (Raghavachar, 1986).

Cytochemistry and cytogenetic analysis were performed by standard methods.

Cell marker analysis: An immunoperoxidase method was used for the analysis of surface and intracellular markers (Frickhofen, 1985). The following monoclonal antibodies were used: HB-3,-35,-41,-75,-76,-77,-102 (anti mouse MHC, ATCC, Rockville, USA), OX3 (anti mouse Ia) and Mac1 (Seralab, UK), Thy1.1, Thy1.2 (Miles). Antibodies against antigens of human lymphocytes, T and B cells, monocytes/macrophages and granulocytic cells were purchased from Becton Dickinson (HLe1, Leu9, Leu1, Leu5, Leu4, LeuM5), Coulter Electronics (My9, My7, My4, B1, B4, J5) and Ortho Diagnostic Systems (OKT6, OKM1, OKT10). Further antibodies were kindly provided by Prof. W. Knapp, Vienna (VIM2, anti granulocyte/monocyte), Dr. P.M. Lansdorp, Amsterdam (C17-27-2, anti platelet gpIIIa and IVB1, anti glycophorin) and Dr. R. Rimmer, London (MCG35, anti mast cell granules). Polyclonal horse anti mouse immunoglobulin (Vector-Lab., USA) was used for the demonstration of mouse immunoglobulin.

RESULTS

Inhibition of hematopoietic progenitor cell growth
During long term culture of bone marrow cells derived form a patient with malignant mastocytosis a rapidly proliferating cell line evolved. Supernatants of this cell line were tested for their effect on the growth of human hematopoietic progenitor cell growth in vitro. Unexpectedly these supernatants were able to completely suppress colony forming cells. CFU-GM were significantly more sensitiv to the inhibitory activity than BFU-E and CFU-E inhibition required about tenfold higher concentrations of the supernatant compared to CFU-GM (Fig.1). Day 7 CFU-GM were suppressed to a similar extent as day 14 CFU-GM. Peripheral blood CFU-GM and BFU-E behaved similar to bone marrow progenitor cells.

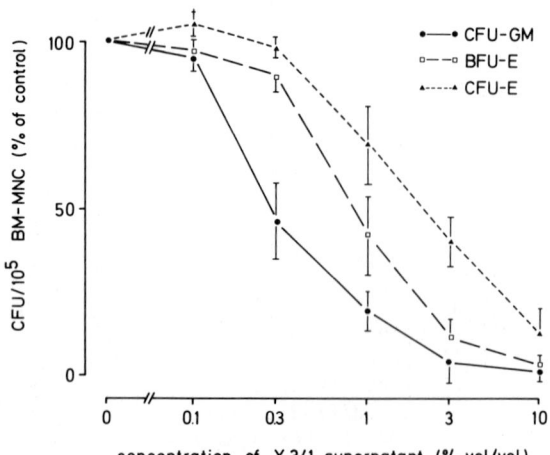

Fig.1: Inhibiton of human hematopoietic progenitor cells by the addition of X431-supernatants (100% = 118+43 CFU-GM, 163+39 CFU-E, 43+24 BFU-E; n=5).

Murine progenitor cells were affected by the inhibitory activity as well but the
sensitivity of the cells was different: Mouse BFU-E suppression was very similar
to human BFU-E (60% inhibition at a concentration of 1%) whereas CFU-GM were
completely resistent to the factor(s) even at a concentration of 10% and there
was no consistent suppression of CFU-E either.

To rule out nonspecific cytotoxicity, active supernatants were added to cultures
of PHA-stimulated peripheral blood lymphocytes and long term cultures of the
cell lines HL-60, K562, CTV2 (myeloid), H9 (T cell) and Whish (human
amnion). There was no significant inhibition of 3H-thymidine uptake by T cells
and the growth of the cell lines was not affected by the addition of inhibitory
supernatants up to 10 days. In line with these data was the observation, that
human CFU-GM growth was not inhibited if X341-supernatants were added to the
cultures on day 5, thereby excluding suppressive effects on more mature
granulopoietic cells.

Characterization of the cell line

X341 cells grew best in alpha medium, supplemented with 5-20% FCS or HS.
Doubling times were 14-22 hrs with a median of 19 hrs. In RPMI 1640 and Iscove's
medium doubling times were significantly longer and cultures were lost after 1-3
weeks. At serum concentrations <2% cells stopped growing and died after 4-7
days. The cell line could be readily cloned in soft agar or methylcellulose with
a cloning efficiency of 65%. Contamination of the cells with mycoplasma or virus
could be excluded by standard culture techniques.

In contrast to the rapid change of the growth pattern, which required daily
feeding of the cultures, there was only a gradual change in cell morphology:
The cells developed large nuclei with prominent nucleoli and there was a gradual
loss of the typical mast cell granules. However after culture for 10 months,
about 15% of the cells retain granules localized to the Golgi zone. They stain
for naphthyl-AS-acetate- and naphthyl-AS-D-chloroacetate esterase and they
exhibit metachromasia on toluidin blue staining.

Cytogenetic analysis unexpectedly showed acrocentric chromosomes and thereby
proved the murine origin of the cells. Surface marker analysis confirmed the
absence of antigens associated with human monocytes/macrophages or granulocytic
cells (My4, LeuM5, OKM1, VIM2, My7, My9), lymphocytes (HLe1; OKT10; Leu9,
OKT6, Leu1, Leu5, Leu4; B4, B1, J5), erythroid cells (IVB1), megakaryocytes
(C17-27-2) and mast cell granules (MCG35). HLA class II antigens (DR, DP,
DQ) were absent but an antibody against a non polymorphic region of human class
I antigens (W6/32) gave repeatedly a strong membrane staining. The cells
reacted with three antibodies against murine class I antigen Dd (HB-75, -76, -
102) compatible with a derivation of the cells from a DBA/2 mouse. Class II
antigens could not be shown with the antibodies available nor did we find T cell
(Thy1) or macrophage antigens (Mac1) or surface immunoglobulin. L1210
leukemia cells gave essentially similar results on marker analysis.

Before the identification of the cells as of murine origin, several attempts
were made to induce a differentiation of the cells. With high concentrations of
DMSO and DMF (0,1%) cell growth was inhibited, the cells aquired a
"macrophage-like" morphology and phagocytosis could be detected. There was no
detectable change of growth pattern or cell morphology during culture in the
presence of hydrocortison, human IFN-gamma, human TNF-alpha, PMA, PHA +/-
peripheral blood MNC, human PHA-LCM, placenta CM or umbilical cord plasma.

DISCUSSION

A cell line "X341", constitutively producing inhibitory factor(s) for
hematopoietic progenitor cells is described. It turned out that X341 derived
from a mouse cell line, which had cross-contaminated long term cultures of bone
marrow cells from a patient with malignant mastocytosis (Frickhofen, 1984).

Surface marker analysis is compatible with a derivation from a CBA/2 mouse. Differentiation antigens of T cells and macrophages or surface membrane immunoglobulin could not be detected. Marker results were identical to those of the mouse B leukemia cell line L1210 (Moore, 1966; Lane, 1982), which could have contaminated the cultures. Although some data are puzzling (toluidin blue positive granules?) a fusion between mouse L1210 cells and human mast cells from the original long term culture seems unlikely. The reactivity of W6/32 (HLA class I) can probably be explained by known cross reactivity of this antibody with murine antigens (A. Woelpl, personal communication).
Growth of mouse as well as human hematopoietic progenitor cells can be completely suppressed by low concentrations of supernatants from the cell line, however the targets cells are different: Human CFU-GM are more sensitive to inhibition than BFU-E and CFU-E whereas only mouse BFU-E are suppressed to a similar extent. Original L1210 leukemia cells do not produce a similar activity consistently (slight inhibition at high concentrations).
The inhibitory activity of X341 supernatants resembles that of a factor isolated from a human T cell line ("CIL"; Trucco, 1984), which likewise primarily inhibits early human progenitor cells. Preliminary data identify the activity in X341 supernatants as a non-dialyzable, trypsin-sensitive factor, which in contrast to CIL is labile to storage even in liquid nitrogen and is rapidly inactivated by heat. We are going to characterize the active principle in more detail since inhibition of early hematopoietic progenitor cells should be an effective mechanism regulating cell growth.

REFERENCES

Frickhofen, N., Heit, W., Czarnetzki, B.M., Carbonell, F., Porzsolt, F., Brudler, O., Heimpel, H. (1984): A case of malignant mastocytosis: unexpected phenotypic properties of the malignant mast cells. Blut 49, 274.

Frickhofen, N., Bross, K.J., Heit, W., Heimpel, H. (1985): Modified immunocytochemical slide technique for demonstrating surface antigens on viable cells. J. Clin. Pathol. 38, 671-676.

Gullberg, U., Nilsson, E., Sarngadharan, G., Olsson, I. (1986): T lymphocyte-derived differentiation-inducing factor inhibits proliferation of leukemic and normal hemopoietic cells. Blood 68, 1333-1338.

Lane, B.C., Bricker, M.D., Cooper, S.M. (1982): Fc receptors of mouse cell lines. II. IgG binding specificity and identification of the Fc receptor on a lymphoid leukemia. J. Immunology 128, 1825-1831.

Metcalf, D. (1986): The molecular biology and functions of the granulocyte-macrophage colony-stimulating factors. Blood 67, 257-267.

Moore, G.E., Sandberg, A.A., Ulrich, K. (1966): Suspension cell culture and in vivo and in vitro chromosome constitution of mouse leukemia L1210. J. Natl. Cancer Inst. 36, 405.

Raghavachar, A., Frickhofen, N., Arnold, R., Schmeiser, T., Porzsolt, F., Heimpel, H. (1986): Hematopoietic colony formation after allogeneic bone marrow transplantation: enhancement by cyclosporin A and anti-gamma-(immune-) interferon antiserum in vitro. Exp. Hematol. 14, 621-625.

Rich, I.N. (1986): Role for the macrophage in normal hemopoiesis. I. Functional capacity of bone marrow derived macrophages to release hemopoietic growth factors. Exp. Hematol. 14, 738-745.

Trucco, M., Rovera, G., Ferrero, D. (1984): A novel human lymphokine that inhibits haematopoietic progenitor cell proliferation. Nature 309, 166-168.

Negative autocrine activity of fractions of leukemic ascites and conditioned medium

Kefu Wu, Li Liu, Jianxin Chu, Jinghua Wan, Yuhua Song and Wenchieh Chen

Institute of Hematology, Chinese Academy of Medical Sciences, Tianjin, China

KEY-WORDS : Autocrine, inhibitory activity, leukemia.

L7811 transplantable mouse ascites leukemia was originated from a leukemic inbred 615 mouse (Staats, 1985) induced by busulfan in our Institute in 1978 (Chu et al. 1983). The cell line of L7811 leukemic cells was established in 1985 and named L7811-85 (Wan et al., 1987).

Inhibitory activity on L7811 leukemic cells and normal 615 bone marrow cells was found from ascites of L7811 mice with ^{3}H-TdR incorporation and morphological observation after cultivation for 6 to 24 hrs. However, the conditioned medium (CM) of L7811-85 cells did not show inhibitory activity. Then, the chromatography analysis of L7811 ascites and L7811-85 CM were performed.

There were stimulative fractions and inhibitory fractions on Sephadex G-150 gel filtrations of L7811 ascites and L7811-85 CM. Since L7811 was a T-cell leukemia (Chu et al., 1983), it might produce some growth factor like IL-2. In this communication, we report the inhibitory activity of them.

Sephadex G-150 gel filtration was performed as described by Wu et al. (1987). The inhibitory activity on L7811, L7811-85 and normal 615 bone marrow (BM) cells was found at range of Mr 70,000 to 80,000 and 40,000 to 50,000 (Figs 1 and 2).

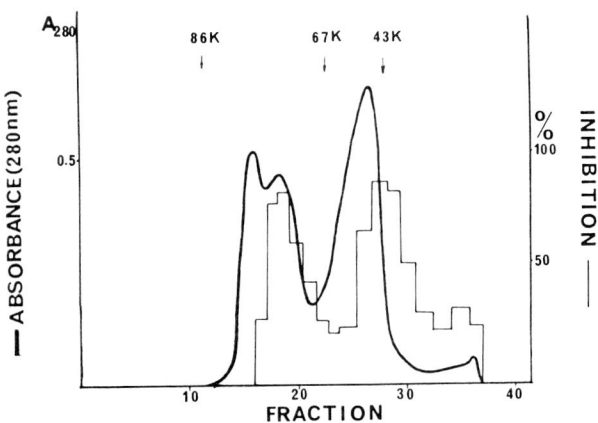

FIG.1   GEL FILTRATION OF L7811 ASCITES WITH SEPHADEX G-150

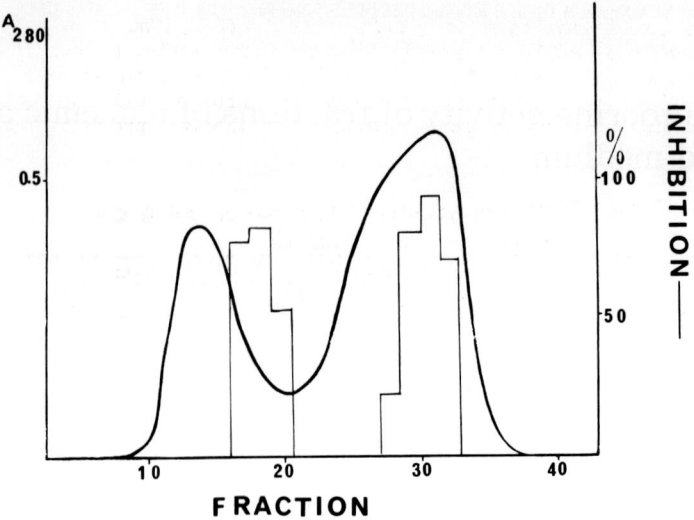

FIG. 2 GEL FILTRATION OF L7811-85 CM WITH SEPHADEX G-150

These fractions could inhibit incorporation of ^{3}H-TdR of L7811 and L7811-85 cells, also normal bone marrow cells and spleen cells cultured 24 hrs and assayed by the method mentioned in previous paper (Wu et al., 1985).

The inhibitory effect appeared in decreasing incorporation of ^{3}H-TdR of these cells and showed morphological and cytochemical damage if 10 % of active fraction was added. Compared with control cell culture, these cells grew poorly and the rate of dead cells were increased after 6 hrs' culture of active fraction. After 24 hrs' culture, the cell number was dramatically decreased and methyl green and pyronin stain was very weak, acid phosphatase, non-specific alfa-naphthol acetate esterase and acid non-specific alfa-naphthol acetate esterase almost disappeared, the PAS reaction of leukemic cells disappeared (Table 1).

Table 1 Cytochemical Effect of Leukemic Cells by Inhibitory active Fraction

	PAS	RNA	ACP	NAE	ANAE
L7811 C	4	5	3	3	3
L7811 Fr.AF	0	1	1	1	1
L7811-85 C	4	5	3	3	3
L7811-85 Fr. CM	0	1	1	1	1

C. Control   Fr. Fraction
AF. Ascitic fluid   CM. Condition medium

These fractions could inhibit L615 (CAMS, 1974), L7212 (Chu et al. 1986) and P388 mouse leukemic cells as well, but not S180 mouse tumor cells, neither human leukemic cell line J6-1 and J6-2 (Wu et al., 1986) and Burkitt's lymphoma cell line Namalva. It seems that L7811 cells could be the most sensitive cells to this inhibitory factor, that means this could be a negative autocrine, produced by L7811 cells (Table 2).

Table 2 Inhibitory Activity of Fractions of AF & CM on Various Cell Cultures

cell	cont. cpm	fr-af cpm	I%	fr-cm cpm	I%
L7811	71072±3546	723±44	99	ND	
L615	49361±3540	17735±1029	64	ND	
L7212	7223±1037	561±44	92	ND	
615-S	1168±122	561±123	52	95±6	92
615-B	1274±391	290±38	77	503±97	61
L7811-85	7977±425	ND		1348±52	83
P388	2190±167	190±7	91	684±204	69
L7212(Cell Line)	9718±479	3072±215	68	3957±193	59
S180	1752±86	1663±93	5	1526±46	13
J6-1	10832±693	9157±278	15	6047±485	44
J6-2	7316±534	9433±650	-29	7063±337	3
Namalva	4057±235	13937±1505	-243	4198±172	-3

$$I\% = (cpm_c - cpm_f)/cpm_c \times 100\%$$

Sephadex X-150 gel filtration of normal 615 mouse's abdomen washing and supernatant of bone marrow did not show inhibitory activity at these molecular weight range.

The purification and characterization and biological significance of this factor remain to be defined.

Projects supported by Nature Science Fund of China.

REFERENCES

Chinese Academy of Medical Science (1974). Laboratory studies on a transplantable mouse leukemia (L614). XI International Cancer Congress (monography).
Chu J.X., Cao S.X., Qian Y.F., Qi S.L., Song Y.H., You S.C. (1983). Establishment of a transplantable mouse ascites leukemia (L7811) and its biological properties. Leuk. Rev. Internat. 1, 82.

Chu J.X., Qi S.L., La J.L. (1981) Establishment of transplantable mouse leukemia (L7212) and its biological properties. Chinese J. Cancer, 3, 287.
Staats J. (1985). Standardized nomenclature for inbred strains of mice: Eight listing. Cancer Res., 45, 945.
Wan J.H., Zhang S.J., Mi J.X., Qi S.L., Wu K.F. (1987). Establishment of murine lymphocytic leukemia cell lines and investigation on their biological characteristics. Acta Academiae Medicinae Sinnicas, 8, (2), (in press) (in Chinese).
Wu K.F., Pope J.H., Ellem K.A.O. (1985). Inhibition of growth of certain human tumour cell lines by a factor derived from human fibroblats-like cell lines. I. Demonstration by mixed culture and use of cell washings. Int. J. Cancer, 35, 477.

Wu K.F., Zhang Y.Q., Song Y.H. (1986). Establishement and characterization of human leukemia cell lines (J6-1, J6-2, J6-3). Proc. CAMS & PUMC, 1, 218.
Wu K.F., Sculley T.B., Pope J.H., Ellen K.A.O. (1987). Partial purification and characterization of the tumor growth inhibitor derived from human fibroblast-like cell lines. Acta Academiae Medicinae Sinicae, 9, 5 (in Chinese with English abstract).

Stimulatory and inhibitory factors of myelopoiesis in normal and AML human serum

Marie-Claude Hofmann, Stefan Arrenbrecht and Christophe Sauter

Division of Oncology, Department of Medicine, University Hospital, Zurich, Switzerland

KEY WORDS

CFU-gm, bone marrow cultures, serum inhibitors and stimulators, AML

INTRODUCTION:

The mechanisms responsible for neutropenia in patients with acute myeloblastic leukemia (AML) are still poorly understood. We tried to determine if the humoral regulation of the hemopoiesis is affected in this disease and if leukemic cells themselves produce an inhibitory factor of the normal hemopoietic stem cell proliferation. We examined the in vitro effects of normal and AML sera or conditioned media (CM) on the proliferation of direct progenitors of granulocytes and macrophages (CFU-gm) and their precursors.

MATERIAL:

Bone marrow cell suspensions were prepared from the spongiosa (illiac crest) of 42 hematologically normal donors. Erythrocytes were eliminated by Dextrane sedimentation. Culture medium was enriched S-MEM supplemented with 20% FCS. As a source for colony stimulating factor (CSF), we used "Human Placenta Conditioned Medium" (HPCM) (5) at a concentration of 10% of the final culture volume. Sera from 22 AML patients, 12 normal donors and mixed AB+ serum were tested for their influence on CFU-gm and CFU-gm precursors proliferation. For conditioned media (CM), blood MNC of 14 AML patients and 8 normal donors were cultured in supplemented S-MEM at 37°C and 5% CO_2. After 3-5 days, the culture supernatants were harvested and filtered (0.45 um pore size) before use.

METHODS:

A. CFU-gm proliferation (soft agar cultures) in normal and leukemic sera or normal and leukemic CM:

Normal bone marrow cells were plated in soft agar after the method of Bruch et al.(1). HPCM (as CSF) and normal or leukemic sera were added both at a final concentration of 10%. After 7 days of incubation (37°C, 5% CO_2) clones (colonies and clusters) were counted. To test the effect of conditioned media, normal bone marrow cells were plated as above with normal or leukemic cell CM to a final concentration of 10%.

B. CFU-gm precursors proliferation (long term liquid cultures) in normal and leukemic sera or normal and leukemic CM:

Normal bone marrow cells were cultured in long term cultures over 7 weeks (method modified after Gartner and Kaplan (3)). Normal or leukemic sera were added during weekly feeding at a final concentration of 10%. The production of CFU-gm by these cultures after 5 weeks was evaluated by the agar test of Bruch et al.(1). To test the effect of conditioned media, normal bone marrow cells were kept in long term cultures as above with normal or leukemic cell CM at a final concentration of 10%.

C. Gel chromatography:

6 normal sera, 6 leukemic sera, 2 normal cell CM and 4 leukemic cell CM were passed over Sephadex G 150. For each sample, 25 fractions were obtained in the MW range from 300 to 13 kD. Fractions were tested for their effects on the proliferation of normal CFU-gm in soft agar and their precursors in long term cultures. Results were compared with the results obtained in control cultures without fractions, containing only HPCM.

RESULTS:

A. Effect of sera and CM on CFU-gm proliferation in agar cultures:

Leukemic sera exert an inhibition of 49.5% on normal clones proliferation and of 85,5% on normal colonies proliferation in comparison with normal sera ($p < 0.01$). However, no inhibition by leukemic cell CM could be detected.

As shown in table 1, gel chromatography results suggest that normal sera contain an inhibitory factor of CFU-gm proliferation with a molecular weight of ~ 210 kD (inhibition of 70%, compared to serum-free cultures; $p < 0.05$) and stimulatory factors with a molecular weight of approximatively 60, 36 and 17 kD. Of the three stimulatory factors, the 36 kD factor is the most prominent, with an average stimulation of 65 % ($p < 0.05$). The 60 and 17 kD serum factors exert a stimulation of ~30%.

Leukemic sera have the same inhibitory factor as normal sera. They lack, however, the stimulatory factors MW = 36 and 60 kD, possessing only the stimulator with MW = 17 kD.

Fractions of media conditioned by normal or leukemic mononuclear cells showed neither stimulatory nor inhibitory activity on normal CFU-gm proliferation.

Table 1: Effect of normal or AML serum fractions (Sephadex G 150) on normal CFU-gm proliferation in agar cultures.

Fraction	V_E/V_0	MW (kD)	CFU-gm/10^5 MNC (% of control) normal *	CFU-gm/10^5 MNC (% of control) AML **	AML/normal (%)
7	1.07	210	28.6 + 29.0	49.5 + 22.1	173.0
15	1.83	60	120.5 + 60.1	80.0 + 45.6	66.3
17	2.01	36	164.4 + 87.1	84.7 + 61.8	51.5
22	2.48	17	139.2 + 18.6	122.4 + 38.7	87.9

* = with 10% fraction of normal serum
** = with 10% fraction of AML serum
results = mean ± standard deviation (n=6)

B. Effect of sera or CM on CFU-gm precursors in long term bone marrow cultures.

Normal sera double the production of CFU-gm by precursors in long term bone marrow cultures in comparison with serum-free cultures. In contrast, leukemic sera exert an inhibition of 78% in comparison with normal sera ($p < 0.01$).

Normal cell CM exert no particular effect on the production of CFU-gm by long term cultures. In contrast, leukemic cell CM exert an inhibition of 99% ($p < 0.01$) in comparison with normal cell CM.

After gel chromatography (table 2), we could demonstrate that normal sera contain an inhibitory factor of CFU-gm production with a molecular weight of 80 kD (inhibition of 80%, $p < 0.01$, in comparison with serum-free fractions). Two stimulating factors were seen, with a MW of 36 kD (stimulation of ~ 20%) and 17 kD respectively (stimulation of ~ 100%).

Leukemic sera contain two inhibitory factors, with MW of 80 kD (inhibition of 75%) and MW > 200 kD (inhibition of 80%, in comparison with serum-free cultures, $p < 0.05$). This latter inhibitory factor was not detected in the corresponding fraction of leukemic CM.

The 36 kD stimulating factor for CFU precursor proliferation present in normal sera was not seen in leukemic sera.

Table 2: Effect of normal or AML serum fractions (Sephadex G 150) on CFU-gm production by precursors (long term cultures of normal bone marrow).

Fraction	V_E/V_0	MW (kD)	CFU-gm/culture (% of control) normal*	CFU-gm/culture (% of control) AML**	AML/normal (%)
7	1.07	200	111.8 + 6.7	19.5 + 7.1	17.4
13	1.59	80	25.2 + 18.9	31.7 + 28.4	125.7
17	2.01	36	114.0 + 31.0	62.0 + 46.8	54.3
22	2.48	17	212.4 + 70.0	75.0***	?

* = cultures with 10% normal fraction
** = cultures with 10% AML fraction
*** = 1 result
results = mean + standard deviation (n = 6)

DISCUSSION :

For simplicity, we assume that chromatographic fractions do not contain multiple factors.

One of the CSFs in normal sera seems to act on CFU-gm as well as on their precursors. Its apparent molecular weight (~ 36 kD) is similar to that of human pluripoietin (4+6).

Leukemic sera lack CFU-gm and CFU-gm precursor stimulating factors of MW = 60 kD and MW = 36 kD seen in normal sera. Reduction of stimulatory factors in leukemic sera might be due to consumption by preponderant leukemic cells. This probably leads to a predominance of normal inhibitory factors and thus to reduced precursor and granulocyte production.

Moreover, leukemic sera probably contain an additional inhibitor of precursor proliferation with MW > 200 kD. Whether this factor causes the strong inhibition seen with leukemic cell CM is unclear at present, since the activity could not be isolated by simple Sephadex chromatography. Similarly, the lack of activity in CM on CFU-gm performance is difficult to assess, since only one set of conditions has been used to produce CM.

Neutropenia in AML would thus be due to reduced levels of stimulators of granulopoiesis and to the appearance of a novel inhibitor of myelopoietic stem cells, not observed in normal serum.

ACKNOWLEDGEMENTS:

This work was supported by the "Schweizerischer Nationalfonds zur Förderung der wissenschaftlichen Forschung".

REFERENCES :

1. Bruch C, Kovacs P, Rüber E (1978): The role of phagocytic cells in human blood leucocyte suspensions for in vitro colony forming cells. Exp. Hematol. 6 : 346-54
2. Francis GE, Berney JJ, Murray VSG, Jackson B, Hoffbrand AV (1980): Adherent cell dependent colony-stimulating activity in human serum: a granulopoietic regulator ? Scand. J. Haematol. 24:13-21
3. Gartner S, Kaplan HS (1980): Long term cultures of human bone marrow cells. Proc. Natl. Acad. Sci. USA 77: 4756-59
4. Metcalf D (1986): The molecular biology and functions of the granulocyte-macrophage colony-stimulating factors. Blood 67: 257-67
5. Schlunk T, Schleyer M (1980): The influence of culture conditions on the production of colony stimulating activity by human placenta. Exp.Hematol. 8 : 179-84
6. Welte K, Platzer E, Lu L, Gabrilove J, Levi E, Mertelsmann R, Moore MAS (1985): Purification and biochemical characterization of human pluripotent hematopoietic colony-stimulating factor. Proc.Natl.Acad.Sci.USA 82: 1526-30

Negative regulation of pluripotent stem cell (CFU-S) proliferation

Inhibiteurs de la prolifération des cellules souches pluripotentes (CFU-S)

Inhibitor of haemopoietic CFU-S proliferation : assays, production sources and regulatory mechanisms

Brian I. Lord [1], Liu Fu-Lu [1,2], Zigmunt Pojda [3] and Elaine Spooncer [1]

[1] Paterson Institute for Cancer Research, Christie Hospital and Holt Radium, Institute, Manchester M20 9BX, England
[2] Institute of Radiation Medicine, Beijing, China
[3] Department of Radiobiology WIHiE, ul Szaserow 128 00-909, Warsaw, Poland

ABSTRACT

Studies have been carried out using an extract of normal adult bone marrow which blocks the entry of spleen colony forming cells (CFC-S) into DNA-synthesis. To develop a direct assay for the inhibitor, CFC-S were highly purified (70%) by centrifugal elutriation and density separation. These, and several IL-3 dependent stem cell lines, all showed satisfactory dose-response relationships with the inhibitor when assayed by tritiated thymidine incorporation. A factor-independent (leukaemic) line was insensitive.

Macrophages, the primary source of inhibitor, from spleen, peritoneum and peripheral blood all produced inhibitor though in degrees which may depend on their functional heterogeneity.

In long-term marrow cultures, the inhibitor can protect the CFU-S population against the cytotoxic effects of ARA-C. It is suggested that comparable studies in human cultures will yield valuable information on the specificy of the inhibitory material.

Finally, two recent pieces of information are reviewed. The first shows that inhibitor production depends on the size of the CFC-S population - a CFC-S feedback factor being required. The second reveals a potentially valuable means of stimulating the immune system.

KEYWORDS

CFU-S, Proliferation, Regulation, Protection, Assays, Macrophages.

Pluripotent haemopoietic progenitor cells or stem cells, measured in the context of this study as spleen colony-forming cells or units (CFC-S or CFU-S, Till & McCulloch, 1961; Siminovitch et al, 1964) normally exhibit a very low proliferative activity. Indeed, although the tissue is turning out extremely large numbers of mature blood cells, the population amplification between the stem and functionally mature cells is such that the stem cell population can readily maintain its numbers with this low rate of turnover. At the same time, the stem cell population has a very large capacity for regeneration and, following damage brought about by such agents as drugs or irradiation, it is found to be proliferating and turning over very rapidly.

In 1972, Gidali and Lajtha disturbed the distribution of haemopoiesis by irradiating a mouse with one hind limb shielded. As a result, they demonstrated that the kinetic behaviour of CFU-S in the shielded limb was independent both of the rest of the skeletal marrow and of the needs of the animal. As a result, they came to the conclusion that CFU-S proliferation was regulated locally within the marrow. Consequently, since CFU-S proliferation in normal marrow is low and the basic cybernetic principle of control is one of negative feedback, a regulatory inhibitor was sought, and found, in normal bone marrow (Lord et al 1976). This inhibitor was shown to be specific for CFU-S proliferation (Lord et al, 1976; Tejero et al, 1984). A comparable inhibitor was also described by Frindel and Guigon (1977) and is discussed elsewhere in this volume by Dr Guigon Differing in detail from the inhibitor described by our group, their inhibitor, however, showed a similar specificity and ability to protect CFU-S from the effects of S-phase cytotoxic agents. The information presented in this report refers to the inhibitor obtained directly from normal adult bone marrow.

The method used for assaying inhibitor has normally been the tritiated thymidine (^{3}HTdR) suicide technique (Becker et al, 1965; Lord et al, 1974). This approach however, is subject to a number of limitations which mean that only relatively small numbers of samples can be handled. The first part of this report, therefore, presents data illustrating the use of purified CFC-S populations or cloned factor-dependent stem cell lines for a more rapid and multisample screening technique for inhibitor activity.

Considerable work has gone into identifying the cells which generate the inhibitor and it now seems certain that some part of the macrophage complex is responsible (Wright, Garland & Lord 1980; Simmons & Lord 1985). The distribution of inhibitor-producing macrophages has, therefore, been investigated and in the second part of this report it will be demonstrated that macrophages from different sites, including the peripheral blood (of both mouse and human) can synthesise the inhibitor.

The obvious use of an inhibitor of stem cell proliferation is its value in protecting stem cells during S-phase cytotoxic chemotherapy and for this it must be effective in vivo. Both the inhibitor reported here and that reported by Dr Guigon have proved to be effective (Guigon et al, 1980; Lord & Wright, 1980). In the third part, it will be shown that long term bone marrow culture may be an effective way of screening an inhibitor for potential in vivo activity.

A variety of stimulators of haemopoietic cells is now known, one of which corresponds to the inhibitor in that it also can be obtained from a macrophage fraction (Wright et al 1982), it acts to trigger DNA-synthesis in quiescent stem cells (Lord et al 1977) and is specific for the CFU-S population (Tejero et al 1984). This stimulator probably corresponds to the stimulatory activity in damaged marrow described by Frindel et al (1976). It has been shown to interrelate with the inhibitor in regulating the proliferative activity of the CFU-S (Lord et al, 1977; Lord & Wright 1982) and in the final

part we will consider their interaction with the bone marrow stem cells and
review some recent data concerning their roles as an integral part of the stem
cells' regulatory microenvironment and in the modulation of the immune response.

A. DIRECT ASSAY OF CFC-S PROLIFERATION INHIBITOR

Since the number of spleen colony forming cells represents about only 0.4% of the
total marrow population, it has not been possible to make direct measurements of
proliferative activity in terms of ^{3}HTdR incorporation. Recent developments in
cell sorting technology, however, have enabled the preparation of very pure
populations of CFC-S. Fluorescence activated cell sorting gave us ~90% pure
CFC-S from normal bone marrow (Lord & Spooncer, 1986) but the yield was of the
order of 10^5 cells only. Furthermore, they were non-proliferating and required
stimulating before using for inhibitor assay. A more appropriate approach has
been found possible using the centrifugal elutriation technique, developed for
CFC-S by Nijhof and Wierenga (1984). This method gave large numbers of cells of
which 50-100% are proliferating CFC-S and which could be explored for use in
inhibitor assays.

Preparation of CFC-S
Male BDF1 mice, 10-12 weeks of age, were implanted, under the loose skin at the
back of the neck, with a dialysis bag containing 300 mg thiamphenicol (Zambon
Chimica-Milan) in 0.5 ml distilled water. Two and three days later, the mice
were bled via the retro-orbital sinus, removing approximately 0.5 ml blood on
each occasion. On the fourth day, the dialysis bag was removed and the mice
allowed to recover for a further 3½ days. At this stage the spleens were grossly
enlarged (~3.5 x 10^8 cells) and, due to the pretreatment, contained large numbers
($\geq 10^5$) of CFU-S. Ten spleens were pooled and the cells suspended in an
elutriation medium (EM) consisting of Fischer's medium supplemented with 0.5 µ/ml
DNA-ase, 0.02% soybean trypsin inhibitor, 5mM napthalenedisulphonic acid and
0.02% bovine serum albumen. The cells were introduced at a flow rate of 12 ml/min
into the elutriation chamber in a Beckman JE-6B rotor spinning at 2000 rpm and
washed through with 200 ml of EM. The flow rate was then increased to 15 ml/min
and the cells now washed through with a further 200 ml of EM were collected.
They were concentrated by centrifugation to about 8 x 10^7 cells/ml and subjected
to a density cut on metrizamide. The low density ($\leq$ 1.079 gm/ml) cells were
retained as the CFU-S-rich fraction.

The cells obtained at each stage of this procedure were assayed for CFU-S by
injecting into mice (10 per group) which had been irradiated with 15.25 Gy ^{60}Co
γ-rays at 0.84 Gy/hr (Lord et al, 1984). Table 1 shows that approximately 70%
of the final cell population was a spleen colony forming cell. In practice,
this number varied from 50 to 100% and about 30% of them were in DNA-synthesis.

Use of elutriated cells to assess inhibitor activity
5 x 10^4 cells obtained by elutriation and density cut were incubated in
triplicate microwell cultures in 0.1ml Fischer's medium supplemented with 10%
WEHI-3b conditioned medium. 50-500 µg inhibitor (a 50-100K dalton extract of
normal pig bone marrow - see Lord et al, 1976) in 0.1ml medium were added to the
cells and incubated for 15 hrs. 20 µl ^{3}HTdR (37kBq) were then added for a
further 4 hrs. Tritiated thymidine incorporation by the cells was then measured
by a cell harvesting process.

Table 1.   Isolation of spleen colony forming cells by centrifugal elutriation and density cut

	Spleen cells from Thiamphenicol treated mice	Cells elutriated at 15 ml/min (E15)	E15 cells subjecte to density cut at $\rho \leqslant 1.079$ gm/ml
CFU-S/10^5 cells injected	30-40	140	900
CFU-S Enrichment	1	4	26
Spleen seeding efficiency (%)	(1.24)*	(1.24)*	1.24
CFC-S (%)	2.8	11.3	73

Figures presented are approximate means from 15 experiments.
* Assumed same as for final cell fraction.

Use of cloned, factor-dependent stem cell lines to assess inhibitor activity
Cloned haemopoietic, IL-3-dependent stem cells lines were also used to assess inhibitor activity. These were the FDCP-Mix cells lines (Spooncer et al 1986) derived from src (MoMuLV) infected long-term bone marrow cultures. They are characterised by their primitive nature, high self-renewal and their capacity for differentiation into all mature myeloid cell types under appropriate culture conditions. Early isolates of FDCP-Mix cell lines also produce spleen colonies when injected into irradiated mice, but the nature of these colonies has not been fully determined yet, eg the long-term life-sparing ability of the injected cells The continued growth in vitro of FDCP-Mix cells is dependent on the continuous presence of interleukin-3 (supplied by WEHI-3b conditioned medium) which is maintained by regular sub-culture of the cells in Fischer's medium supplemented with horse serum (20% v/v) and WEHI-3b CM (10% v/v).

Cells from a variety of the FDCP-Mix clones were taken, one day after refeeding, and their response to inhibitor was assayed as described for the elutriated cells.

Results
The incorporation of ^{3}HTdR by elutriated spleen colony-forming cells (EC) was reduced by 40% using 50 µg/ml of inhibitor and this was further reduced to about 20% of control with 250 µg/ml (Fig. 1). The factor-dependent cells were less sensitive to inhibitor but still showed clear dose-related responses. FDCP-Mix clones 1(15S) and 9(17S) decreased to about 30% and A4 to 50% of control respectively with 500 µg/ml of inhibitor.

A derivative F19 (17H) of clone FDCP-Mix9 which spontaneously became factor-independent and leukaemic, was largely insensitive to inhibitor (Fig. 1).

These results indicate that elutriated and factor-dependent cells may be useful for the rapid assessment of multiple samples of putative inhibitors. The doses required appear relatively high but this may relate to the fact that virtually all the cells are potential responders whereas in the spleen colony assay, <0.5% of the cells are sensitive targets.

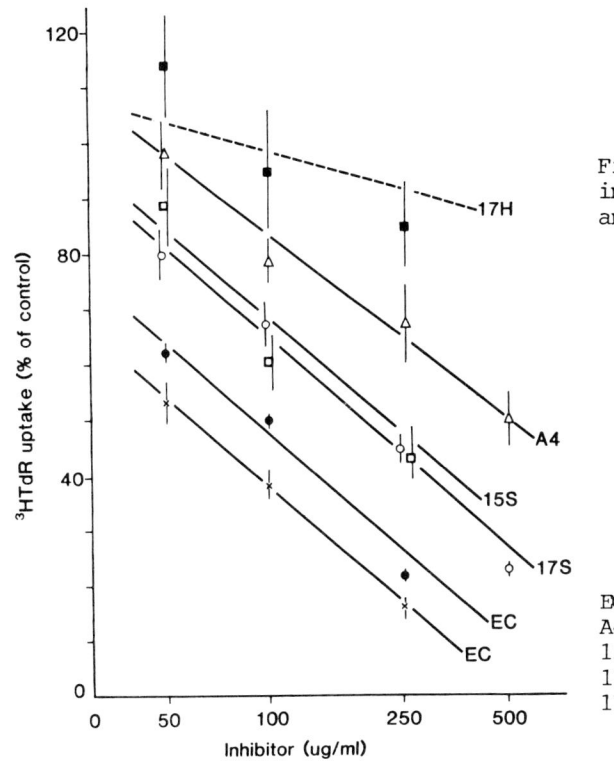

Fig 1. Inhibition of ^{3}HTdR incoporation in Elutriated CFC-S and IL-3 dependent cell lines

EC-elutriated CFC-S
A4
15S (FDCP-1)   IL-3 dependent
17S (FDCP-9)   cells
17H (FI-9)   IL-3 independent cells

The fact that clone FI9 is insensitive suggests that the inhibitor remains non-cytotoxic as in the spleen colony assay (Lord et al, 1976) at these larger doses. If this clone is indeed equivalent to a leukaemic clone, then its lack of sensitivity also extends the value of such an inhibitor as protection for the normal stem cell populations during cytotoxic chemotherapy. It also supports the observation that CML cells do not respond to the normal inhibitory cycle apparent in human (and mouse) long-term marrow cultures (Eaves et al, 1986).

B SOURCES OF INHIBITOR

In view of the properties of the inhibitor-producing cells; adherence to plastic radio-resistance, low density, Fc^+ and $Thy\ 1,2^-$, it was concluded that macrophages were involved (Wright et al 1980). Subsequently we used a macrophage specific monoclonal antibody, F4/80, to collect macrophages by fluorescence activated cell sorting and were able to demonstrate their ability to produce inhibitor. Furthermore, these macrophages could be cultured over extended periods, continuously synthesising inhibitor (Simmon & Lord, 1986). More recently, we have used macrophages selected from marrow by culture as a source of inhibitor. In order to investigate the ubiquity of inhibitor production, other sources of macrophages have now been investigated.

Sources and preparation of macrophages

Full details of the preparation of macrophage cultures are to be published elsewhere (Pojda et al, 1987). Briefly, mouse or human blood, the latter obtained after informed consent had been received, was fractionated on lymphocyte separating medium and the light density mononuclear fraction cultured with a source of macrophage colony stimulating factor, CSF-1, until a good healthy layer of adherent macrophages was obtained. Bone marrow, peritoneal cells and mononuclear spleen cells (after fractionation on lymphocyte separating medium) were subjected to a two-stage culture system. The cells were cultured with CSF-1 for 4 days. The supernatants, including suspended cells were then transferred to a second culture flask and incubated for a further 3 days. The macrophages adhering to the plastic in this second stage were used as putative producers of inhibitor.

When satisfactorily established, the cultures were washed and fresh medium (without CSF-1) was added. After 3 days further culture, the conditioned supernatant media were collected and assayed for inhibitory activity by the ^{3}HTdR suicide technique.

For bone marrow, spleen and peritoneal cells, conditioned media were also collected from the cells before starting the culture processes.

Results

CFU-S from phenylhydrazine-treated bone marrow were rapidly proliferating and 32.5% were found to be in DNA-synthesis (table 2). Extracts from fresh and cultured marrow macrophages reduced this to non-significant levels as previously demonstrated. Extracts from fresh splenic and peritoneal cells were only mildly inhibitory and while macrophages from spleen increased their activity on prolonged culture, those from the peritoneum did not (table 2).

Table 2. Inhibitor activity of Cultured Macrophages

Extract from:	^{3}HTdR kill (% in DNA-S)
Control	32.5 ± 2.5*
Fresh bone marrow	5.2 ± 8.8
Cultured bone marrow Mϕ	0.8 ± 3.6
Fresh splenic mononuclears	22.6 ± 2.7
Cultured splenic Mϕ	12.7 ± 1.8
Fresh peritoneal Mϕ	18.8 ± 3.1
Cultured splenic Mϕ	22.8 ± 2.9

* Standard error of the mean of 3-6 experiments

Macrophages from peripheral blood had the capacity at least to develop inhibitory activity. Mouse blood reduced the ^{3}HTdR kill to 12.8% while only one in 6 samples of normal human blood proved inactive (table 3). Interestingly, macrophages from a single patient with polycythaemia vera were not active (table 3).

Table 3.   Inhibitory Activity of Cultured Blood Macrophages

Extract from:		^{3}HTdR kill (% in DNA-S)
Control		32.5 ± 2.5*
Mouse		12.8 ± 2.3
Human	(i)	12.3 ± 5.2
	(ii)	30.4 ± 7.7
	(iii)	1.7 ± 5.6
	(iv)	11.7 ± 4.5
	(v)	4.4 ± 4.6
	(vi)	10.0 ± 4.1
	Mean	11.7 ± 4.1

NB Patient with polycythaemia vera 25.8 ± 5.2

* Standard error of the mean of 3-6 experiments

Clearly, macrophages are a primary souce of inhibitor and while they are widespread in the body, those in the marrow are the most active. The variable degress of inhibitory and potential inhibitory activity in other sites probably indicate a large degree of functional heterogeneity in the macrophage population which may be associated with development and increased specialisation.

While not in itself significant, it is interesting that macrophages from the single polycythaemia vera patient did not produce inhibitor. Eaves et al (1986) have documented preliminary studies showing that long-term marrow cultures of polycythaemia vera do not respond to the inhibitory cycle seen for normal marrow stem cells (Tóksoz et al, 1980; Cashman et al 1985). It seems likely, therefore, that lack of response to, or production of, proliferation inhibitor may be a major defect in cases of polycythaemia vera.

C. USE OF LONG-TERM BONE MARROW CULTURES

Long-term bone marrow culture is, in many respects, an extremely good model for marrow performance in vivo. It depends for its continued growth on an adherent layer containing a variety of cell types (including macrophages) which form an effective microenvironment in which haemopoiesis can develop (Allen & Dexter, 1983). This microenvironment appears to provide all the factors necessary to generate haemopoiesis since the culture is maintained simply by replacing old growth medium with fresh medium (containing no growth factors) at weekly intervals (Spooncer & Dexter, 1984) and indeed, the culture can recover completely to normal within a few days after replacing all the supernatant medium (and cells) with fresh medium (Dexter et al, 1983). It has been shown also that the cultures respond to (Tóksoz et al, 1980) and produce inhibitor spontaneously and that the pluripotent progenitor cells undergo cyclic proliferation in relation to the feeding programme (Tóksoz et al, 1980; Cashman et al, 1985).

We have attempted to exploit this system in assessing the capacity of the inhibitor to protect CFU-S against the S-phase cytotoxicity of cytosine arabinoside, ARA-C. In a series of preliminary experiments, mouse bone marrow cultures were established over a period of 4 weeks in the standard manner with refeeding every 7d by replacing half the supernatant medium (and cells) with fresh medium (Spooncer & Dexter, 1984). They were then divided into four groups, washed completely free of supernatant medium and cells and treated as shown in table 4.

Table 4. Protocol for treatment of marrow cultures

Group	t = 0hrs	t = 4hrs	t = 5hrs
1	2ml medium	-	wash and reestablish full culture with 10 ml medium
2	2ml medium with 100 µgI	-	
3	2ml medium	200µl ARA-C (100 or 200 µg)	
4	2ml medium with 100 µgI	200 µg ARA-C (100 or 200 µg)	

I = inhibitor   ARA-C = cytosine arabinoside

After 7, 14 and 21 days, the cultures were again refed by total replacement of the medium and supernatant cells. The cellularity of the cultures was measured at 1 to 3 day intervals. At certain points, cultures were sacrificed. Supernatant cells were pooled with adherent layer cells (harvested using a silicone rubber policeman) and assayed for CFU-S and granulocyte/macrophage CFC.

Results
Treatment with 200 µg ARA-C for just one hour resulted in a long-term depression in the ability of the cultures to grow. In the third week of culture, cellularity was reaching only 6×10^6 cells/3 flasks compared with 24×10^6 in the controls (fig. 2). By contrast, those pretreated with inhibitor before adding ARA-C were performing at least as well as the controls. Surprisingly, those treated with inhibitor alone performed even better and this was a repeated feature in all experiments and for each weekly refeeding cycle. Recovery from the lower dose of ARA-C was complete by the third week of culture (fig. 3). Those treated with inhibitor, however, produced nearly 50% more cells than those treated with ARA-C alone.

Figure 4 shows a series of results for CFU-S, GM-CFC and the cellularities of culture over 3-4 weeks of growth. Inhibitor initially delayed the growth of the CFU-S population so that on the third day of the first week, the population was only 60% of control. However, it gave significant protection against the effect of ARA-C. This led to a 'knock on' effect in the GM-CFC and maturing cell compartments where, due to population amplification, growth was considerably better than in the ARA-C only treated group and by the second week was producing normal numbers of cells.

Figs 2 & 3 Protection of LTBMC from the cytotoxic effect of ARA-C by Inhibitor

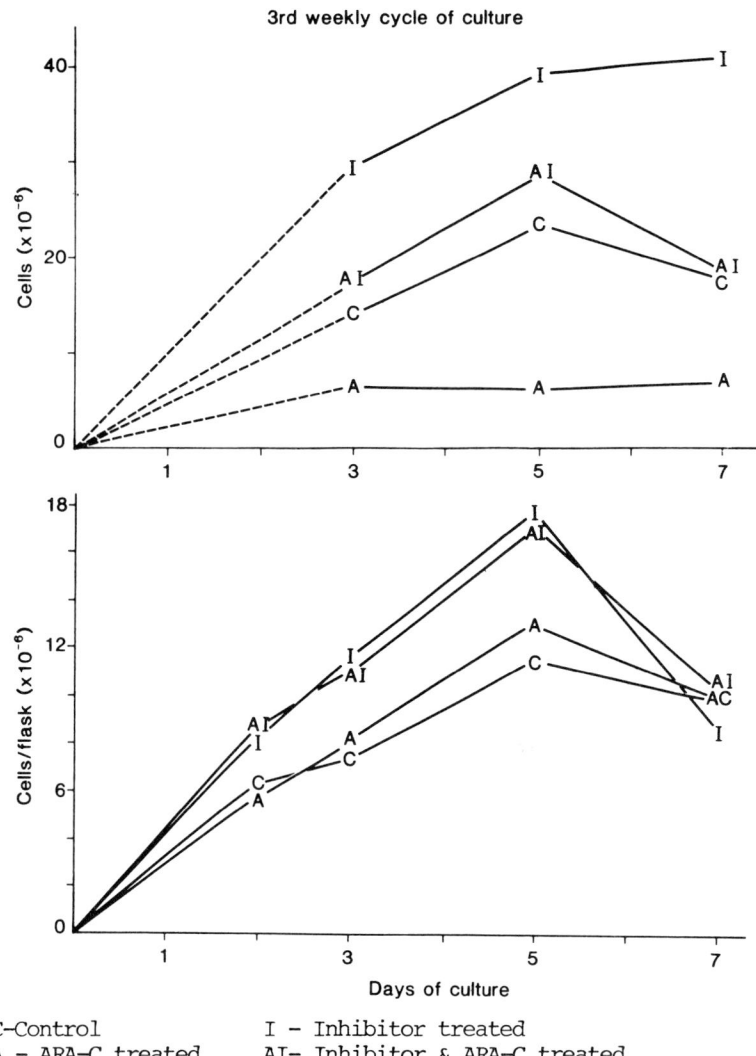

C - Control
A - ARA-C treated
I - Inhibitor treated
AI - Inhibitor & ARA-C treated

It has previously been demonstrated that this inhibitor can protect CFU-S in vivo from the cytotoxic effects of ^{3}HTdR hydroxyurea (Lord & Wright, 1980) and ARA-C (Lord - unpublished results). It is clearly effective also in this 'simulated' in vivo situation.

A problem which has occasionally arisen with haemopoietic regulatory factors is while factors of human origin are effective on mouse cells, the reverse is not always true. Since the inhibitor is not reliably effective against the CFC-Mix (they are the limit of its specificity - Tejero et al, 1984) a direct observation of this kind is not possible. Human long-term cultures, presumably containing the human equivalent of the CFU-S, however, are possible and we are currently carrying out these experiments to ascertain appropriate species specificities an potential in vivo applicability.

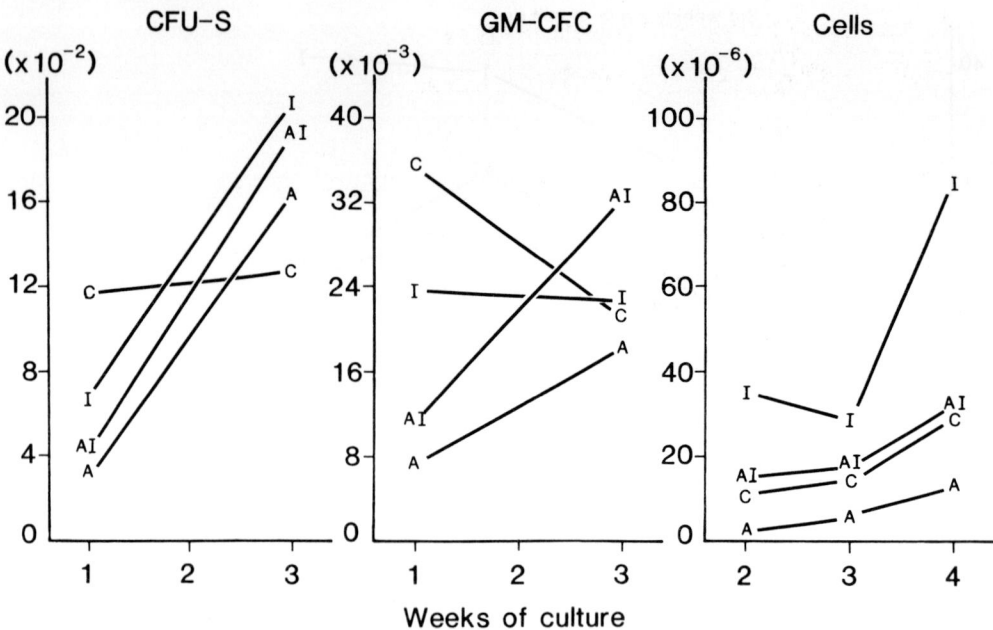

Fig. 4 CFU-S, GM-CFC & Cell Numbers in LTBMC treated with Inhibitor and/or Cytosine Arabinoside

D. THE CFC-S IN ITS MICROENVIRONMENT - INTERACTION WITH INHIBITOR AND STIMULATOR

Stem Cell Feedback Factor

The inhibitor appears to work by directing cells into the out-of-cycle, resting Go phase. Restimulation, however, is not effected simply by removing the inhibitor. Go CFU-S must be triggered into DNA-synthesis by a stimulator, i.e. an ON/OFF proliferation switch is operated. The stimulator, like the inhibitor, appears to be synthesised by macrophages - though different ones from the inhibitor-producers (Wright et al, 1982). Since the adherent layer of a long-term culture provides all the necessary ingredients it has been assumed that the inhibitor and stimulator producing cells are part of the microenvironment which regulates and maintains the integrity of the stem cell population.

In vivo, normal marrow knows to produce inhibitor, damaged marrow with a depleted CFU-S population knows to produce stimulator. The question then arose as to how did the marrow recognise what was needed? This must clearly be by a negative feedback signal from the CFU-S population telling the regulatory compartment its size. i.e. it presents a feedback factor which is proportional to the size of the population. This factor was found by sorting a 'pure' population of CFC-S (Lord & Spooncer, 1986) and using it to complement the depleted CFU-S population in phenylhydrazine-damaged bone marrow, or using an extract from it equivalent to the additional CFU-S required to complement its own population. Under these conditions, the production of stimulator by the phenylhydrazine-damaged marrow was switched over to production of inhibitor. By prior removal of the inhibitor-producing cells, it was further shown that the feedback signal acted by switching off stimulator production (Lord, 1986).

Thus, an appropriate feedback loop appears to act between the stem cell and its regulatory microenvironment.

Interaction with the immune system

The distributions of inhibitor and stimulator producing cells within the marrow are in accordance with the distribution of proliferative activity and the maturational age of the CFU-S population (Lord & Schofield, 1980; Lord & Wright, 1984) and, since it appears that the lymphoid populations are an early offshoot from the stem cell it is pertinent to refer briefly to some recent observations on the influence of inhibitor and stimulator on the immune system.

In a new publication by Kozlov et al (1987) it is demonstrated, using the method of adoptive transfer, that the addition of inhibitor to marrow containing proliferating CFU-S augments the generation of immune plaque-forming cells (PFC). The addition of stimulator to marrow containing non-proliferating CFU-S reduces the generation of PFC. Table 5 summarises their results.

Table 5. Generation of Plaque Forming Cells after Adoptive Transfer

Cells Injected	% CFU-S in DNA-synthesis	PFC
Normal bone marrow	<10	700 (350)[++]
Normal bone marrow + stimulator	35	200 (120)[++]
Testosterone propionate bone marrow	40	600
Testosterone propionate bone marrow + Inhibitor	0	1500
Lymph Node Cells	*	15000
Lymph Node Cells + Inhibitor	*	20000

* No CFU-S
+ Difference from lymph node cells not significant
++ Figures in parentheses - cells transferred without additional thymocytes
  (from Kozlov et al, 1987).

Conversely, erythroid differentiation was suppressed by inhibitor and augmented by stimulator. These results were tentatively discussed in the light of CFU-S population shifts, due to interaction with inhibitor or stimulator, to the relatively immature and mature ends of the CFU-S age spectrum. More work will obviously be necessary to elucidate the mechanism of this effect but modulation of the immune system in this way is clearly an important facet to be borne in mind and, possibly, exploited during future in vivo use of these factors.

ACKNOWLEDGEMENTS

This work was supported by a grant from the Cancer Research Campaign.

REFERENCES

Allen, T.D., & Dexter, T.M. (1983): Long-term bone marrow cultures: An ultrastructural review. Scan. Elect. Micros 4, 1851-1866.

Becker, A.J., McCulloch, E.A., Siminovitch, L. & Till, J.E. (1965): The effect of differing demands for blood cell production on DNA synthesis by hemopoietic colony forming cells of mice. Blood 26, 296-308.

Cashman, J., Eaves, A.C., & Eaves, C.J. (1985): Regulated proliferation of primitive hematopoietic progenitor cells in long-term human marrow cultures. Blood 66: 1002-1005.

Dexter, T.M., Spooncer, E., Varga, J., Allen, T.D. & Lanotte, M. (1983): Stromal cells and diffusible factors in the regulation of haemopoietic cell development in Haemop. Stem Cells Alfred Benzon Symp 18 (Ed S.A. Killman, E.P. Cronkite, C.N. Muller-Berat). Munksgaard, Copenhagen 303-322.

Eaves, A.C., Cashman, J.D., Gaboury, L.A., Kalonsek, D.K. & Eaves, C.J. (1986). Unregulated proliferation of primitive chronic myeloid leukaemia progenitors in the presence of normal marrow adherent cells. Proc Natl. Acad Sci USA. 83, 5306-5310. Frindel, E., Croizat, H. &

Vassort, F. (1976): Stimulating factors liberated by treated bone marrow: in vitro effect on CFU kinetics. Exp Hemat 4, 56-61.

Frindel, E., Guigon, M. (1977): Inhibition of CFU entry into cycle by a bone marrow extract. Exp. Hemat. 5, 74-76.

Gidali, J. & Lajtha, L.G. (1972): Regulation of haemopoietic stem cell turnover in partially irradiated mice. Cell Tissue Kinet 5, 147-159.

Guigon, M., Enouf, J. & Frindel, E. (1980): Effects of CFU-S inhibitors on murine bone marrow during ARA-C treatment-1. Effects on stem cells. Leuk Res. 4, 385-91.

Kozlov, V.A., Orlovskaya, I.A. & Tsyrlova, I.G. (1987): Haemopoietic stem cells with different proliferative activity in the formation of the immune response. Cell Tissue Kinet, in press.

Lord, B.I., Lajtha, L.G. & Gidali, J. (1974): Measurement of the kinetic status of bone marrow precusor cells: three cautionary tales. Cell Tissue Kinet, 7, 505-515.

Lord, B.I., Mori, K.J., Wright, E.G. & Lajtha, L.G. (1976): An inhibitor of stem cell proliferation in normal bone marrow. Brit. J. Haematol 34, 441-445.

Lord, B.I., Mori, K.J. & Wright, E.G. (1977): A stimulator of stem cell proliferation in regenerating bone marrow. Biomed Exp 27, 223-226.

Lord, B.I. & Schofield, R. (1980): Some observations on the kinetics of haemopoietic stem cells and their relationship to the spatial cellular organisation of the tissue. Lect. Notes Biomaths 38, 9-22.

Lord, B.I. & Wright, E.G. (1980): Sources of haemopoietic stem cell proliferation: stimulators and inhibitors. Blood Cells 6, 581-593.

Lord, B.I. & Wright, E.G. (1982): Interaction of inhibitor and stimulator in the regulation of CFU-S proliferation. Leuk Res. 6, 541-551.

Lord, B.I., Hendry, J.H., Keene, J.P., Hodgson, B.W., Xu, C.X., Rezvani, M. & Jordan, T.J. (1984): A comparison of low and high dose-rate radiation for recipient mice in spleen-colony studies. Cell & Tissue Kinet 17, 323-334.

Lord, B.I. & Wright, E.G. (1984): Spatial organisation of CFU-S proliferation regulators in the mouse femur. Leuk Res 8, 1073-1083.

Lord, B.I. (1986): Interactions of regulatory factors in the control of haemopoietic stem cell proliferation. In: Biological Regulation of Cell Proliferation (eds. R. Baserga, P. Foa, D. Metcalf, E.E. Polli) Serono Symposia Publications 34, 167-171.

Lord, B.I. & Spooncer, E. (1986): Isolation of haemopoietic spleen colony forming cells. Lymphokine Res 5, 59-72.

Nijhof, W. & Wierenga, P.K. (1984): Isolation of homopoietic pluripotent stem cells from spleens of thiamphenicol-pretreated mice. Exp. Cell Res. 155, 583-587.

Pojda, Z., Dexter, T.M. & Lord, B.I. (1987): Production of a multipotential cell (CFU-S) proliferation inhibitor by various populations of mouse and human macrophages. Submitted for publication.

Siminovitch, L., McCulloch, E.A. & Till, J.E. (1963): Distribution of colony-forming cells among spleen cells. J. Cell Comp. Physiol. 62, 327-336.

Simmons, P.J. & Lord, B.I. (1985). Enrichment of CFU-S proliferation inhibitor-producing cells based on their identification by the monoclonal antibody F4/80. J. Cell Sci 78, 117-131.

Spooncer, E. & Dexter, T.M. (1984): Long-term bone marrow cultures. Biblthca Haemat. 40, 366-383.

Spooncer, E., Heyworth, C.M., Dunn, A. & Dexter, T.M. (1986): Self renewal and differentiation of interleukin-3-dependent multipotent stem cells are modulated by stromal cells and serum factors. Differentiation 31, 111-118.

Tejero, C., Testa, N.G. & Lord, B.I. (1984): The cellular specificity of haemopoietic stem cell proliferation regulators. Brit. J. Canc. 50, 335-341.

Till, J.E. & McCulloch, E.A. (1961): A direct measurement of the radiation sensitivity of normal mouse bone marrow cells. Radiat. Res. 14, 213-222.

Töksoz, D., Dexter, T.M., Lord, B.I., Wright, E.G. & Lajtha, L.G. (1980): The regulation of hemopoiesis in long-term bone marrow cultures. II Stimulation and inhibition of stem cell proliferation. Blood 55, 931-936

Wright, E.G., Garland, J.M. & Lord, B.I. (1980): Specific inhibition of haemopoietic stem cell proliferation: characteristics of the inhibitor producing cells. Leuk. Res. 4, 537-545.

Wright, E.G., Ali, A.M. Riches, A.C. & Lord, B.I. (1982): Stimulation of haemopoietic stem cell proliferation: characteristics of the stimulator producing cells. Leuk Res. 6, (531-539).

Résumé

Les auteurs rapportent les études qu'ils ont réalisées avec un extrait de moelle adulte normale qui bloque l'entrée en cycle des cellules pluripotentes de la souris (CFU-S). Pour permettre une étude directe de l'inhibiteur les CFU-S ont été séparées par élutriation et gradient de densité (enrichissement à 70 %). Ces cellules, ainsi que plusieurs lignées cellules IL3 dépendantes, sont sensibles à l'inhibiteur de manière dose dépendante dans un test basé sur l'incorporation de thymidine tritiée. Une lignée leucémique qui peut se développer sans facteur de croissance n'est pas sensible à l'effet de l'inhibiteur.

Les macrophages, source de l'inhibiteur, quelque soit leur origine : rate, péritoine ou sang, produisent cet inhibiteur de manière variable, en fonction peut être de leur hétérogénéité fonctionnelle.

En cultures à long terme, l'inhibiteur protège les CFU-S contre les effets cytotoxiques de l'Ara-C. Des études comparables sur des cultures de cellules humaines devraient apporter des données intéressantes sur la spécificité de l'inhibiteur. Les auteurs montrent que la production de l'inhibiteur dépend de la taille de la population des CFU-S ; un facteur de retro-contrôle des CFU-S est nécessaire. Ils soulignent aussi la possibilité de stimuler par l'inhibiteur des CFU-S le système immun.

Biological properties of low molecular weight pluripotent stem cell (CFU-S) inhibitors

Martine Guigon

Laboratoire d'Hématologie, Faculté de Médecine St-Antoine, Paris, France

ABSTRACT

This paper is a review of the experimental data leading to the present knowledge of some biological properties of small peptidic inhibitors of pluripotent stem cell (CFU-S) proliferation. A partially purified material obtained from fetal calf bone marrow has been shown to inhibit CFU-S entry into DNA synthesis after irradiation or chemotherapy. It can protect CFU-S from the damage caused by repeated doses of Cytosine Arabinoside (Ara-C) and increase the proportion of mice surviving quasi-lethal protocols comprising successive high doses of Ara-C. The same increase in animal survival was obtained by the injection of the synthetic hemoregulatory peptide HP5b, known to be a strong inhibitor of myelopoiesis. The increase in animal survival appears to be due to marrow protection, since a similar protective effect was obtained by a bone marrow graft following the same Ara-C protocol. Besides its effect on CFU-S kinetics and on normal mouse resistance to the toxicity of high doses of Ara-C, the CFU-S inhibitor does not modify the growth of a murine solid tumor (EMT6), nor its response to Ara-C treatment. The CFU-S inhibitor is not specific for CFU-S, as it also inhibits GM-CFC growth and hepatocyte proliferation. Its species-specificity and its possible relation with other suppressive molecules are discussed. If all these results are confirmed by the use of the pure molecule, by the studies of other drugs and of other tumors, the CFU-S inhibitor could be a promising tool to protect bone marrow in patients during cancer treatment.

KEY WORDS

Inhibitor, CFU-S, Ara-C, EMT6 tumor, Marrow graft, Hemoregulatory peptide.

The data presented in this article resulted from a collaborative work. It was initiated by E. Frindel and carried out under her direction at Institut Gustave Roussy (INSERM U.66, U.250, Villejuif, France), with the participation of M. Lenfant (ICSN, CNRS, Gif/Yvette, France), J. Wdzieczak-Bakala (INSERM U.250, Villejuif, France) and J.Y. Mary (INSERM U.263, Paris, France). Selected experiments resulted from collaborations with H. Izumi-Hisha (Japan), W.R. Paukovits (Austria), O.D. Laerum (Norway), M-N. Lombard and C. Nadal (INSERM U.22, Orsay, France).

INTRODUCTION

The study of murine hemopoiesis has provided evidence of the existence in hemopoietic tissues, mainly the bone marrow, of a small compartment of pluripotent stem cells that, due to their properties of self renewal and differentiation, are at the origin of all hemopoietic cells. Despite numerous attempts to purify this population using sophisticated technologies (cell sorting, monoclonal antibodies,...), these cells cannot be morphologically recognized. The spleen colony assay devised by Till & McCulloch some twenty five years ago (1961) allows the study of colony forming units in the spleen (CFU-S), which were proven to derive from pluripotent stem cells. At present, it remains the only available technique for their study. Although the pluripotent nature of CFU-S in some particular experimental conditions (day 8 colonies) has been questioned and debated recently (Magli et al., 1982; Blackett et al., 1986), this technique has been widely used and has been helpful in providing essential information on CFU-S properties and kinetics.

The majority of CFU-S are quiescent in normal healthy mice. Various agents, such as X-rays, cytostatics or antigens (see review in Tubiana & Frindel, 1982) trigger them into cell cycle. For example, the administration of a dose of Cytosine Arabinoside (Ara-C)*, a S-phase specific agent, does not kill quiescent CFU-S, but provokes their recruitment into DNA synthesis (Frindel et al., 1978). The cycling CFU-S become vulnerable to further Ara-C injections and their destruction will prevent the recovery of hemopoietic cells, which may lead to irreversible aplasia. It would be therefore of major importance to maintain CFU-S in the quiescent state at the time of drug administration during a sequential protocol.

It was Frindel's pioneer idea that endogenous inhibitors of CFU-S recruitment into cycle could protect these cells during cancer treatment and therefore prevent the occurence of severe and long lasting aplasia, which is one of the main limiting factors in the use of potent cytostatics.

The purpose of this paper is 1) to review the biological properties of low molecular weight CFU-S inhibitors, pointing out their efficiency in protecting normal and tumor-bearing mice during cytostatic treatment and 2) to discuss their specificity and possible relationship with other hemopoietic inhibitors.

DO ENDOGENOUS CFU-S INHIBITOR(S) EXIST?

When this work started in 1972, few data had been published on hemopoietic inhibitors and they concerned the inhibition of progenitor and/or differentiated hemopoietic cells only (Rytömaa, 1967; Paukovits, 1971). These studies were based on the concept of "chalone" introduced by Bullough (1962), who assumed that tissue proliferation -at least epidermis- was under the control of a negative factor, tissue-specific, but devoid of species-specificity. This definition, which was later shown to be too restrictive, has been very useful in promoting research on inhibitory substances. Therefore, CFU-S inhibitors should be found in hemopoietic tissues, particularly in the bone marrow, even if it not the primary and only source. It appeared likely that, as it is the case for other regulatory molecules and hormones, the concentration in tissues would be very low. Isolation and identification would thus require very large amounts of starting material.

First attempts were done using fetal calf bone marrow, which is highly hemopoietic, not ossified and not too rich in lipid. Crude extracts were injected into mice, the CFU-S of which were triggered into DNA synthesis by whole body irradiation. The number of CFU-S (day 9 colonies) was determined by the spleen colony assay (Till & McCulloch, 1961) and their percentage in DNA synthesis was assessed by the thymidine suicide technique (Becker et al., 1965). The lack of reproducibility of

* Aracytine Upjohn

these techniques needed repetitive experiments and careful check of their significance (Znojil & Necas, 1987).

Fetal calf marrow was shown to contain a dialysable substance which inhibited CFU-S entry into DNA synthesis after irradiation (Frindel & Guigon, 1977). This was the first evidence of the existence of an inhibitor of CFU-S recruitment.

The purification was undertaken and since 1980, has been carried out by Lenfant and Wdzieczak-Bakala. They developed a large scale extraction procedure, which resulted in supplying great quantities of a partially purified inhibitory material from fetal calf marrow (Wdzieczak-Bakala et al., 1984) and from fetal calf liver (Guigon et al., 1984). Unless stated, the results presented in this paper were obtained with a partially purified factor isolated by ultrafiltration (<10 kDa) and gel chromatography on BioGel P-2 as previously published (Wdzieczak-Bakala et al., 1983). The dose of 4µg/mouse was found to have an inhibitory effect.

EFFECTS ON CFU-S KINETICS

The CFU-S inhibitor was shown to prevent CFU-S entry into DNA synthesis after Ara-C (Guigon & Frindel, 1978). These early results, obtained by in vitro assays using a crude inhibitor, were then confirmed by in vivo assays carried out with the partially purified material from fetal calf marrow (Wdzieczak-Bakala et al., 1983) and fetal liver (Guigon et al., 1984). As shown on figure 1, a single injection of 500 mg/kg bw of Ara-C triggers quiescent CFU-S into DNA synthesis 8 hours later. The administration of a partially purified CFU-S inhibitor obtained either from fetal calf marrow or liver (5 µg/mouse i.p.) given 6 hours after Ara-C injection, prevents CFU-S entry into DNA synthesis.

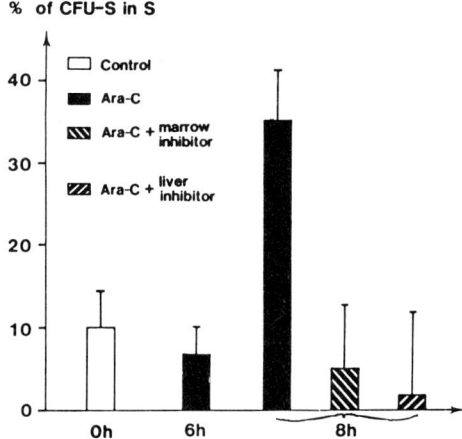

Fig.1: Effects of the administration of a partially purified CFU-S inhibitor obtained from fetal calf marrow or liver on CFU-S entry into DNA synthesis after a single dose of Ara-C (500 mg/kg bw).

No change was observed in the number of CFU-S, as previously reported (Guigon et al., 1984). Although these inhibitors strongly prevented CFU-S recruitment after Ara-C, they did not switch off already cycling CFU-S either in vivo or in vitro.

The inhibitor extracted from fetal calf marrow is also capable of protecting marrow CFU-S against the killing effect of repeated doses of Ara-C (Guigon et al., 1980). In these experiments, mice received 3 injections of 500 mg/kg bw of Ara-C at 8 hour-intervals, with or without a simultaneous injection of a few milligrams of a crude inhibitor. Eight hours later, the percentage of remaining CFU-S was of 11% after Ara-C alone and of 25% after Ara-C+the inhibitor ($p<0.001$). These findings suggested that this factor might be useful to protect marrow against the toxicity of S-phase specific agents. In view of possible clinical applications, it was of importance to know whether, by protecting CFU-S with the inhibitor, it was possible to increase the tolerance of animals to higher doses of cytostatics, which would be more efficient against malignant cells.

SURVIVAL EXPERIMENTS

It has been Frindel's idea to devise an assay using animal survival as endpoint. Groups of 10 specific-pathogen-free (SPF) mice of the same strain, sex, age and weight received repeated injections of high doses of Ara-C (500 to 1000 mg/kg bw/injection), which caused the death of a high proportion of the animals within 10 days. In the first series of experiments, an increase in animal survival was obtained through repeated injections of several milligrams of a rather crude inhibitor (Guigon et al., 1982). These results were confirmed by using the partially purified inhibitor from fetal calf marrow at a dose of 4µg/mouse/injection (Wdzieczak-Bakala et al., 1983). Moreover, the study of CFU-S kinetics in response to a sequential protocol of high doses of Ara-C permitted to optimize the timing of the inhibitor administration and to reduce the necessary amount to a single dose. Figure 2 shows the percentage of CFU-S in S phase at the time of each injection of 900 mg/kg bw of Ara-C. Based on the percentage in S phase, only a small number of CFU-S are killed by the two first injections of Ara-C. Conversely, at the time of the third and fourth drug injections, the majority of CFU-S are cycling and therefore sensitive to these Ara-C injections.

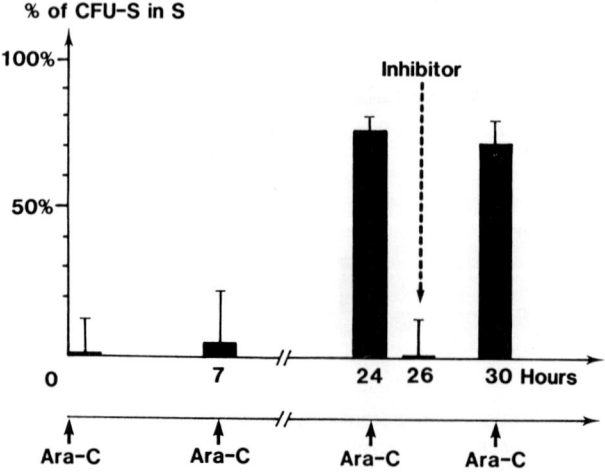

Fig. 2 : Percentage of CFU-S in S phase at different times during a protocol of Ara-C treatment (900 mg/kg bw/injection).

Thus, it was tested whether the protection of a few remaining CFU-S by maintaining them quiescent with the inhibitor was sufficient to allow animal survival. Consequently, the inhibitor was given between the two last drug injections. On table I, are presented results of experiments performed with different doses of Ara-C and/or inhibitor. Animal lethality depended on the dose of Ara-C, but in all cases, the administration of the inhibitor increased by a factor of about 2 the proportion of survivors. Furthermore, it appeared that a single dose of inhibitor was sufficient if given 4 hours before the last Ara-C injection, at which time almost no CFU-S were in S phase.

Table I : Percentage of mice surviving high doses of Ara-C without or with the administration of different doses of the partially purified CFU-S inhibitor from fetal calf marrow (4 µg/mouse). Mice received 3 injections of Ara-C at 0, 8 and 24 hrs and two successive injections of inhibitor at 6 and 22 hours or 4 injections at 0, 7, 24 and 30 hrs with injection of the inhibitor either at 26 and 28 hrs or at 26 hrs only.

ARA-C Dose/injection mg/kg bw	CFU-S INHIBITOR Number of injections	Number of experiments	PERCENTAGE (%) OF SURVIVING MICE		
			ARA-C	ARA-C+INHIBITOR	Significance
1500 x 3	2	3	15	45	p<0.01
625 x 4	2	4	40	70	p<0.01
625 x 4	1	4	37	70	p<0.01
900 x 4	1	4	25	50	p<0.02

These results were obtained with the partially purified inhibitor from fetal calf marrow. When assayed in similar protocols, at the same dose range, the partially purified inhibitor from fetal calf liver did not increase animal survival. This might be due either to the effect of unspecific components still present in the semi-purified material or to unappropriate dosage.

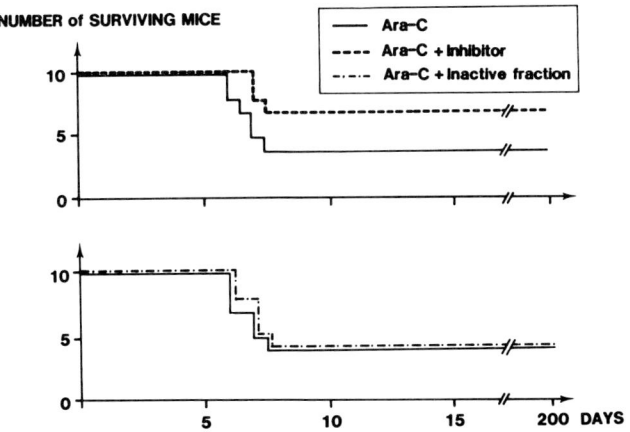

Fig. 3 : Upper curve : Number of surviving mice given 4 injections of high doses of Ara-C, without (——) or with (----) an injection of 4 µg/mouse of the CFU-S inhibitor 4 hours before the last Ara-C injection. Lower curve : Same protocol with the injection of a non inhibitory fraction (—.—.).

Figure 3 represents an example of survival curves. The majority of Ara-C treated mice died within 10 days, while a large number of mice given the inhibitor survived at day 10 and were still alive at day 200.

It is worth mentioning that in surviving mice after inhibitor administration, the number of CFU-S, which was drastically decreased soon after Ara-C treatment, returned to almost normal values after one month and remained unchanged for the ensuing six months (Table II). Therefore, it seems that, when given with the inhibitory factor, very high doses of Ara-C did not induce long lasting impairment of murine hemopoiesis (Guigon et al., in preparation).

Table II : Number of CFU-S per leg in mice surviving high doses of Ara-C and one injection of CFU-S inhibitor as compared to age-matched untreated controls (mean of 3-4 experiments).

	time after treatment		
	1 Day	1 Month	6 Months
ARA-C + INHIBITOR	5 ± 2.6	2970 ± 1729	3240 ± 825
CONTROL	2744 ± 186	2904 ± 1602	3239 ± 884

Of interest is the fact that another stem cell inhibitor, the hemoregulatory peptide, which was shown to be a strong inhibitor of granulomacrophagic colony (GM-CFC) growth in vitro (Paukovits & Laerum, 1982) and to decrease the number of GM-CFC and CFU-S in vivo (Laerum & Paukovits, 1984), also protected mice against high doses of Ara-C. When administered into mice treated with Ara-C (4 injections of 900 mg/kg bw), as shown on figure 2, the synthetic hemoregulatory peptide HP5b increased significantly and durably the proportion of surviving animals : Ara-C 24%, Ara-C + HP5b 51% $p< 0.02$ (Paukovits & al., 1986). This effect, resulting from the injection of 600 ng/mouse of the synthetic pentapeptide 4 hours before the last Ara-C injection, appeared to be similar to that obtained with the partially purified CFU-S inhibitor: Ara-C 25%, Ara-C + CFU-S inhibitor 50%, $p< 0.02$ (Table 1).

It is not yet clear whether the protection of mice caused by either hemopoietic stem cell inhibitor is due to its known effects on hemopoietic cells or/and to some effects at other levels. To determine if the increased animal survival could be due to marrow protection, mice receiving the same protocol of Ara-C were grafted with normal marrow 2 hours after the last drug injection: this resulted also in an increase in animal survival (Guigon et al., 1985). Moreover, as shown on Table III, the increase in animal survival was of the same order of magnitude (about 2) when Ara-C treated mice received either stem cell inhibitor or a bone marrow graft .

Although no direct comparison could be done between the effect of the administration of stem cell inhibitors and of the bone marrow graft due to difference in the protocols, the observations strongly suggest that the increase in animal survival is related to marrow protection.

Table III : Increase in Ara-C-treated mouse survival due to the administration of stem cell inhibitors or to bone marrow grafting (each line corresponds to the mean of 4 experiments). The CFU-S inhibitor (4 µg/mouse) or the synthetic hemoregulatory peptide HP5b (0.6 µg/mouse) were administered to Ara-C treated mice (4 doses of 900mg/kg bw), scheduled as described on fig.2. Normal marrow cells (10^7 cells) were grafted 2 hours after the last Ara-C injection.

	SURVIVAL (%) IN ARA-C TREATED GROUP (CONTROL)	INCREASE IN MOUSE SURVIVAL DUE TO TREATMENT AS COMPARED TO CONTROL	
		Mean	95 % Confidence Interval
CFU-S INHIBITOR	25	2.0	1.1 - 3.8
MARROW GRAFT	26	2.8	1.7 - 4.7
HEMOREGULATORY PEPTIDE	24	2.1	1.1 - 3.9
MARROW GRAFT		2.7	1.6 - 4.7

Therefore, at present, the protection of normal mice against Ara-C toxicity can be obtained by a single application of a small amount of either stem cell inhibitor: a partially purified low molecular weight factor isolated from fetal calf marrow (MW<1000) and a synthetic pentapeptide (p-Glu-Glu-Asp-Cys-Lys), which is an analog of the natural inhibitor of myelopoiesis associated with mature human granulocytes (Paukovits & Laerum, 1984). Besides, the protection induced by these molecules appears to be as effective as that resulting from marrow grafting. Nevertheless, neither treatment can save all animals. It might be due to non optimal timing and/or doses, or more likely to the toxicity of high doses of Ara-C for non-hematological tissues.

DO CFU-S INHIBITORS HAVE ANY EFFECT ON TUMOR GROWTH AND ON THE RESPONSE OF TUMOR CELLS TO CHEMOTHERAPY?

These points are of importance before envisaging clinical applications. The effect of the partially purified CFU-S inhibitor from fetal calf marrow was checked by Izumi-Hisha using a solid murine tumor, the EMT6 tumor (Rockwell et al., 1972). A single dose of the CFU-S inhibitor did not modify the tumor growth when injected to EMT6 tumor-bearing mice, which confirmed earlier results obtained in vitro with the crude extract (Guigon et al., 1981). When tumor-bearing mice received a similar protocol of Ara-C treatment as described for normal mice, an arrest of tumor growth was observed during the first week, indicating that this tumor was responsive to high doses of Ara-C. However, most animals died within a week. The administration of a single dose of inhibitor did not modify the response of the tumor to Ara-C and significantly increased the number of survivors : Ara-C 32%, Ara-C + inhibitor 57% $p< 0.05$ (Guigon et al., 1986a; 1986b). These results are encouraging, but they must be confirmed by the study of other solid tumors and leukemias as well, by using other cytostatics and the pure inhibitor.

CELL SPECIFICITY

The recent knowledge of the colony-stimulating factors has provided evidence that the concept of cell specificity must be reassessed (Metcalf et al., 1986). It is now well documented that the cell specificity of many regulatory molecules depends on several parameters, such as dose-range, assays, cellular environment. Besides, it must be emphasized that no valuable conclusion about the actual specificity of a regulator can be drawn until the pure molecule has been obtained and assayed on the pure target cells. Target cells could be different from producing cells and cells of unrelated organs could also be responsive. Such eventuality should be considered before envisaging human applications in order to avoid the occurence of harmful side-effects.

The relevance of some of these remarks can be illustrated by the following data, resulting from studies of the effects of CFU-S inhibitors on other hemopoietic and non hemopoietic cells.

- The CFU-S inhibitor has previously been reported to be rather specific for CFU-S, as it did not modify the number and the percentage in DNA synthesis of GM-CFC (Guigon et al., 1984). This resulted from in vivo experiments in which the partially purified inhibitors from fetal calf marrow and liver were administered to normal mice; GM-CFC proliferation was assessed 2 hours after injection of the inhibitor by the hydroxyurea suicide technique. But, when assayed in vitro by the preincubation technique used in the studies of the hemoregulatory peptide (Laerum & Paukovits, 1984), it significantly decreased the number of GM-CFC, within the dose-range active for CFU-S. These preliminary results must be confirmed and other progenitors studied as well. Nevertheless, such findings underline the fact that in some cases, the absence of observed effects should be interpreted with caution.

- When injected to normal baby rats, the CFU-S inhibitor was shown to inhibit hepatocyte G1-S transition, as reported by Lombard et al. in an accompagnying paper.

In view of these data, it appears that the CFU-S inhibitor, at least the partially purified substance isolated from fetal calf marrow which protected mice against Ara-C toxicity, cannot be considered as specific for CFU-S any longer. It is perhaps an inhibitor of cell recruitment, as suggested by its effect on hepatocyte kinetics; however, its effect on GM-CFC, which are a cycling population, is not in favor of this hypothesis. It is hoped that such questions will be answered in the future by using the pure molecule.

SPECIES SPECIFICITY

The fact that the inhibitory molecules extracted from fetal calf tissues acted on murine cells (mouse and rat) indicates that they are not species-specific. Effects on human cells are now being investigated, but these studies are up to now restricted to in vitro cultures of already committed stem cells. Hopefully, recent techniques of blast colony culture (Leary & Ogawa, 1987; Rowley et al., 1987) may allow the approach of more primitive stem cells. At present, preliminary results indicate that inhibitors of murine CFU-S are not cytotoxic for normal human blood and marrow progenitor cells.

RELATION WITH OTHER HEMOPOIETIC INHIBITORS

Some of the presented data indicate that the CFU-S inhibitor obtained from fetal calf marrow shared some properties with a synthetic pentapeptide, which is an analog of the myelopoietic inhibitor present in human granulocytes. Both molecules are small peptides, but it is not yet known whether they are chemically different. Conversely, it appeared to be different from CFU-S inhibitors reported by other groups (Lord et al., 1976; Wright et al.,1980a; Cork et al., 1981), which have a high molecular weight and which swith off cycling CFU-S. It might be that the differences between these inhibitors is due to their origin, one being extracted from fetal hemopoietic tissues, the others being released from adult tissues, probably macrophages (Wright et al., 1980b). However, no conclusion can be drawn until pure inhibitors are available.

The CFU-S inhibitor is not chemically related to other well defined molecules suppressing hemopoiesis, such as interferons, tumor-necrosis factor, tumor-growth factor, prostaglandin E2, lactoferrin... But it cannot be ecxluded that its inhibitory effect on CFU-S recruitment is mediated via a cascade of events involving the participation of one or several of these substances, directly and/or via accessory cells.

CONCLUSION

It appears of major importance now to answer the following questions : 1) does the pure CFU-S inhibitor keep the protective property as just reported with the partially purified material and can this effect be obtained when using other drugs and other tumors? 2) what is the effect of the pure CFU-S inhibitor on human normal and malignant hemopoietic and non hemopoietic cells? The eventual clinical applications of this factor will depend on the responses to these main points.

ACKNOWLEDGMENTS

This work was supported by Institut National de la Recherche et de la Santé Médicale, by Association pour la Recherche sur le Cancer and by Institut Gustave Roussy.

I want to express my gratitude to Pr. M. Tubiana (Institut Gustave Roussy, Villejuif) for his constant help and interest in this work, to Pr. C. Parmentier (Institut Gustave Roussy, Villejuif) and to Pr. A. Najman (Hopital St Antoine, Paris) who gave me the possibility to start studies on human material. I also thank C. Lacout, F. Lepesteur and M.F. Frey (INSERM U250) for their excellent technical assistance throughout this work and the scientists of this laboratory for fruitful discussions.

REFERENCES

Becker A.J., McCulloch E.A., Siminovitch L. & Till J.E. (1965) : The effect of differing demands for blood cell production on DNA synthesis by hemopoietic colony forming cells of mice. Blood 26, 296-308.

Blackett N., Necas E. & Frindel E. (1986) : Diversity of haemopoitic stem cell growth from a uniform population of cells. Nature Lond 322, 289-290.

Bullough W.S. (1962) : The control of mitotic activity in adult mammalian tissues. Biol.Rev. 37, 307-342.

Cork M., Anderson I., Thomas D.B. & Riches A. (1981): Regulation of the growth fraction of CFU-S by an inhibitor produced by bone marrow. Leukemia Res. 5, 101-105.

Frindel E. & Guigon M. (1977) : Inhibition of CFU entry into cycle by a bone marrow extract. Exp. Hematol. 5 , 74-76.

Frindel E., Guigon M., Dumenil D. , and Fache M-P. (1978) : Stimulating factors and cell recruitment in murine bone marrow stem cells and EMT6 tumors. Cell Tissue Kinet. 11, 393-403.

Guigon M., Enouf J., & Frindel E., (1980) : Effects of CFU-S inhibitors on murine bone marrow during Ara-C treatment. Leukemia Res. 4, 385-391.

Guigon M. & Frindel E. (1978): Inhibition of CFU-S entry into cycle after irradiation and drug treatment. Biomedecine 29, 176-178.

Guigon M. & Frindel E. (1981) : Inhibiteurs de la prolifération des cellules souches hématopoiétiques. Perspectives d'utilisation en chimiothérapie. Bull.Cancer 68 , 150-153.

Guigon M., Mary J.Y., Enouf J. & Frindel E.(1982) : Protection of mice against lethal doses of 1-ß--D-arabinofuranosylcytosine by pluripotent stem cell inhibitors. Cancer Res. 42, 638-642.

Guigon M., Wdzieczak-Bakala J., Mary J.Y., & Lenfant M., (1984) : A convenient source of CFU-S inhibitors : the fetal calf liver. Cell Tissue Kinet. 17, 49-54.

Guigon M., Wdzieczak-Bakala J., Lenfant M., Mary J.Y., & Frindel E., (1985) : Comparative effect of pluripotent stem cell (CFU-S) inhibitors and bone marrow grafting on mouse survival during lethal Cytosine Arabinoside treatment. Exp.Hematol.13, 337.

Guigon M., Wdzieczak-Bakala J., Izumi H., Lenfant M. & Frindel E.(1986a) : Low molecular weight hemopoietic pluripotent stem cell (CFU-S) inhibitors. In Biological regulation of cell proliferation, eds R. Baserga, P. Foa, D. Metcalf and E.E. Polli, pp 159-166. New-York : Raven Press.

Guigon M., Wdzieczak-Bakala J., Izumi H., Lenfant M. & Frindel E. (1986b) : Protection de la souris normale ou porteuse de tumeur par un inhibiteur des CFU-S au cours d'un protocole letal de chimiothérapie (Ara-C)(1986b), Bull.Cancer 73, 409.

Laerum O.D. & Paukovits W.R. (1984) : Inhibitory effects of a synthetic pentapeptide on hemopoietic stem cells in vitro and in vivo. Exp.Hematol. 12 , 7-17.

Leary A.G. & Ogawa M. (1987) : Blast cell colony assay for umbilical cord blood and adult bone marrow progenitors. Blood 69, 953-956.

Lombard M.N., Wdzieczak-Bakala J., Sotty D., Lenfant M., Nadal C. & Guigon M. (1987) : Effect on hepatocyte G1-S transition of an inhibitor of CFU-S entry into DNA synthesis : this book .

Lord B.I., Mori K.J., Wright E.G. & Lajtha L.G.(1976) : An inhibitor of stem cell proliferation in normal bone marrow. Brit.J. Haematol. 34, 441-445.

Magli M.C., Iscove N.N. & Odartchenko N. (1982) : Transient nature of early haemopoietic spleen colonies. Nature 295, 527 - 529.

Metcalf D., Burgess A.W., Johnson G.R., Nicola N.A., Nice E.C., Delamarter J., Thatcher D.R., & Mermod J.J. (1986) : In vitro actions on hemopoietic cells of recombinant murine GM-CSF purified after production in Escherichia coli : comparison with purified native GM-CSF. J. Cell.Physiol. 128, 421-431.

Paukovits W.R. (1971) : Control of granulocyte production : separation and chemical identification of a specific inhibitor (chalone). Cell Tissue Kinet. 4, 539-547.

Paukovits W.R. & Laerum O.D. (1982) : Isolation and synthesis of a hemoregulatory peptide. Z. Naturforsch. 37c, 1297-1300.

Paukovits W.R. & Laerum O.D. (1984) : Structural investigation on a peptide regulating hemopoiesis in vitro and in vivo.Hoppe Seyler's Z. Physiol. Chemie 365, 303-311.

Paukovits W.R., Laerum O.D. & Guigon M. (1986) : Isolation, characterization and synthesis of a chalone-like hemoregulatory peptide. In : Biological regulation of cell proliferation, eds R. Baserga, P. Foa, D.Metcalf and P.P. Polli, pp 111-120. New-York : Raven Press.

Rockwell S., Kallman R.F. & Fajardo L.F.(1972) : Characteristics of a serially transplanted mammary tumor and its tissue-culture adapted derivative. Nat. Cancer Inst. 49, 735-747.

Rowley S.D., Sharkis S.J., Hattenburg C. & Sensenbrenner L.L. (1987) : Culture from human bone marrow of blast progenitor cells with an extensive proliferative capacity. Blood 69, 804-808.

Rytömaa T. (1967) : Regulation system of blood cell production. In Control of cellular growth in the adult organisms. Teir ed, 106-138. New-York : Academic Press.

Till J.E. & Mc Culloch E.A. (1961) A direct measurement of the radiation sensitivity of normal mouse bone marrow cells. Radiat. Res.14, 213-222.

Tubiana M. & Frindel E. (1982) : Regulation of pluripotent stem cell proliferation and differentiation : the role of long-range humoral factors. J. Cell. Physiol. suppl.1, 13-21.

Wdzieczak-Bakala J., Guigon M., Lenfant M. & Frindel E. (1983) : Further purification of a CFU-S inhibitor: in vivo effects after Cytosine Arabinoside treatment. Biomedicine & Pharmacoth. 37,467-471.

Wdzieczak-Bakala J., Guigon M., & Lenfant M., (1984) : Purification and biochemical characterization of a CFU-S proliferation inhibitor : preliminary results. IRCS Med. Sci. Biochem.12, 868-869.

Wright E.G., Sheridan P., & Moore M.A. (1980a) : An inhibitor of murine stem cell proliferation produced by normal human bone marrow. Leukemia Res. 4, 309-314.

Wright E.G., Garland J.M. & Lord B.I. (1980b) : Specific inhibition of haemopoietic stem cell proliferation : characteristics of the inhibitor producing-cells. Leukemia Res. 4, 537-545.

Znojil V., & Necas E. (1987) : The spleen colony technique. I. Correction for the overlap effect and sources of errors in CFU-S determination. II. Errors in estimation of the fraction of CFU-S synthesizing DNA. Cell Tissue Kinet. in press.

Résumé

Cet article est une revue des données expérimentales qui ont conduit à la connaissance actuelle de certaines propriétés biologiques de petits peptides qui inhibent la prolifération des cellules souches pluripotentes hématopoïétiques (CFU-S). Il a été montré qu'une substance partiellement purifiée, extraite de moelle de foetus de veau, inhibe l'entrée des CFU-S en synthèse d'ADN après irradiation ou chimiothérapie. Cet inhibiteur protège les CFU-S au cours d'un traitement séquentiel par la Cytosine Arabinoside (Ara-C) et augmente la proportion de souris survivant à l'injection de doses quasi-létales de ce médicament. Un effet protecteur du même ordre a été obtenu par l'injection d'un peptide hémorégulateur synthétique (HP5b), qui est un inhibiteur puissant de la myélopoïèse. Bien qu'aucune preuve directe ne puisse en être donnée, il semble que l'augmentation du nombre de souris survivant à des doses quasi-létales d'Ara-C observée après administration de l'un de ces inhibiteurs soit liée à la protection de la moelle : en effet, un effect protecteur du même ordre a été obtenu par une greffe de moelle realisée après la fin du traitement par l'Ara-C. En outre, l'inhibiteur des CFU-S n'a pas d'effet sur la croissance d'une tumeur solide murine (tumeur EMT6), ni sur sa réponse au traitement par l'Ara-C; par contre, il augmente de façon significative la survie des souris porteuses de tumeur traitées par des fortes doses d'Ara-C. L'inhibiteur des CFU-S ne semble pas spécifique, puisque il inhibe aussi la croissance des progéniteurs granulomacrophagiques (GM-CFC) et la prolifération des hépatocytes normaux. Sa spécificité d'espèce ainsi que ses relations avec d'autres molécules suppressives de l'hématopoïèse sont discutées. Si l'ensemble de ces résultats sont confirmés en utilisant la molécule pure, d'autres médicaments et d'autres tumeurs, l'inhibiteur des CFU-S pourrait être un outil précieux pour protéger la moelle chez l'homme au cours de traitements anticancéreux.

Discussion

Chairpersons/*Modérateurs*: W.R. Paukovits (Autriche)
J. Breton-Gorius (France)

Discussion

M. AGLIETTA to B. LORD : What can you tell us about the chemical structure and the origin of your inhibitory molecule, its relationship with other inhibitory molecules I mean? Do you know anything about it?

B. LORD : In a word, nothing. We are presently making a serious attempt to solve this problem and I hope that in the near future, I will be able to give a better answer to that question.

P. BAINES : I didn't see very much data on primary tumor cells. You have read that the inhibitors once inhibited cycling of those sorts of populations. I didn't see much data on the effects of the inhibitors for all the speakers on primary tumor cells and I am wondering if the inhibitor won't in fact protect tumor populations.

B. LORD : No, I don't think we have got that sort of information yet. A primary tumor tissue is scarce ; inhibitor for large scale in vivo experiments is in short supply. You may be right, but you will recall that the stem cell line that spontaneously became leukemic was insensitive to inhibitor. If a leukemic tumor population is insensitive, I remain optimistic that tumor populations, particularly of other tissues, will also be insensitive.

M. GUIGON : We have already shown that the marrow extracted low molecular weight CFU-S inhibitor had no effect on the growth on a solid murine tumor, the EMT6 tumor and that it did not modify the response of tumoral cells to Ara-C. We are now carrying out experiments on other types of tumors and on leukemias, and especially on human malignancies.

N.C. GORIN : My question is for Dr Guigon. It is somewhat related to the previous intervenant. I mean you suggested that using this inhibitor could do as well as a transplant, which is really something aggressive and could be very interesting in the future. And also you may use a stimulating agent after chemotherapy. So you know, we could skip transplantation but then, questions arise. I have two questions : the first one is the same one : what about the tumor? and you said that you are starting studying the tumor but of course we would be very much interested in tumors with low proportion of cycling cells, I guess. And what do you know about the L1210, I mean the leukemia model? Do you think that this inhibitor would also put brakes on the tumor cell proliferation? That is the first question. And the second question would be about the antigenicity of these molecules. If we are going to try to use that in human, then, I guess, we may have problems.

M. GUIGON : I think I have already answered the first part of your question. Of course, we will do experiments on other tumors, including leukemias. Experiments on the L1210 leukemia are already in progress.

M. LENFANT : We have some negative results which suggest that the homogenous molecule is not antigenic. In Dr Frindel's group, Dr Miyanomae tried to raise antibody against the inhibitory factor and so far he has not succeeded.

M. GUIGON : I may add something on this point. I don't think that the molecule, at least the partially purified and probably this will be also true for the pure molecule, has any antigenic effect because, as you know, when you inject some antigen into mice, you will get an increased proliferation of CFU-S and we just have observed the reverse. When injected to normal mice, this inhibitor did not stimulate CFU-S proliferation although these cells are very sensitive to antigenic stimulation.

A.C. RICHES : Dr Guigon, is your inhibitor cell population specific? Does it act only on the CFU-S population or does it act on other progenitor populations?

M. GUIGON : My last slide showed that this molecule has also an inhibitory effect on GM-CFC growth. We have not enough results on BFU-E, but these studies are now in progress. Anyway I think that it is better to use a pure molecule to answer the question of specificity and not only a pure molecule, but also purified progenitor populations. This is one of our purposes. And also, I have forgotten to mention it, but this is presented in the poster of Marie-Noelle Lombard and colleagues, this inhibitor also affects normal hepatocyte proliferation. So its specificity should be checked now very carefully.

M. LENFANT : Maybe it will be a matter of doses which might be active more precisely on CFU-S or on other progenitors or on hepatocytes. Actually we haven't carried out all these experiments with pure compound and dose effects are not available.

M. GUIGON : May I add something to answer your question, Dr Riches. I think that we are maybe facing the same problem with the inhibitor as with the CSFs, that Dr Metcalf pointed out yesterday : it means that according to the dose, the concentration, the environment..., we can have a specific effect, an unspecific effect and perhaps a non-physiological effect on other progenitors or other cells.

E. FRINDEL : I would like to propose another hypothesis about the specificity of the inhibitor. Maybe it is not specific for a special tissue or a cell, but specific for a certain phase of the cell cycle. We have some preliminary experiments to show that, at least for the CFU-S, the cell has to be in Go to respond to the inhibitor. Even if it is in G1, when it is still not in DNA synthesis, it can no longer be inhibited. So, it is possible that once the mitotic machinery is triggered, it is irreversible. Very few tissues are normally in Go and capable of entering the cell cycle under various conditions. Therefore we were perhaps misled into thinking that the inhibitor is tissue specific.

M. AGLIETTA : May I ask you something I didn't understand well. So you mean that your inhibitor is just keeping cells in Go. As soon as they are entered in G1 phase, that is the reverse part. What about growth factors? Can you counteract the action of inhibitors with specific growth factors? Do you have any data, for example with IL-3?

E. FRINDEL : I have no data to show that the inhibitory effect can be reversed by any specific growth factor, but perhaps, this is what happens when cells enter into cycle.

B. LORD : Could I just add the same point? This is where are the major differences between inhibitors I have talked about and those Dr Frindel's group has talked about. We are not talking specifically about preventing recruitment. We are talking about switching cells off that are already proliferating. In this situation we cannot do what you are asking with the stimulators. We have previously reported that we can switch CFU-S off with inhibitors and switch them back on with stimulators. These are active changes in the cell cycle status, as least as far as is measured by thymidine suicide experiments. It is not specifically a restriction of triggering into cell cycle.

M. LENFANT : Actually I am evaluating in Dr Mori's laboratory (Kyoto University) the activity of the inhibitor on Interleukin-3 dependent cell lines and mast cell colony formation. Results are too preliminary to be reported.

W. PAUKOVITS : I have a question to Dr Lord. Your inhibitor comes from macrophages exclusively?

B. LORD : No, I can't say that specifically, because we have not looked at other populations in detail. However, all the evidence that we have got for this is that any cell population we have seen to produce inhibitor carries many of the characteristics of a macrophage population. Finally, we used a macrophage-specific monoclonal antibody, F4/80 and fluorescence activated cell sorting to select a macrophage population which, after several weeks of culture, was still synthesizing inhibitor. I don't know whether Dr Wright has any more information on this point.

E. WRIGHT : The early experiment which was published a few years ago was essentially based upon density adherence. But more recently, and this is now in press, we have actually been able to grow macrophages using CSF-1, IL-3 and various other growth factors and to show that the cultured macrophages can be taken. Using such a protocol, Dr Lord has been describing, we can use these cultured macrophages to produce both inhibitors and stimulators which have got all the characteristics of material obtained either from bone marrow or from the cell separated macrophages.

J. BRETON-GORIUS : On the same subject, did you have the opportunity to test some monocytic human cell lines?

B. LORD : I think it is possible that some of these cell lines may produce inhibitor. We are actually looking at a number of activated macrophage lines. I did the experiments last week, but these were provided from someone else completely blind, so I don't know whether my answers were correct or not. It is in process.

W. PAUKOVITS : Dr Lenfant, on the first reverse column you need 60% of methanol to get it off, or maybe 50%, anyway a rather high percentage. On the second reverse phase column, it comes off with 4.5% acetonitrile. Of course methanol and acetonitrile are not the same, but the difference of the percentages is very large. How do you explain that?

M. LENFANT : Semi preparative purification was carried out without addition of counter-ions, whereas 0,1% trifluoroacetic acid was added to the Methanol/Water elution solvent at the analytical separation step.

D. ROODMAN : Macrophages can be activated or they can be at resting stage : do you have any information whether your inhibitors are made by the activated fraction of macrophages or by resting macrophages?

B. LORD : It is a question that we cannot specifically answer directly. The situation is that, no matter what the status of the cycling of CFU-S in a particular tissue is, there are present enough macrophages that are capable of producing an inhibitor and there are enough macrophages which are capable of producing a stimulator. They don't do it at the same time. But, on the other hand, if you do a density separation on the density gradient, you can separate two distinct density fractions, one of which, on its own, will produce inhibitor, the other one, on its own, will produce stimulator and that is irrespective of whether it is from normal bone marrow which is normally producing inhibitor or whether it is from regenerating marrow which is normally producing stimulator. Fetal liver which is normally producing stimulator is the same. So if it happens that activation brings about a physical change in the density of the cells, then maybe that is the case. But at the same time you have also got to say OK, the high density fraction is perhaps the activated form and is producing stimulator. In normal marrow, it is there, but it is not producing stimulator, although it is activated and vice versa for the inhibitor in regenerating marrow or fetal liver. I think probably they are separate populations and that it is not a straight question of activation or inactivation of the population. That is, I think, the best answer I can give at this stage.

A.C. RICHES : Talking about macrophages, one also has to realize that, within an activated macrophage cell population, there are subpopulations which can perform specific functions. So although we talk about activated or inflammatory macrophages and resident macrophages, there are subpopulations within those two levels which can perform specific functions.

B. LORD : Macrophages represent an extremly complicated set of cells, very heterogeneous. It has many functions and produces many factors. It is going to be very difficult to untangle all the possible functions of the macrophage population.

E. FRINDEL : I would like to suggest some other food for thought. We have always talked about the direct effect of inhibitors on the target cells, which in our case are CFU-S. What if the effect of the inhibitor was to inhibit the secretion of stimulators and that one of the main target cells would be the cell secreting the stimulating factors. We have some very preliminary data which are in favour of this possibility.

B. LORD : I think that by the fact that we have been able to sort out pure populations of stem cells, we can conclude that the effect is direct. Whether it is by FACS sorting, when you need to handle cells a little bit more and stimulate them, before assaying for response to inhibitor or whether it is by centrifugal elutriation when the CFU-S are truly proliferating, we have seen a direct effect on them with inhibitor. With regard to stimulator, a variety of agents obviously will stimulate the same CFU-S population. A number of years ago, we studied CFU-S stimulation by testosterone hemisuccinate. If we separated out again by density the stimulated or producing cell population cuts, we were then unable to stimulate the stem cells with the testosterone hemisuccinate. So my tentative conclusion, at least from that, was that the stimulation was probably taking place directly by the stimulator producing cells, while the testosterone hemisuccinate acted via the stimulator-producing cells. Yes, I like to believe that we are talking about direct effects on the stem cells.

A. AXELRAD : Just a small point of clarification, please. My understanding is that the Manchester inhibitor is strictly specific to CFU-S. The Villejuif inhibitor can effect GM-CFC as well as CFU-S. Is that correct?

B. LORD : With the Manchester inhibitor, we have, I believe, shown that it is completely specific for CFU-S. We have found no effect on GM-CFC, no effect on BFUE, early BFU-E or later ones. The mixed colony formers, there is a slight ambivalance about that population. I think it is right on the edge of sensitivity and right on the edge of the CFU-S population. So I would say yes, it is restricted to CFU-S.

M. GUIGON : I don't agree with Brian Lord, because, I think, we cannot say that a molecule is specific as long we have not assayed it on many tissues. Have you looked at the effect of your inhibitor for example, on liver cells, kidney cells or other non-hematopoietic cells? Because one point is the specificity for the hematopoietic tissue and the other point is the specificity for different organs. If you want to do in vivo experiments, you have to bear in mind that a mouse is a complex organism and that a human person is even more complicated, so you have to check maybe more carefully the effect on other tissues, normal and tumoral as well. To answer the second part of your question, Dr Axelrad, the French group has shown that the inhibitor they are studying cannot be considered as cell specific.

M. LENFANT : I appreciate your answer. I want to ask you if you have studied these effects at the same dose, on the various cell lines, the various progenitors, same dose of inhibitors, same concentration. It is very interesting because I am very interested in the specificity and dose response curve.

B. LORD : Yes, I am just going to add to this point. We always tended in looking at the specificity of inhibitors in hematopoietic tissue. Quite to say that the more closely related the cell population, the better chance we have to define specificity. It is better to define specificity within the tissue itself rather than against unrelated cells. It would be nice, however, to think that perhaps we were dealing with a stem cell inhibitor rather than specifically a hematopoeitic stem cell inhibitor. I have tried to look at this in the gut stem cells. In the crypt, I don't know whether I say this, but I thought there might have been a little effect there as well. We should think of it a little more, it is a possibility.

M.N. LOMBARD : Just a small point. How hopeful and how far are we on the way of labelling these factors?

M. LENFANT : The synthesis of a labelled molecule will be undertaken as soon as the structure of the inhibitory molecule is completely established.

Brief reports

Communications brèves

Importance of the measurement error in demonstrating inhibition of stem cell proliferation

Emanuel Nečas and Vladimír Znojil

Faculty of Medicine, Charles University, 128 53 Prague, Czechoslovakia

Key words: stem cell, CFU-S, S-phase, statistics

INTRODUCTION

The proliferation rate of stem cells detected by the spleen colony technique (CFU-S) is measured by calculating the fraction of CFU-S synthesizing DNA (CFU-S in S-phase) using the ratio of CFU-S numbers from two samples (Becker et al., 1965). The CFU-S assay should be strictly linear and CFU-S numbers should be determined with a negligible error to obtain correct values of the fraction of CFU-S in S-phase. Neither of these two assumptions are fulfilled. Therefore, a correction procedure for the non-linearity of the CFU-S assay has been developed. Furthermore, average errors can be calculated both for CFU-S numbers and the fraction of CFU-S in S-phase. The measurement error which should be expected in experiments designed to measure the CFU-S proliferation rate has been derived.

MATERIAL AND METHODS

A large and heterogeneous data set was used for statistical analysis. Several hunderds of measurements were performed on three inbred strains of mice (CBA/Ca, C57Bl/10, DBA 2/J), on F_1 hybrids of two strains (CBF_1), and on ICR randombred mice.

Specialized measurements were performed to complete the analysis where necessary. Linearity of the CFU-S assay was tested by injecting five to six groups of lethally irradiated mice with increasing numbers of bone marrow cells from normal donors (1.0×10^4 to 2.4×10^5 cells per a mouse). Curves were plotted through the results by the least square fit. The difference of these curves from that corresponding to the function derived by Armitage (1946) for the number of overlapping objects was statistically tested.

The statistical homogeneity of the measurements was studied by means of the distribution function of the normalized variances of the CFU-S numbers, $\alpha^2 C/C$, where C is the mean CFU-S number and $\alpha^2 C$ its variance. The distribution function should be the modified $\tilde{\chi}^2$ distribution (after the linear transformation of the x axis) with degrees of freedom equal to the number of mice used for CFU-S determination minus one. The correspondence of these two distribution functions was tested.

RESULTS

Eight separate determinations of linearity of the CFU-S assay were performed. Four persons counted the spleen colonies independently. The curves plotted through these data did not reveal significant differences when compared to the theoretical function derived by Armitage (1946). Consequently, this function can be used to correct spleen colony counts for colony overlap. The correct number of spleen colonies N is

$$N = M \ln(M/(M - C)) - P \qquad (1)$$

where C is the mean colony number from a group of mice, P are the endogenous colonies (background of the assay), and M is the maximum distinguishable colony number (Znojil and Nečas, in press a). The M value is calculated by fitting the curve corresponding to the equation of Armitage to the colony numbers obtained in measurements of the CFU-S assay linearity.

The distribution of the normalized variances of CFU-S numbers ($\alpha^2 C/C$) was not significantly different from the $\tilde{\chi}^2$ distribution. Therefore, the average variance of the CFU-S numbers could be defined as

$$\alpha^2 C = ((M-C)/M)^2 (N + \alpha^2 P + N^2 (\alpha K/K)^2 + (\exp((P+N)/M) - (M+P+N)/M)^2 \alpha^2 M) \qquad (2)$$

where the meaning of C, N, P and M is the same as in equation (1), and K is frequency of CFU-S among nucleated cells injected to the lethally irradiated mouse recipients, i.e. the product of the colony forming cells (CFC-S) and the fraction seeding the spleen (the f-factor). The $\alpha^2 P$, $\alpha^2 M$ and $(\alpha K/K)^2$ are variances of the P, M and K values (Znojil and Nečas, in press a).

The equations (1) and (2) were used to define the ratio of two CFU-S numbers, R, and its standard error $\tilde{\delta}R$. The ratio is calculated when the proliferation rate of the CFU-S is estimated from the fraction of CFU-S in S-phase. The functions derived were tested on results from repeated measurements where R was fixed at a certain known value by an appropriate dilution of the bone marrow cell suspension in one of the two samples compared (Znojil and Nečas, in press b).

Figure 1 shows size of the standard error, $\tilde{\delta}R$, for the case that groups of 10 mice are used for CFU-S determination. The size is shown for various mean spleen colony numbers (in the control sample), C, and for two extreme cases which can occur in the experiments studying inhibition of the CFU-S proliferation, i.e. for the fraction

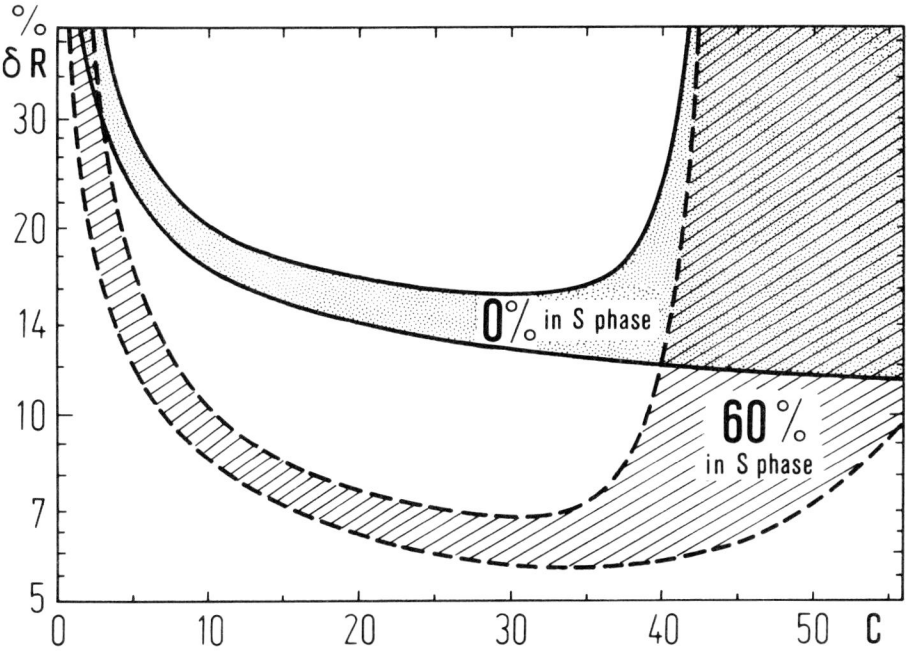

Fig. 1 The standard error (δR) of the fraction of CFU-S in S-phase. The range obtained for three inbred mouse strains and CBF mice is shown. C is the mean spleen colony count in the control group.

in S-phase either 0 or 60 per cent.

DISCUSSION

The proliferation rate of CFU-S can be underestimated due to non-linearity of the CFU-S assay. The correction procedure which has been derived should, therefore, be used in studies of CFU-S proliferation inhibition.

The measurement error (coefficient of variation) is relatively large in the CFU-S assay. Furthermore, its size fluctuates due to relatively small sampling size represented by groups of about ten mice. This decreases quality of the results considerably. The equation enabling to calculate the average error of a single measurement brings a further improvement to this widely used method.

The logarithmically transformed values of the CFU-S numbers and of the CFU-S ratio should be used when summing-up results from several independent experiments. Their distribution is closer to the normal distribution than that of non-transformed original values.

CONCLUSION

More correct and more reliable data can be obtained from experiments based on the spleen colony technique by the procedures briefly described here. This is of a special importance in studies where inhibition of the CFU-S proliferation is investigated.

REFERENCES

Armitage, P. (1946): An overlap problem arising in particle counting. Biometrica 36, 257-269.
Becker, A.J., McCulloch, E.A., Siminovitch, L., Till, J.E. (1965): The effect of differing demands for blood cell production on DNA synthesis by hemopoietic colony-forming cells. Blood 26, 269-308.
Znojil, V., Nečas, E. (in press a): The spleen colony technique I: correction for the overlap effect and sources of error in CFU-s determination. Cell Tissue Kinet.
Znojil, V., Nečas, E. (in press b): The spleen colony technique II: errors in estimation of the fraction of CFU-s synthesizing DNA. Cell Tissue Kinet.

The effect of a CFU-S proliferation inhibitor on stimulator production *in vitro* by human foetal liver cells

Andrew Riches and Michael Cork

Department of Anatomy and Experimental Pathology, University of St-Andrews, St-Andrews, Scotland

KEYWORDS

Human fetal liver, proliferation regulation, CFU-S, stem cell regulation.

INTRODUCTION

The proliferation of haematopoietic stem cells appears to be regulated at the local level by the production of factors which switch stem cells into cell cycle or out of cell cycle. The CFU-S proliferation stimulator is found in the 30-50K dalton fraction separated by Amicon filtration of supernatants produced from cell suspensions of haematopoietic tissues containing a high proportion of CFU-S in cell cycle (e.g. fetal liver, regenerating bone marrow). Whereas the CFU-S proliferation inhibitor is found in the 50-100K dalton fraction separated by Amicon filtration of supernatants produced from cell suspensions of haematopoietic tissues containing a low proportion of CFU-S in cell cycle (e.g. normal adult bone marrow). The local concentrations of these factors varies depending on the kinetic state of the CFU-S population. The production of these regulatory activities by human haematopoietic tissues has been assayed using murine CFU-S. Normal adult human bone marrow produces a CFU-S proliferation inhibitor (Wright et al, 1980) whereas human fetal liver cells produce a CFU-S proliferation stimulator (Cork et al, 1982; 1986). The production of these regulators in vitro by murine haematopoietic tissues is influenced by the presence of the opposing acting regulator. Thus in the presence of stimulator the inhibitor is not produced and in the presence of inhibitor the stimulator is not produced (Lord & Wright, 1982). The effects of the CFU-S proliferation inhibitor on the production of CFU-S and GM-CFC proliferation stimulators produced by human fetal liver cells in vitro will be studied.

MATERIALS AND METHODS

Human fetal tissue was obtained following therapeutic prostaglandin inductions. Cell suspensions were prepared in Fischer's medium plus 15% horse serum, incubated for 2 hours and cells removed by centrifugation. The cells were resuspended in fresh medium and divided into three aliquots. To the first suspension was added 100 μgm/ml of the 50-100K dalton fraction from normal murine bone marrow (Lord, 1983). To the second suspension 100 μgm/ml of the 50-100K dalton fraction from regenerating murine bone marrow (3 days post 4.5 Gy X-irradiation) and the third acted as a control. The cells were incubated for 5 hours and all the supernatants were passed through an Amicon Diaflo membrane (50K cut off; XM50) to remove the 50-100K activity and allow the stimulators (30-50K) to pass through. The CFU-S and GM-CFC stimulator activities were then measured using normal bone marrow CFU-S or quiescent normal bone marrow GM-CFC (Cork et al, 1982; 1986). The response was monitored by measuring the % of CFU-S or GM-CFC in S phase following incubation.

RESULTS AND DISCUSSION

The CFU-S proliferation stimulator is removed following washing of human fetal liver cells. After incubation of the cells for 5 hours in vitro, the activity can be recovered again after Amicon filtration (Fig. 1; Table 1A).

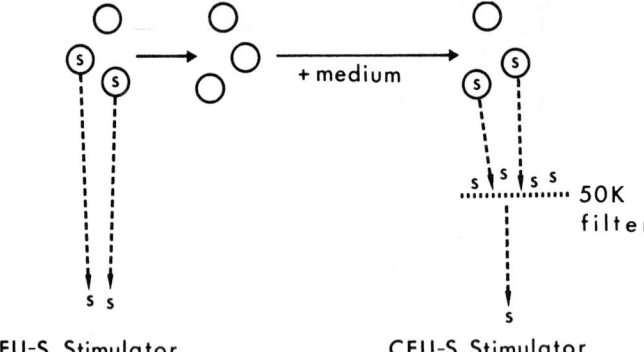

Fig. 1. Production of the CFU-S proliferation stimulator from human fetal liver cell suspensions following 5 hours incubation in vitro.

If the washed cells are incubated in the presence of 100 µgm/ml of the 50-100K dalton fraction from normal bone marrow (CFU-S proliferation inhibitor), then following Amicon filtration to remove the inhibitor, no CFU-S proliferation stimulator activity (30-50K fraction) is produced (Fig. 2; Table 1A).

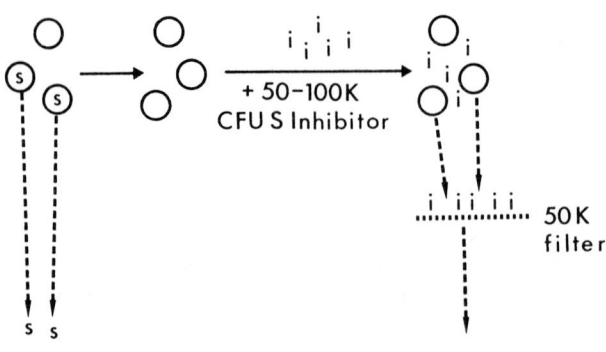

Fig. 2. Production of the CFU-S proliferation stimulator from human fetal liver cell suspensions incubated in the presence of the 50-100K dalton fraction from normal murine bone marrow (CFU-S proliferation inhibitor).

Similarly if the washed cells are incubated in the presence of 100 µgm/ml of the 50-100K dalton fraction from regenerating bone marrow (non-specific control), then following Amicon filtration to remove the 50-100K fraction, the CFU-S stimulator activity is produced (Fig. 3; Table 1A).

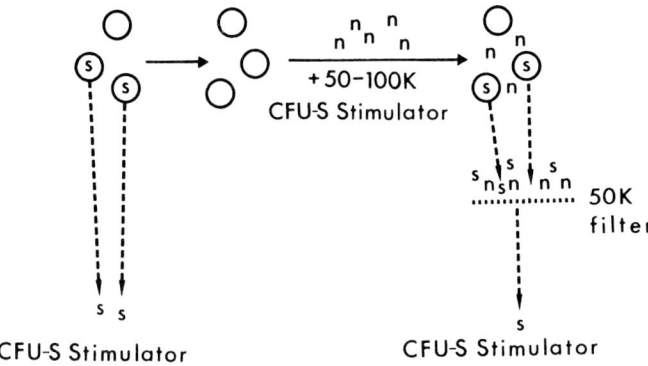

Fig. 3. Production of the CFU-S proliferation stimulator from human fetal liver cell suspensions incubated in the presence of the 50-100K dalton fraction from regenerating murine bone marrow (non-specific control).

When the different supernatants were tested for activity on quiescent GM-CFC from normal murine bone marrow, in all cases they induced proliferation (Table 1B).

Table 1. The effect of a CFU-S proliferation inhibitor on the production of proliferation regulators by human fetal liver cells in vitro.

(A) CFU-S proliferation stimulator

	% CFU-S in S phase
NBM + medium	3.0 ± 0.9
NBM + HFL (control)	29.3 ± 0.9
NBM + HFL (50-100K Inhibitor)	1.0 ± 3.9
NBM + HFL (50-100K Stimulator)	28.9 ± 2.9

(B) GM-CFC proliferation stimulator

	% GM-CFC in S phase
NBM + medium	56.6 ± 2.6
NBM + HFL (GMI)	1.4 ± 0.7
NBM + HFL (GMI) + HFL (control)	54.0 ± 4.0
NBM + HFL (GMI) + HFL (50-100K Inhibitor)	53.1 ± 2.0
NBM + HFL (GMI) + HFL (50-100K Stimulator)	53.9 ± 2.7

NBM normal murine bone marrow cells
HFL (control) supernatant from a 15 week gestational age human fetal liver suspension containing a CFU-S and a GM-CFC proliferation stimulator.
HFL (50-100K Inhibitor) supernatant from a 15 week gestational age human fetal liver cell suspension following incubation with the 50-100K dalton fraction from normal bone marrow (CFU-S proliferation inhibitor)
HFL (50-100K Stimulator) supernatant from a 15 week gestational age human fetal liver cell suspension following incubation with the 50-100K dalton fraction from regenerating bone marrow (non-specific control; contains no detectable inhibitor)
HFL (GMI) supernatant from a 13 week gestational age human fetal liver cell suspension containing a GM-CFC proliferation inhibitor (switches GM-CFC into a quiescent state; Cork et al, 1982)

These results support the view that the 30-50K CFU-S stimulator and 50-100K CFU-S inhibitor have a role in regulating CFU-S proliferation. Further these factors also seem to interact in the regulation of the production of these factors by the producer cells. The CFU-S inhibitor seems to regulate stimulator production by human fetal liver cells in a similar way to that described using murine producer cells (Lord & Wright, 1982). Not only are the CFU-S proliferation regulators non-species specific in their target activity i.e. human regulators act on murine CFU-S but also in their regulatory activity of factor production i.e. murine regulators act on human producer cells.

CONCLUSIONS

Cell suspensions of human fetal liver cells produce factors in vitro which switch murine CFU-S and GM-CFC into cycle. In the presence of the 50-100K dalton fraction from normal murine bone marrow (CFU-S proliferation inhibitor) the production of the 30-50K dalton CFU-S proliferation stimulator is suppressed but not the activity stimulating GM-CFC proliferation. In the presence of the 50-100K dalton fraction from regenerating murine bone marrow (contains no detectable CFU-S inhibitor) all the factors are produced. There thus appears to be a specific regulatory role for the CFU-S proliferation inhibitor on the production of the CFU-S proliferation stimulator by human fetal liver cells in vitro. (Supported by the Scottish Home and Health Department.)

REFERENCES

Cork, M.J., Wright, E.G. & Riches, A.C. (1982): Regulation of murine granulocyte-macrophage progenitor cells and haemopoietic stem cell proliferation by factors produced in human fetal liver. Leukaemia Research 6, 553-565.

Cork, M.J., Riches, A.C. & Wright, E.G. (1986): A stimulator of murine haemopoietic stem cell proliferation produced by human fetal liver cells. British Journal of Haematology 63, 775-783.

Lord, B.I. & Wright, E.G. (1982): Interaction of inhibitor and stimulator in the regulation of CFU-S proliferation Leukaemia Research 6, 541-551.

Lord, B.I. (1983): Haemopoietic stem cells. In Stem Cells: Their Identification and Characterisation, ed. C.S. Potten, pp 118-154. Edinburgh: Churchill-Livingstone.

Wright, E.G., Sheridan, P. & Moore, M.A.S. (1980): An inhibitor of murine stem cell proliferation produced by normal human bone marrow. Leukaemia Research 4, 309-319.

Effect on hepatocyte G1-S transition of an inhibitor of CFU-S entry into DNA synthesis

Marie-Noëlle Lombard, Joanna Wdzieczak-Bakala *, Dominique Sotty **, Maryse Lenfant **, Claude Nadal and Martine Guigon ***

    Institut Curie Biol. Bat. 110, Centre Universitaire 91405 Orsay Cedex, France
  * INSERM U250, Institut Gustave-Roussy, 94805 Villejuif, France
 ** Institut de Chimie des Substances Naturelles, CNRS, 91190 Gif-sur-Yvette, France
*** Laboratoire d'Hématologie, Faculté de Médecine St-Antoine, 27, rue de Chaligny, 75012 Paris, France

KEY-WORDS

CFU-S inhibitor, in vivo synchronized hepatocytes, hepatocyte-derived cell line.

INTRODUCTION

A low molecular weight factor inhibiting in vivo haemopoietic pluripotent stem cell (CFU-S) entry into DNA synthesis was extracted from foetal calf bone marrow (Frindel & Guigon 1977). After purification by Biogel P2 chromatography (Wdzieczak et al. 1983), it was shown to be active at dose-ranges of 5-10 µg/mouse. A further purification by HPLC in reverse phase resulted in a 100-fold increase in efficiency (M. Lenfant et al. in preparation). In the present work, we investigated the specificity of its action by testing its effect on synchronized rat hepatocytes either in an in vivo system (Nadal 1973) or in vitro (Lombard et al. 1985, 1987). Experiments were performed in two series:
- a) with the partially purified factor.
- b) with the factor purified to homogeneity.

MATERIALS AND METHODS

An in vivo synchronous wave of hepatocytes in S phase was obtained 18 hr after a single injection of alkaline casein solution (1 mg/g body weight) to ~ 10 day - old rats. CFU-S inhibitory fractions were injected at a time t = 12 hr when ~ 20 % of the hepatocytes were in G1 (Fig. 1). Injected dose-ranges were similar to the ones active on mice: series a) = 5 µg/rat (25 controls, 25 treated), series b) = 50, 100 or 200 ng/rat (19 controls, 33 treated). In each a) and b) series, five litters were used. Controls were treated with equal doses of a simultaneously collected inactive fraction or with saline alone. Six hours later (t = 18 hr), the animals received a pulse (6-^{3}H)-thymidine injection (1 µCi/g body weight). Labelling indexes (LI) were recorded after autoradiographic processing.

In vitro synchronization of HAB liver cells was obtained after a serum starvation period (~10 hr), followed by re-feeding with serum-enriched medium (12 %), resulting in L.I. changes from ~ 4 % to 38 % in the next 20 hr. Inhibiting fractions to be tested were added in the medium when ~ 30 % of the cells were in G1 (Fig. 1).

In parallel experimental series, cultures were treated either with CFU-S inhibiting fraction or with a glycopeptide isolated from human plasma which was shown to inhibit G1-S transition of hepatocytes in vivo (Auger et al. 1983).

Fig. 1 - Schematic organigramme summarizing in vivo and in vitro experimental procedures.

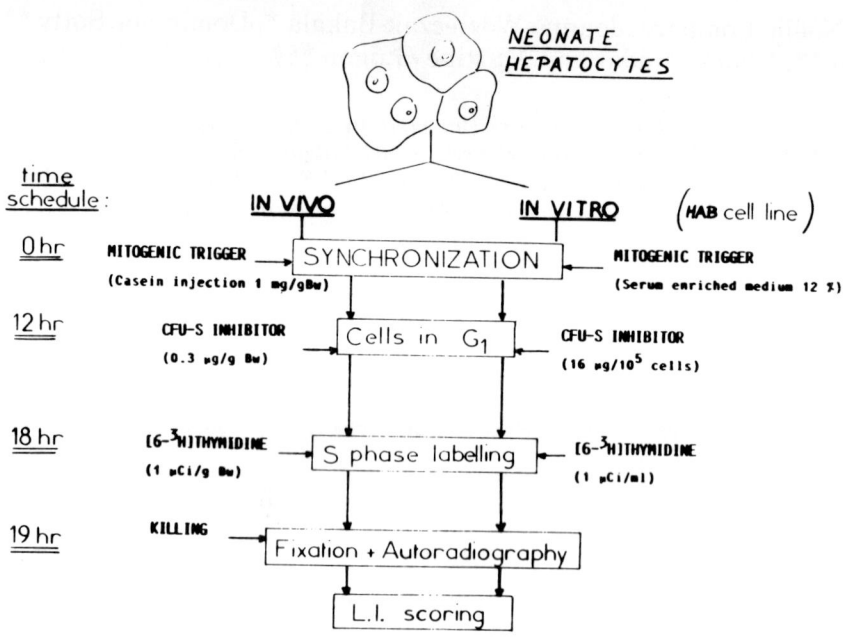

RESULTS

Our first set of experiments testing the activity of the partially purified CFU-S-inhibiting factor showed that in vivo synchronized hepatocytes were responsive: 26-30 % of the induced cell cycles were blocked before entering S phase in both male and female treated baby rats (Fig. 2). No comparable effect could be obtained on hepatocyte-derived cells synchronized in vitro: under our culture conditions, the G1-S transition of the cells was impaired under the action of a glycopeptide blocking hepatocyte proliferation in vivo (Fig. 3), whereas no such response could be detected with the tested CFU-S inhibitor.

The results we obtained in vivo, indicating an activity of the CFU-S inhibitory factor in a heterologous system could be related to some non-specific properties of co-migrating fractions remaining in the partially purified factor. A second set of experiments was therefore carried out with the purified molecule. The results showed that within dose-ranges similar to the ones active on mice, 40-50 % of the in vivo synchronized hepatocytes were blocked in G1 after a single injection of CFU-S inhibitor (Fig. 4).

Fig. 2 - In vivo synchronized hepatocytes. Labelling indexes (LI) 18 hr after synchronizing stimulation. C = controls, T = treated with CFU-S partially purified inhibitor (Statistical significance according to Student's t test: $0.01 < p < 0.02$) N = non-synchronized hepatocyte population).

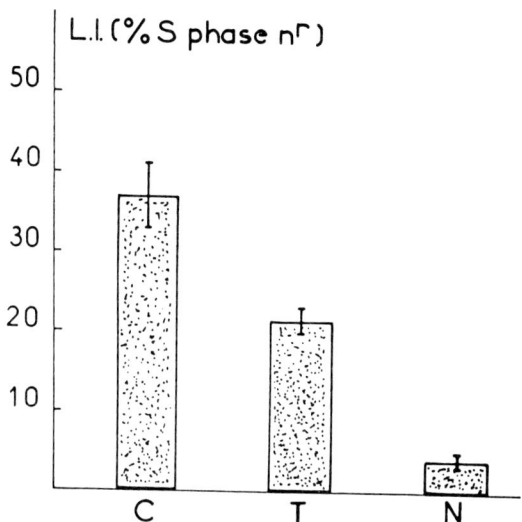

Fig. 3 - In vitro L.I. of HAB liver cells ■---■ (see details in materials and methods). Effect of two different inhibitory fractions added to the medium (arrow): 1) a small peptide extracted from foetal calf bone marrow, active on in vivo mouse CFU-S (crosses: x   x) 2) a low molecular weight glycopeptide obtained from purified human plasma, active in vivo on rat hepatocytes (empty squares : □).
Reported values are averages from 3 - 6 cultures. ** = Statistical significance according to Student's t test: $0.01 > p > 0.005$.

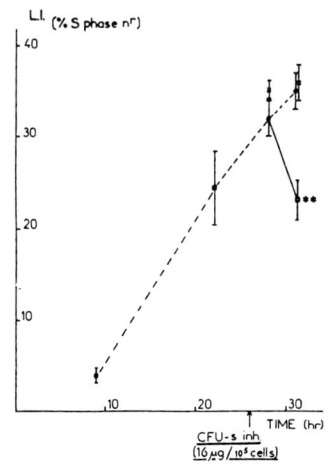

273

Fig. 4 - <u>In vivo</u> synchronized hepatocytes treated with 50, 100 or 200 ng/rat of CFU-S inhibiting factor purified to homogeneity. C = controls (0.001 > p > 0.01).

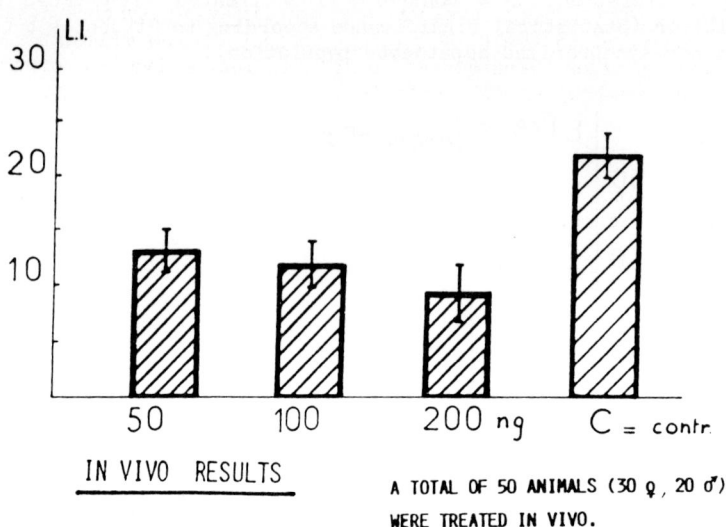

IN VIVO RESULTS

A TOTAL OF 50 ANIMALS (30 ♀, 20 ♂) WERE TREATED <u>IN VIVO</u>.

DISCUSSION & CONCLUSION

This is a first report on the effect on rat hepatocyte proliferation of purified inhibitory factors highly efficient on mouse CFU-S entry into DNA synthesis.
Our data show that a factor inhibiting CFU-S recruitment into the cell cycle can be active in a system which is heterologous with respect to both species and target cell type. This could indicate a certain receptivity of synchronized mammalian epithelial cells corresponding to early steps of the cell cycle as a counterpart to the well-established common refractoriness of cells having entered the S phase. From the results obtained in both series of <u>in vivo</u> experiments, it can be concluded that hepatocytes in G1 are responsive to CFU-S inhibitors. We do not know yet whether the measured decrease in L.I. results from a G1-S block sending the cells back to G_0 or from a slowing down of the transit due to an induced prolongation of the G1 period. There was, however, no cell death to be seen nor any sign of cytotoxicity throughout histological examinations of tissue sections from the 58 livers of sucklings treated with CFU-S inhibitory fractions.

The possibility that post-natal hepatocytes throughout differentiation steps have kept common receptivity properties with foetal haemopoietic populations of the liver is one among many hypotheses. <u>In vitro</u>, this receptivity seems to have disappeared.

Significant also, in connection with our results, is the fact that cells of the foetal liver were shown to produce a CFU-S inhibiting factor (Guigon et al. 1984). From the present data, it appears that the factor purified to homogeneity as well as the partially purified one are actively inhibiting in vivo hepatocyte G_1-S transition. This indicates that the effect of the latter is not simply due to co-migrating components. The former, however, was already significantly active at doses of 50 ng/baby-rat.

In conclusion, it appears at present that an in vivo synchronized hepatocyte population is receptive to the G1-S blocking effect of low molecular weight CFU-S inhibitors. Whether such hepatocytes would be responsive to other stem cell inhibitory factors remains an open question.

ACKNOWLEDGMENTS

The skilfull assistance of M.F. Gournay, J. Samouël and N. Barat throughout this study is greatfully acknowledged.

REFERENCES

Auger G., BlanoT D., Van Heijenoort J., Nadal C. & Gournay M-F. (1983). Partial purification of rat and human serum factors inhibiting the G-S transition in rat hepatocytes. Eur. J. Biochem. 133, 363-369.
Frindel E. & Guigon M. (1977). Inhibition of CFU-S entry into cycle by a bone marrow extract. Exp. Hemat., 5: 74-76.
Guigon M., Wdzieczak-Bakala J., Mary J-Y. & Lenfant M. (1984). A convenient source of CFU-S inhibitors: the foetal calf liver. Cell & Tissue Kinet., 17, 49-55.
Lombard M-N., Houssais J-F. & Zilberfarb D. (1985). In: "Hormones & cell regulations", Ed Dumont et al. Elsevier Sci. Publ. 9, 410-411.
Lombard M-N. et al. (1987). In vitro effect of a glycopeptide acting in vivo on hepatocyte. Submitted to Cell & Tissue Kinet.
Nadal C. (1973). Synchronization of baby rat hepatocytes: a new test for the detection of factors controlling DNA synthesis in rat hepatic cells. Cell & Tissue Kinet., 6, 437-446.
Wdzieczak-Bakala J., Guigon M., Lenfant M. and Frindel E. (1983). Further purification of a CFU-S inhibitor: in vivo effects after cytosine arabinoside treatment. Biomedicine Pharmacoth., 37, 467-471.

Lymphokines in the negative regulation of hematopoiesis

Rôle des lymphokines dans l'inhibition de l'hématopoïèse

Interferons and other lymphokines in bone marrow suppression *in vitro* and *in vivo* : implications for the pathogenesis of aplastic anemia

Neal Stuart Young

Cell Biology Section, Clinical Hematology Branch, National Heart, Lung and Blood Institute, ACRF 7C103, Bethesda, Maryland 20892, USA

ABSTRACT

Clinical and laboratory data have suggested a relationship between the immune system and bone marrow failure. Using an in vitro system to generate a suppressor activity for hematopoietic colony formation, we have identified gamma-interferon as an inhibitory molecule produced by lectin-stimulated normal lymphocytes. Interferon production in aplastic anemia is abnormal, as high amounts are produced by stimulated cells and spontaneously. Some patients have interferon in their circulation, and colony formation by aplastic bone marrow in vitro can be improved in the presence of anti-interferon antibodies. The cell that produces gamma-interferon in aplastic anemia is an activated suppressor lymphocyte, which may be the target of immunosuppressive therapy.

Interferons are potent suppressors of hematopoiesis in vitro. Observations from cell culture bear on the role of soluble inhibitors in bone marrow failure. First, interferon interacts with other factors in its effect on cell proliferation in vitro. Synergy between gamma-interferon and alpha-interferon, tumor necrosis factor, and lymphotoxin is profound, and, conversely, interferon's negative effects can be overcome by positive growth factors in vitro. Second, in vitro, endogenous interferon is more active than added interferon in cell cultures, emphasizing that the local mileu of effector and target cells is probably as important for inhibitors as it is for growth factors. Third, interferon production is complexly regulated. For example, hematologic toxicity with IL-2 administration is probably mediated by interferon and lymphotoxin, mimicking increased production of IL-2 and interferon in aplastic anemia.

Lymphokine and lymphocyte abnormalities in aplastic anemia may be manifestations of an underlying viral etiology. Three examples are discussed: Epstein-Barr virus-associated aplastic anemia, for which the virus has been demonstrated in bone marrow; B19 parvovirus bone marrow failure, in which the virus is cytotoxic to marrow cells; and HIV-induced neutropenia, in which antibodies to the envelope of the virus may mediate suppression of myelopoiesis.

KEY WORDS

interferon, lymphokines, interleukins, hematopoiesis, growth factors, T cells, monocytes, aplastic anemia, bone marrow, parvovirus, Epstein-Barr virus

INTRODUCTION

The hematopoietic and immune systems are linked: they share origin in a common stem cell; they have parallel patterns of development and organization; cells and factors from each system have complex interactions in vitro and in the organism. The specific role of the immune system in bone marrow failure was suggested by clinical observations. The laboratory studies that followed showed interaction of immune cells and factors with bone marrow, but cell culture experiments were necessarily selective, driven by the clinical relevance of the results and the availability of appropriate assays. While the path from the initial observations of co-culture inhibition of hematopoiesis to definition of suppressive cells and factors is clear, the biological significance of many phenomena remains uncertain.

Mathe first described the recovery of autologous bone marrow function in patients whose transplants had failed to engraft, and he interpreted cure of aplastic anemia in these cases to the immunosuppressive conditioning regimen. Subsequently, anti-thymocyte globulin, anti-lymphocyte globulin, high doses of corticosteroids, cyclophosphamide, and cyclosporine A have all shown the ability to restore hematopoiesis in aplastic patients, implicating the immune system in the pathogenesis of bone marrow failure. Although aplastic anemia has been usually viewed as a heterogeneous disorder, very high recovery rates with some regimens including anti-thoracic duct lymphocyte globulin have suggested that an immune mechanism may be a common final pathway of marrow suppression in this disease.

HEMATOPOIETIC SUPPRESSION IN VITRO

The mechanism of hematopoietic suppression in aplastic anemia was inferred from laboratory experiments in which colony formation, indicative of hematopoietic progenitor function, was measured in the presence or absence of putative suppressor cells (Kagan et al., 1976; Hoffman et al., 1977). Colony formation by aplastic bone marrow and peripheral blood is poor. Mixing of aplastic marrow or blood cells and normal marrow or blood also resulted in poor colony formation. Conversely, removal of a subpopulation of cells from aplastic marrow (usually T cells but in some cases monocytes or B cells) enhanced colony number. These results were consistent with a suppressor cell activity present in aplastic blood and bone marrow.

Coculture experiments were criticized because they 1) frequently employed cells from multiply transfused patients who were alloimmunized and, 2) effector (aplastic) and target (normal) cells were not matched for HLA antigen identity. In retrospect, alloimmunization was probably not an explanation for the phenomena observed, as the immune system of multiply transfused patients is hypo- rather than hyperactive in vitro (Gascon et al., 1986). Nevertheless, some investigators were unable to reproduce coculture inhibition using untransfused aplastic tissue and HLA-matched culture conditions (Singer et al., 1978, 1979; Torok-Storb et al., 1979).

LYMPHOKINES THAT INHIBIT HEMATOPOIESIS

Hematopoietic Inhibitors in Culture Systems

An experimental approach that avoided the problems of alloimmunization and the difficulties of obtaining compatible samples for in vitro studies--and ultimately the dependence on aplastic patient samples--was developed by Bacigalupo and colleagues (1981, 1982). They found that the culture supernatants of aplastic patients' bone marrow and (E-rosetted) peripheral blood suppressed normal colony formation. The presence of inhibitory activity in supernatants of cell cultures circumvented the practical problems of alloimmunization and pointed to a soluble molecule as a mediator of hematopoietic suppression in aplastic anemia. In the same experiments, similar inhibitory activity was obtained with lectin stimulation of normal peripheral blood cells. The authors interpreted their data as showing a hierarchy of lymphocyte function, with aplastic bone marrow cells fully activated, aplastic peripheral blood cells activated by the rosette procedure, and normal cells requiring mitogen exposure for activation. Using normal cells under well-defined culture conditions also represented a major advantage over reliance on tissue samples from clinically unstable patients.

Lymphocytes exposed to lectins are stimulated to produce a number of lymphokines, some of which are known to be general inhibitors of cell proliferation, including hematopoietic colony formation. Zoumbos and colleagues (1984) measured interferon in the supernatants of phytohemagglutinin-stimulated mononuclear cells isolated from normal blood. Interferon activity, determined in a virus-protection bioassay, peaked at days 2-3, coincident with the appearance of colony inhibitory activity. The suppressive activity for hematopoiesis could be neutralized by addition of antiserum or monoclonal antibody to gamma-interferon. Antiserum to lymphotoxin also abrogates colony inhibitory activity (Murphy et al., 1986), consistent with the coordinated production of lymphokines by activated cells (reviewed in Zoumbos et al., 1985a).

Similar to growth factors, the cloning of negative regulators has confounded, at least temporarily, the original, simplistic view of the biologic role of these molecules, usually derived from a single assay system using relatively impure factors (reviewed in Zoumbos et al., 1985a). The interferon molecules were defined by their ability to protect cells from virus-induced cytotoxicity. However, first purified natural interferons and later recombinant interferons have been shown to be potent inhibitors of cell proliferation, including hematopoietic colony formation in vitro (Broxmeyer et al., 1983; Raefsky, et al., 1985; Hosoi, et al., 1985; Rigby et al., 1985) and hematopoiesis in animals and human patients (Talpaz et al., 1986). Conversely, the lymphotoxins, defined by tumoricidal activity, also inhibit cell proliferation and are active in virus protection systems (Wong and Goeddel, 1986). In addition, the interaction of these molecules with each other (synergy) and with growth factors (antagonism), discussed below, complicates the interpretation of experiments.

<u>Hematopoietic Inhibitors in Patients</u>

Production of gamma-interferon and lymphotoxin by lymphocytes likely explained the majority of reported coculture and depletion experiments employing aplastic blood and bone marrow cells. That lymphokines might have a role in the pathogenesis of aplastic anemia and other bone marrow failure syndromes was inferred from a number of observations (Zoumbos et al., 1985b). First, repetition of Bacigalupo's experiments with substitution of interferon for colony inhibition assays showed that aplastic cells behaved abnormally: blood mononuclear cells stimulated by lectin produced very large quantitites of interferon compared to normal, and cells from some aplastic patients produced interferon spontaneously. Some patients' marrow cells also produced large amounts of interferon, and incubation of aplastic bone marrow with antibodies to interferons improved colony formation. Interferon was detected in the circulation of about 30% of patients using the virus protection bioassay; by radioimmunoassay, 10-20% of sera have detectable gamma-interferon (Torok-Storb et al., 1987 and Platanias and Young, unpublished data). The difference is likely due to the presence of lymphotoxin, which has intrinsic activity in the bioassay and also amplifies the effect of gamma-interferon synergistically.

An extensive survey of circulating cells in aplastic anemia, using one and two color flow cytometry, identified a cell that produced interferon (Zoumbos et al., 1985c). The majority of patients examined had elevated numbers of activated suppressor T cells, defined by the presence of surface antigens for Leu 2, HLA-DR, and the interleukin 2 receptor (Tac). The soluble form of the interleukin 2 receptor, which is detected by radioimmunoassay, is also markedly elevated in the majority of aplastic anemia patients (Nelson and Young, unpublished data). When obtained from patients or when produced in vitro by lectin stimulation, it is the Tac$^+$ cell population that makes interferon. Inhibition in autologous coculture by activated lymphocytes is linear with the number of Tac$^+$ cells and mediated through gamma-interferon production (Young et al., 1987). Spontaneous production of interferon by bone marrow in one aplastic patient was mediated by HLA-DR$^+$ and Leu 2$^+$ lymphocytes (Zoumbos et al., 1985c).

Although generally reproducible (Laver et al., 1985; Pochart et al., 1986; Herrmann et al., 1986; Hinterberger et al., 1986), these in vitro studies do not give direct evidence for a pathogenic role of interferon and other lymphokines in bone marrow failure states. In support of such a role, anti-thymocyte globulin therapy selectively reduces the number of Tac$^+$ lymphocytes, and patients who recover always have normal numbers of these cells (Platanias et al., 1987). However, treatment of patients with large numbers of circulating activated lymphocytes with anti-Tac monoclonal antibodies had no hematologic effect, probably because this antibody fixes human complement poorly (Tac$^+$ cells were either liganded in vivo with antibody and not removed from the circulation or decreased transiently) (Young and Waldmann, unpublished data). Interferon therapy of patients with chronic myelogenous leukemia or cancer does suppress peripheral blood counts and hematopoietic colony number (Talpaz et al., 1986), although the doses of administerd interferon are usually very high compared to the amounts measured in aplastic serum. Interleukin 2 therapy of cancer patients is profoundly hematosuppressive, possibly through production of gamma-interferon and lymphotoxin by the patients' own activated lymphocytes (Ettinghausen et al., 1987).

MECHANISMS OF LYMPHOKINE ACTIVITY

Synergy and Antagonism

As general anti-proliferative agents, inhibitory factors can interact synergistically (see Zoumbos et al., 1985a for review). Synergy of these factors in hematopoietic cell culture can have quantitatively striking results. For example, addition of small amounts (5 u) of gamma-interferon to varying concentrations of alpha-interferon can increase the inhibitory effect compared to alpha-interferon alone ten fold (Raefsky et al., 1986). Gamma-interferon also acts synergistically with tumor necrosis factor (Broxmeyer et al., 1986; Young et al., 1987) and lymphotoxin (Murphy et al., 1986) on hematopoietic colony formation.

Positive growth factors can antagonize the activity of negative agents. The effects of recombinant interferons on hematopoietic colony formation can be partially overcome by addition of growth factors as fetal calf serum or conditioned medium (Raefsky et al., 1986). Similarly, the inhibitory action of tumor necrosis factor on myeloid colony formation is dependent on the source of colony stimulating factor (Munker and Koeffler, 1987).

Interactive Regulation

In cell culture, factors can induce the production of oppositely acting molecules. IL-2 induction of lymphocytes leads to release of gamma-interferon. In an hematopoietic cell line, tumor necrosis factor causes production of GM-CSF

(Munker et al., 1986). These regulative interactions may explain inverted effects with dose of some factors (stimulation at low concentrations, inhibition at high concentrations) and variably stimulatory or inhibitory effects of factors on different cell types. In one well-studied example, tumor necrosis factor, which acts as a growth factor on a fibroblast line, induced cells to transcribe interferon mRNA and produce beta-interferon, resulting in a state of proliferative homeostasis (Kohase et al, 1987).

A further level of complexity involves the interaction of monocytes and their factors. IL-1 is a monokine with broad effects on the immune and other systems. IL-1 induces GM-CSF release by endothelial cells (Fibbe et al., 1986; Zucali et al., 1986; Sieff, et al., 1986). IL-1 also appears to be equivalent to H-1, a murine growth factor that acts on primitive stem cells only in the presence of other proliferation factors (Stanley et al., 1986). IL-1 production by monocytes is abnormally low in aplastic anemia and abnormally elevated in myelofibrosis (Gascon, Scala, and Young, unpublished data); IL-1 production thus broadly correlates with circulating hematopoietic progenitor number in these two diseases.

Endogenous Factor Production

Positive growth factors like colony stimulating factor were defined in clonal prognitor cell assays; the physiologic importance of these factors has been questioned because they are difficult to detect in Dexter flasks and their addition does not alter growth of long-term bone marrow cultures. Nevertheless, administration of sufficient quantities of recombinant factors like GM-CSF and G-CSF hsd dramatic effects on hematopoiesis in intact animals. Under normal conditions, these factors probably act locally in both long-term culture and in bone marrow, either because they are membrane-associated (Daniak et al., 1984) or concentrated in the extracellular matrix (Gordon et al., 1987).

Negative factors may also act locally. Human bone marrow stromal cells can produce beta-interferon (Shah et al., 1983) and addition of anti-beta-interferon antibodies to clonal cultures can improve plating efficiency (Young et al., 1987). A non-gamma interferon has also been measured in normal bone marrow sera (Zoumbos et al., 1985b). Local action may be inferred from experiments with the a monocytoid cell line that produces beta-interferon (Resnitsky et al., 1986). Beta-interferon that is produced on induction of U937 cells to differentiate is much more active than exogenous beta-interferon in its effects on HLA-DR, oligo-synthetase, and c-myc expression.

Implications of Mechanism for the Study of Aplastic Anemia

Synergy between factors implies that even low concentrations of factors may have profound effects on cell proliferation. Synergy of pathologically produced factors (gamma-interferon and possibly lymphotoxin among others) with endogenous factors (beta-interferon) may exaggerate the effect of small numbers of negative molecules. Locally concentrated factors may be much more potent inhibitors of marrow cell production than comparable quantities of exogenous factors.

However, the interaction of negative and positive factors complicates extrapolation to disease pathogenesis from in vitro assay systems. Colony stimulating factor/burst promoting activity is elevated in most patients with aplastic anemia, and the increased production of inhibitory molecules may be only a secondary phenomenon, perhaps resulting from a normal homeostatic regulatory mechanism. Indeed one possible mechanism of action of anti-thymocyte globulin may be its ability to stimulate mononuclear cells to produce colony stimulating factors (Gascon et al., 1985). Deficiency of in vitro IL-1 production by monocytes from aplastic anemia patients suggests the possibility

of fundamentally dysregulated growth factors. Treatment of patients with
GM-CSF, IL-3, and IL-1 will be revealing as to the underlying mechanism and the
role of the immune system in the pathogenesis of bone marrow failure.

VIRUSES IN BONE MARROW FAILURE

Many of the abnormalities of the immune system described in aplastic anemia
commonly occur in acute and chronic virus infections: dysregulated lymphokine
production, inverted helper/supppressor lymphocyte ratios, lymphocyte
activation, decreased natural killer cell activity (reviewed in Zoumbos et al.,
1985a). Regardless of their direct role on the bone marrow, these immunologic
abnormalities have suggested the presence of viruses in aplastic anemia.
Clinically, a relationship between viruses and bone marrow failure has been
inferred from the history of preceding infectious mononucleosis or hepatitis in
some patients with aplasia (Young and Mortimer, 1984).

Epstein-Barr Virus

Epstein-Barr virus is the etiologic agent of infectious mononucleosis.
Recently, we have detected this virus in the bone marrow of three patients with
aplastic anemia using immunofluorescence for nuclear antigen, in situ
hybridization for viral RNA/DNA within individual cells, and Southern analysis
for viral DNA (Baranski et al., 1987). Epstein-Barr virus was not detected by
these methods either in the circulating mononuclear cells of the aplastic
patients or in the bone marrow a a patient with acute infectious mononucleosis.
The index case had had an episode of infectious mononucleosis one month prior to
the onset of severe marrow failure; a second case had suffered pharyngitis,
cervical adenopathy, and viral pneumonia about six weeks prior to the
development of aplastic anemia. In the third case, however, who was identified
retrospectively, there was no striking history of a viral illness. The first
two patients died despite therapy with anti-thymocyte globulin and Acyclovir.
The third patient had had a complete hematologic remission with anti-thymocyte
globulin treatment alone. Concurrent in vitro studies have suggested that
Epstein-Barr virus can directly infect hematopoietic progenitors (Baranski et
al., 1987). In some patients with aplastic anemia, it seems likely that direct
virus infection of stem cells and/or the immune response provoked by infection
(Tosato et al., 1979) interact to contribute to marrow failure.

B19 Parvovirus

B19 parvovirus is the etiologic agent of fifth disease, a common childhood
exanthem and polyarthralgia syndrome in adults (Anderson, et al., 1986). In
persons with underlying hemolysis, B19 parvovirus infection results in transient
aplastic crisis, an abrupt failue of erythopoiesis that leads to severe anemia
in the setting of high peripheral demand for red blood cells. Studies from our
laboratory have shown that the B19 parvovirus infects and kills erythroid
progenitor cells (reviewed in Young and Mortimer, 1984); the virus may be highly
selective for bone marrow cells, as it has only been successfully propagated in
human erythroid bone marrow cultures (Ozawa, et al., 1986).

With acute natural and experimental infection (Anderson et al., 1985), B19
parvovirus causes reticulocytopenia and also leucopenia and thrombocytopenia.
The virus is usually present in the circulation for only a brief period of time,
probably due to the host's production of neutralizing antibodies. However, in a
child with defective antibody production (Nezelof's syndrome), the clinical
result of parvovirus infection was chronic bone marrow failure (Kurtzman et al.,
1987). B19 parvovirus persisted in the circulation for five months and has
recurred twice; onset of viremia and reappearance of virus have been associated
with transfusion-dependent anemia, absolute reticulocytopenia, and severe

neutropenia. Virus has been demonstrated in the bone marrow by immunofluorescence for capsid proteins, in situ hybridization for single stranded DNA and RNA, and by Southern analysis for replicative forms. In this patient, chronic parvoviremia was not associated with the usual signs of a viral illness like fever, skin eruption, or joint pain. The appearance of the bone arrow in this chronically infected patients and (in acute infection) was similar to the appearance of infected bone marrow cultures (Ozawa et al., 1987): numerous giant pronormoblasts, a paucity of other erythroid precursor cells, and "maturation arrest" in the myeloid differentiation. Without the suspicion of underlying viral etiology and in the absence of specific probes for the B19 parvovirus, chronic bone marrow failure in this child would have been credited to his underlying immunologic abnormalities.

A third model of virus interaction with immune cells is the neutropenia of acquired immunodeficiency syndrome. Hematopoietic cultures of bone marrow from patients with this syndrome are inhibited in the presence of sera containing antibody to the HIV envelope glycoprotein (Donahue et al., 1987). For the case of this retrovirus, infection of granulocytic progenitors is not lytic, and the clinically manifest neutropenia is dependent on development of a specific humoral immune response in an infected patient.

REFERENCES

Anderson, M.J., Higgins, P.G., Davis, L.R., Willman, J.S., Jones, S.E., Kidd, I.M., Pattison, J.R., Tyrell, D.A. (1985): Experimental parvoviral infection in humans. J. Inf. Dis. 152, 257-265.

Anderson, M.J., Pattison, J.R., Young, N.S. (1986): Pathogenesis of parvovirus-induced disease in humans. In Concepts in Viral Pathogenesis II, ed A.L. Notkins, M.B.A. Oldstone, pp. 261-268. New York: Springer-Verlag.

Bacigalupo, A., Podesta, M., Mingari, M.C., Moretta, L., Van Lint, M.T., Marmont, A. (1981a): Immune suppression of hematopoiesis in aplastic anemia: activity of T-γ lymphocytes. J. Immunol. 125, 1449-1453.

Bacigalupo, A., Podesta, M., Mingari, M.C., Moretta, L., Piaggio, G., Van Lint, M.T., Durando, A., Marmont, A. (1981b): Generation of CFU-C/suppressor T cells in vitro: an experimental model for immune-mediated marrow failure. Blood 57, 491-496.

Baranski, B., Moore, J., Armstrong, G., Magrath, I., Young, N. (1987): Epstein-Barr virus associated aplastic anemia: demonstration of virus in clinical bone marrow samples and in hematopoietic progenitor cells infected in vitro. Clin. Res. 35, 419a.

Broxmeyer, H.E., Lu, L., Platzer, E., Feit, C., Juliano, L., and Rubin, B.Y. (1983): Comparative analysis of the influences of human $\gamma\psi$ α, and β interferons on human multipotential (CFU-GEMM), erythroid, (BFU-E), and granulocyte-macrophage (CFU-GM) progenitor cells. J. Immunol. 131, 1300-1309.

Broxmeyer, H.E., Williams, D.E., Lu, L., Cooper, S., Anderson, S.L., Beyer, G.S., Hoffman, R., Rubin, B.Y. (1986): The suppressive influences of human tumor necrosis factors on bone marrow hematopoietic progenitor cells from normal donors and patients with leukemia: synergism of tumor necrosis factor and interferon-γ. J. Immunol. 136, 4487-4495.

Dainiak, N., Feldman, L., Cohen, C.M. (1985): Neutralization of erythroid burst promoting activity in vitro with antimembrane antibodies. Blood 65, 877-885.

Donahue, R.E., Johnson, M.M., Zon, L.I., Clark, S.C., Groopman, J.E. (1987): Suppression of in vitro hematopoiesis following human immunodeficiency syndrome. Nature 326, 200-203.

Ettinghausen, S.B., Moore, J.G., Platanias, L., Young, N., Rosenberg, S.A. (1987): Hematologic toxicity of lymphokine activated killer (LAK) cells and recombinant interleukin 2 in cancer patients. Blood, in press.

Fibbe, W.W., van Damme, J., Billiau, A., Voogt, P.J., Duinkerken, N., Kluck, P.M., Falkenburg, J.H. (1986): Interleukin-1 (22-K factor) induces release of granulocyte-macrophage colony-stimulating activity from human mononuclear phagocytes. Blood 68, 1316-1321.

Gascon, P., Zoumbos, N.C., Scala, G., Djeu, J., Moore, J.G., Young, N.S. (1985): Lymphokine abnormalities in aplastic anemia: implications for the mechanism of action of antithymocyte globulin. Blood 65, 407-412.

Gascon, P., Scala, G., Djeu, J., Young, N. (1986): Decreased lymphokine and monokine production in hypertransfused patients. Blood 68 (suppl. 1), 297a.

Herrmann, F., Griffin, J.D., Meuer, S.G., Zum Buschenfelde, K.-H.M. (1986): Establishment of an interluekin-2 dependent T cell line derived from a patient with aplastic anemia, which inihibits in vitro hematopoiesis. J. Immunol. 136, 1629-1634.

Hinterberger, W., Adolph, G., Koller, U., Knapp, W., Bettleheim, P., Geissler, K., Godner, A., Volc-Platzer, B., Lechner, K. (1986): The significance of interferon in patients with aplastic anemia. J Cell Biochem 10D (suppl.), 243.

Hoffman, R.A., Zanjani, E.D., Lutton, J.D., Zalusky, R., Wasserman, L.R. (1977): Suppression of erythroid-colony formation by lymphocytes from patients with aplastic anemia. N. Engl. J. Med. 296, 10-14.

Hosoi, T., Ozawa, K., Urabe, A., Takaku, F. (1985): Effects of recombinant interferons on the clonogenic growth of leukemic cells and normal hemopoietic progenitors. Int. J. Cell Clon. 3, 304-309.

Kagan, W.A., Ascensao, J.A., Pahwa, R.N., Hansen, J.A., Goldstein, G., Valera, E.B., Incefy, G.S., Moore, M.A.S., Good, R.A. (1976): Aplastic anemia: presence in the human bone marrow of cells that suppress hematopoiesis. Proc. Nat. Acad. Sci. USA 73, 2890-2894.

Kohase, M., Henrikson-DeStefano, D., May, L.T., Vilcek, J. (1986): Induction of β_2-interferon by tumor necrosis factor: a homeostatic mechanism in the control of cell proliferation. Cell 45, 659-666.

Kurtzman, G.J., Ozawa, K., Cohen, B., Hanson, G., Oseas, R., Young, N.S. (1987): Chronic bone marrow failure due to persistent B19 parvovirus infection. N. Engl. J. Med., in press.

Laver, J., Kernana, N.A., Levick, J., Moore, M.A.S., O'Reilly, R.J., Castro-Malaspina, H. (1985): Gamma interferon-mediated myelosuppression in severe aplastic anemia: in vitro studies identifying potential responders to ATG. Blood 66 (suppl 1), 121a.

Mamus, S., Beck-Schroeder, S., Zanjani, E.D. (1985): Suppression of normal human erythropoiesis by gamma interferon in vitro. Role of monocytes and T lymphocytes. J. Clin. Invest. 75, 1496-1503.

Munker, R., Gasson, J., Ogawa, M., Koeffler, H.P. (1986): Recombinant human TNF induces production of granulocyte-macrophage colony stimulating factor. Nature 323, 79-82.

Munker, R., Koeffler, P. (1987): In vitro action of tumor necrosis factor on myeloid leukemic cells. Blood 69, 1102-1108.

Murphy, M., Loudon, R., Kobayashi, M., Trinchieri, G. (1986): γ-Interferon and lymphotoxin released by activated T cells synergize to inhibit granulocyte/monocyte colony formation. J. Exp. Med. 164, 263-279.

Ozawa, K., Kurtzman, G., Young, N. (1986): Replication of the B19 parvovirus in human bone marrow cell cultures. Science 233, 883-886.

Ozawa, K., Kurtzman, G., Young, N. (1987): Productive infection by B19 parvovirus of human erythroid bone marrow cells in vitro. Blood, in press.

Pochart, F., Rhodes-Feuillette, A., Devergie, A., Vilmer, E., Gluckman, E. (1986): The interferon compartment in aplastic anemia. Exp. Hematol. 14, 435.

Platanias, L., Gascon, P., Bielory, L., Griffith, P., Nienhuis, A., Young, N. (1987): Lymphocyte subsets and lymphokines following anti-thymocyte globulin therapy for aplastic anemia. Brit. J. Haematol., in press.

Raefsky, E., Platanias, L., Zoumbos, N., Young, N. (1986): Studies of interferon as a regulator of hematopoietic proliferation. J. Immunol. 135, 2507-2512.

Resnitsky, D., Yarden, A., Zipori, D., Kimchi, A. (1986): Autocrine β-related interferon controls c-myc suppression and growth arrest during hematopoietic cell differentiation. Cell 46, 31-40.

Rigby, W.G.C., Ball, E.D., Guyre, P.M., Fanger, M.W. (1985): The effects of recombinant-DNA-derived interferons on the growth of myeloid progenitor cells. Blood 65, 858-861.

Shah, G., Dexter, M., Lanotte, M. (1983): Interferon production by human marrow stromal cells. Brit. J. Haematol. 54, 365-372.

Sieff, C.A., Tsai, S., Faller, D.V. (1986): Interleukin 1 induces cultured human endothelial cell production of granulocyte-macrophage colony stimulating factor. J. Clin. Invest. 79, 48-51.

Singer, J.W., Brown, J.E., James, M.C., Doney, K., Warren, R.P., Storb, R., Thomas, E.D. (1978): Effect of peripheral blood lymphocytes from patients with aplastic anemia on granulocytic colony growth from HLA-matched and -mismatched marrows: effect of transfusion sensitization. Blood 52, 37-46.

Singer, J.W., Doney, K.C., Thomas, E.D. (1979): Coculture studies of 16 untransfused patients with aplastic anemia. Blood 54, 180-185.

Stanley, R., Bartocci, A., Patinkin, D., Rosendaal, M., Bradley, T.R. (1986): Regulation of very primitive, multipotent hemopoetic cells by hemopoietin-1. Cell 45, 667-674.

Talpaz, M., Spitzer, G., Hittelman, W., Kantsrjian, H., Gutterman, J. (1986): Changes in granulocyte-macrophage colony-forming cells among leukocyte-interferon-treated chronic myelogenous leukemia patients. Exp. Hematol. 14, 668-671.

Torok-Storb, B. Sieff, C., Storb, R., Adamson, J., Thomas, E.D. (1980): In vitro tests for distinguishing possible immune-mediated aplastic anemia from transfusion-induced sensitization. Blood 55, 211-215.

Torok-Storb, B., Johnson G., Bowden R., Storb R. (1987): Gamma-interferon in aplastic anemia: inability to detect significant levels in sera or demonstrate hematopoietic suppressing activity. Blood 69, 629-633.

Tosato, G., Magrath, I., Koski, I., Dooley, N., Blaese, M. (1979): Activation of suppressor T cells during Epstein-Barr virus-induced infectious mononucleosis. N. Engl. J. Med. 301, 1133-1137.

Wong, G.H.W., Goeddel, D.V. (1986): Tumour necrosis factors α and β inhibit virus replication and synergize with interferons. Nature 323, 819-822.

Young, N., Mortimer, P.P. (1984): Viruses and bone marrow failure. Blood 63, 729-737.

Young, N.S., Leonard, E., Platanias, L. (1987): Lymphocytes and lymphokines in aplastic anemia: pathogenic role and implications for pathogenesis. Blood Cells, in press.

Zoumbos, N.C., Djeu, J.Y., Young, N.S. (1984): Interferon is the inhibitor of hematopoiesis generated by stimulated lymphocytes in vitro. J. Immunol. 133, 769-774.

Zoumbos, N., Raefsky, E., Young, N. (1986): Lymphokines and hematopoiesis. Prog. Hematol. 14, 201-227.

Zoumbos, N.C., Gascon, P., Djeu, J., Young, N.S. (1985b) Interferon is the inhibitor of hematopoietic suppression in aplastic anemia in vitro and possibly in vivo. Proc. Nat. Acad. Sci. USA 82, 188-192.

Zoumbos, N.C., Gascon, P., Djeu, J., Trost, S.R., Young, N.S. (1985c): Circulating activated suppressor T lymphocytes in aplastic anemia. N. Engl. J. Med. 312, 257-265.

Zucali, J.R., Broxmeyer, H.E., Dinarello, C.A., Gross, M.A., Weiner, R.S. (1987): Regulation of early hematopoietic (BFU-E and CFU-GEMM) progenitor cells in vitro by interleukin 1-induced fibroblast-conditioned medium. Blood 69, 33-37.

Résumé

Les données cliniques et biologiques suggèrent une relation entre le système immunitaire et l'insuffisance médullaire. A l'aide d'un modèle permettant in vitro d'obtenir une activité inhibitrice pour la formation de colonies, nous avons montré que l'interféron gamma était une molécule inhibitrice produite par des lymphocytes normaux stimulés par une lectine. La production d'interféron dans l'aplasie médullaire est anormale : des quantités importantes sont produites par les cellules stimulées et spontanément certains malades ont de l'interféron dans leur circulation et la formation de colonies in vitro à partir de moelle osseuse en aplasie peut être améliorée par des anticorps anti interféron. La cellule qui produit l'interféron gamma dans l'aplasie médullaire est un lymphocyte activé de type suppresseur, qui peut être la cible du traitement immunosuppresseur. Les interférons sont de puissants inhibiteurs de l'hématopoïèse in vitro. Les observations faites en culture soulignent le rôle d'inhibiteurs solubles dans l'insuffisance médullaire. L'effet de l'interféron sur la prolifération cellulaire in vitro est modifié par d'autres facteurs. La synergie entre l'interféron gamma et l'interféron alpha, le "tumor necrosis factor" et la lymphotoxine est très importante : à l'inverse les effets négatifs de l'interféron peuvent être dépassés par des facteurs de croissance à action positive in vitro. Par ailleurs, in vitro l'interféron endogène est plus actif que l'interféron ajouté en culture, ce qui souligne que le micro-environnement des cellules cibles et effectrices est probablement aussi important pour les inhibiteurs qu'il l'est pour les facteurs de croissance. Enfin, la production d'interféron est réglée de manière complexe. Par exemple, la toxicité hématologique lors de l'administration d'IL2 est peut être due à l'interféron et à la lymphotoxine, ce qui ressemble à la production augmentée d'IL2 et d'interféron dans l'aplasie médullaire. Les anomalies des lymphokines et des lymphocytes dans l'aplasie médullaire peuvent être des manifestations d'une maladie virale sous jacente. Trois exemples sont présentés : l'aplasie médullaire associée au virus Epstein Barr, où la présence du virus a pu être démontré dans la moelle osseuse ; l'insuffisance médullaire due au parvovirus B19 où on a pu montrer que le virus est toxique pour les cellules de la moelle et la neutropénie induite par le virus HIV, au cours de laquelle des anticorps dirigés contre l'enveloppe du virus peuvent favoriser l'inhibition de la myélopoïèse.

Suppression of hematopoietic cells by TNF-Alpha

G. David Roodman

Audie L. Murphy Veterans Administration Hospital and University of Texas Health Science Center, San Antonio, TX, USA

ABSTRACT

We have examined the effects of tumor necrosis factor-alpha, a product of activated macrophages, on human erythroid progenitors (CFU-E, BFU-E) and the hematopoietic cell lines, K562, HL60 and HEL cells. Tumor necrosis factor (TNF) significantly inhibited CFU-E and BFU-E growth at concentrations as low as 10^{-11} to 10^{-12}M (0.2 u/ml), although erythroid colony and burst formation were not totally ablated. Preincubation of marrow samples with TNF for 15 minutes was sufficient to suppress erythroid colony and burst formation. Addition of TNF after the start of culture inhibited CFU-E and BFU-E derived colony formation if TNF was added within the first 48 hours of culture. Additionally, TNF inhibited the growth of highly purified erythroid progenitors harvested from day 5 BFU-E. The colonies which formed in cultures treated with TNF were significantly smaller than those formed in control cultures. TNF (10^{-8}-10^{-10}M) also suppressed the growth of the hematopoietic cell lines K562, HL60 and HEL cells, with 40-60% of the cells being sensitive to TNF. Preincubation of HL60 cells with TNF for 15 minutes significantly inhibited their growth. K562, HL60, and HEL cells expressed high affinity receptors for TNF in low numbers (6,000-10,000 receptors per cell). Fluorescence activated cell sorter analysis of TNF binding to HEL cells demonstrated that the majority of these cells expressed TNF receptors. Additionally, differentiation of K562 cells with hemin resulted in loss of TNF sensitivity and decreased expression of TNF receptors on their surface. These data suggest that TNF is a rapid irreversible and extremely potent inhibitor of hematopoietic cells and acting directly on a subpopulation of early erythroid cells predominately CFU-E, BFU-E and possibly proerythroblasts.

KEY WORDS

Tumor necrosis factor, CFU-E, BFU-E, K562, HL60, HEL cells, differentiation, hemin, erythroid progenitors

Cell to cell interaction in the regulation of hematopoiesis has been a topic of intensive investigation over the past decade. In particular, monocyte-macrophages have been shown to be important regulators of erythroid progenitor cell growth. Resting macrophages have been shown to stimulate the growth of CFU-E and BFU-E (Zuckerman, 1981; Kurland, et al, 1980) while activated macrophages have been shown to suppress the growth of CFU-E and BFU-E (Roodman, et al, 1983; Zanjani, et al, 1982).

Macrophages secrete a variety of cellular products when activated by endotoxin or antigens, including tumor necrosis factor-alpha (Cohn, 1978; Wang, et al, 1985). This monokine inhibits the growth of a variety of tumor cells (Old, 1985) and have multiple effects on normal tissues (Bertolini, et al, 1986; Silberstein, et al, 1986).

TNF appears to be a prime candidate as a negative regulator of erythropoiesis: 1) It is not produced by resting macrophages but is produced by activated macrophages (Pennica, et al, 1984). 2) Endotoxin and BCG, products associated with chronic disease processes, stimulate its production (Old, 1985). 3) TNF production by monocytes in patients with malignancies has been correlated with the severity of their disease (Aderka, 1985) and 4) TNF has been reported to be an inhibitor of BFU-E, CFU-GEMM and CFU-GM (Broxmeyer, et al, 1986).

Therefore, we have undertaken studies to examine the effects of TNF-alpha on CFU-E and BFU-E and to delineate its mechanism of action on erythroid progenitors. In addition, we have used the hematopoietic cell lines K562, HL60 and HEL cells as model systems to investigate the mechanism of action of TNF-alpha on hematopoietic cells.

METHODS

Bone marrow samples: After obtaining informed consent, bone marrow samples (2 to 3 mls), were aspirated under xylocaine anesthesia from the posterior iliac crests of normal volunteers into syringes containing 1,000 units per ml of preservative free heparin in alpha Minimum Essential Media (a-MEM). Marrow samples were diluted with two volumes of α-MEM, and the marrow mononuclear cell preparation prepared after density gradient centrifugation over Hypaque ficoll. The cells were then washed x 3 with αMEM and the nucleated cells counted in a hemocytometer. The cells were then resuspended in αMEM-5% fetal calf serum.

Assay of CFU-E and BFU-E: Marrow mononuclear cells were cultured in plasma clots (Tepperman, et al, 1974). Marrow cells were cultured in 0.1 ml plasma clots with 1 Iu/ml of human urinary erythropoietin (Toyoba) in the presence or absence of highly purified human recombinant tumor necrosis factor-α (Cetus Corp., Emeryville, CA, 1.44×10^7 units/mg protein). In selected experiments, marrow cells were preincubated with varying concentrations of TNF or αMEM for 0, 5, 10, 15, 30, 45, 60 or 120 minutes then washed extensively or treated with a neutralizing polyclonal antibody to TNF (Lot# 57-030685, Cetus Corp., 1:100 dilution neutralizes 4,000u of TNF). The antibody completely blocked the effects of TNF on erythroid progenitors (for BFU-E: control = 99 ± 6, TNF=3 ± 3, TNF+anti TNF=84 ± 40: for CFU-E: control=194 ± 2, TNF= 3 ± 1, TNF+anti TNF=193 ± 9). The cells were then cultured as described above. In other experiments marrow cells were cultured in the absence of TNF for varying periods of time and then 10^{-8}M TNF in 0.01 ml of αMEM or media were added to each clot (final concentrations of TNF=10^{-9}M) to determine the effects of adding TNF after the start of culture. The number of cells per colony was also determined microscopically.

Preparation of highly purified erythroid progenitors: Marrow mononuclear cells were cultured in methylcellulose (Iscove, et al, 1974) at 4×10^4 cells/ml in the presence of 1 Iu/ml of human urinary erythropoietin. Cells were cultured in 100 mm tissue culture plates in a final volume of 10 mls/per plate. Cultures were allowed to incubate as described above for 5 days and then nonhemoglobinized colonies with the characteristic morphology of BFU-E were harvested with a finely drawn pipette (Young, et al, 1984). The cells were washed x1 with αMEM and a single cell suspension prepared. The cells were then divided equally and half the cells were cultured in the presence of 10^{-8}M TNF and the other half were cultured in plasma clots for 7 days in the presence of vehicle. Clots were harvested, stained, and small single hemoglobinized colonies with the characteristics of CFU-E were scored microscopically.

Hematopoietic cell lines: K562 (provided by Dr. Linda Marshall), HL60 (provided by Dr. D. Von Hoff) and HEL cell, (provided by Dr. Rodger McEver) were used in these studies. Cells (2×10^4 to 1×10^5 cells/ml) were cultured in RPMI 1640 media with 10% fetal calf serum (Hyclone). Varying concentrations of TNF were added to cells in log phase, and the effects of TNF on cell viability were determined by Trypan blue exclusion. Viable cell numbers were determined 0, 24, 48 and 72 hours after the start of cultures. In selected experiments HL60 cells were preincubated with TNF (10^{-8}M) for 5, 10, 15, 30, 60 or 120 minutes, neutralizing antibody to TNF added and the cells washed extensively and cultured in the absence of TNF. The cells were then counted at 24 and 48 hours after the start of culture.

Detection of TNF receptors on hematopoietic cell lines: High specific activity radioiodinated TNF was prepared by incubating 10 μg of TNF with 2 mC of ^{125}Iodine and 2 iodobeads. Iodinated TNF was then separated from unreacted iodine by gel filtration chromatography on G25 Sephadex (Pharmacia) and used for receptor studies. ^{125}I-TNF had approximately 80% of the activity of unlabeled TNF in a TNF bioassay and had a specific activity of 153,010 cpm/pmole. In preliminary experiments K562, HL60 and HEL cells were incubated with 10^{-9}M radioiodinated TNF for 1, 2, 3 or 4 hours at 4°C in the presence or absence of 100-fold excess of unlabeled TNF, washed extensively and counted in a gamma counter. Specific binding of TNF was determined by subtracting the amount of radioactivity bound to cells in the presence of excess unlabeled TNF (10^{-7}M) from the radioactivity in the presence of ^{125}I labeled TNF alone. Nonspecific binding was 10-15% of specific binding. In this manner the kinetics of TNF binding to its receptor were determined. K562, HL60 or HEL cells (4×10^6/ml) were then incubated with 6×10^{-9} - 2×10^{-10}M radioiodinated TNF for 3 or 4 hours at 4°C in the presence or absence 100-fold excess of unlabeled TNF, transferred to glass fiber discs, the discs washed extensively and then counted in a gamma counter. Receptor concentrations per cell and dissociation constants for binding of TNF to its receptor were determined by Scatchard analysis (Scatchard, 1949). The distribution of TNF receptors on cells was determined as follows: Cells were washed with ice cold phosphate buffered saline (PBS) and resuspended at 3×10^6 cells/ml into each of four tubes (1 ml/tube). TNF (10^{-6}M) or vehicle was added (0.01 ml) and the cells incubated on ice for 3 hours. Media or anti-TNF was then added and the cells incubated for 1 hr on ice. The cells were washed x3 with ice cold PBS and resuspended in 0.25 ml ice cold PBS. FITC conjugated goat anti-rabbit IgG F(ab)$_2$ fragment (Cappel) (0.005ml) was then added to the cells, and the cells incubated for 30 min on ice. Cells were washed x3 with ice cold PBS and then analyzed on an Epics C Fluorescence Activated Cell Sorter (Coulter).

In selected experiments K562 cells were cultured in the presence of 0.05 mM freshly prepared hemin for at least 7 days. Cells became hemoglobinized after 72 hours.

<u>Statistical methods</u>: Cultures were done in quadruplicate and the mean $\pm$ standard error of the mean calculated. Results were compared using a 2 way analysis of variance for repeated measures and a Newman Keuls range test. Results were considered significantly different for $p<.05$.

RESULTS AND DISCUSSION

As seen in Table 1, 10^{-7} to 10^{-12}M TNF significantly inhibited erythroid colony and burst formation in a dose-dependent fashion. This concentration of TNF (10^{-12}M) corresponds to 0.2 units of TNF per ml. Different marrow samples displayed different sensitivities to TNF with some not affected by TNF concentrations less than 10^{-10}M and others inhibited by TNF concentrations as low as 10^{-13}M. Similar results were seen in 8 separate experiments. Interestingly not all CFU-E and BFU-E were inhibited by TNF regardless of the concentration tested.

TABLE 1. TNF-α Suppresses the Growth of CFU-E and BFU-E

TNF (M)	CFU-E	BFU-E
	(Colonies per 10^5 cells plated)	
Control	449±22	98±5
10^{-8}	69±9*	10±4*
10^{-9}	177±32*	24±4*
10^{-10}	198±31*	49±3*
10^{-11}	241±17*	57±6*
10^{-12}	243±11*	61±7*
10^{-13}	419±43	90±6

Results represent the mean ± SEM of four determinations. CFU-E and BFU-E were cultured as described in Methods.
*$p<.05$ compared control.

We then examined the time requirements for TNF to suppress CFU-E and BFU-E. As seen in Table 2, preincubation of marrow cells with 10^{-10}M TNF for 10-15 minutes was sufficient to inhibit CFU-E and 30 min was sufficient to inhibit BFU-E growth.

TABLE 2. Effects of Preincubating Marrow Cells with TNF

TNF Exposure Time (Min)	CFU-E	BFU-E
	Colonies Per 10^5 Cells Plated	
Control	311±10	95±4
5	243±38	93±5
10	194±25*	85±6
15	199±29*	95±5
30	135±25*	62±5*
45	150±16*	61±7*

Marrow cells were incubated with 10^{-10}M TNF for varying periods of time and then excess of anti-TNF was added. Control cultures contained only anti-TNF. Results represent the mean ± SEM for four determinations. *$p<.05$

Addition of TNF after the start of the cultures (Table 3) showed that TNF could only suppress erythroid colony or burst formation if added within 24 (CFU-E) or 48 hours (BFU-E) of culture.

TABLE 3. Effects of Adding TNF After the Start of Culture

Time TNF Added After Start of Culture	CFU-E	BFU-E
	Colonies per 10^5 Cells Plated	
Control	136±4	84±8
0 hr	59±9*	49±3*
24 hr	46±7*	46±3*
48 hr	116±11	46±3*
72 hr	164±15	65±3
96 hr	102±23	71±5

TNF (10^{-9}M) was added to plasma clot cultures of marrow cells after the clots had formed at the times specified. Results represent the mean ± SEM for four determinations. *p<.05 compared to control cultures.

Interestingly TNF significantly decreased the size of erythroid colonies and bursts that formed if added within 48 hours after the start of cultures (Table 4).

TABLE 4. Effects of TNF on Erythroid Colonies

	% Small Colonies	p-Value
Control	29.7±3.5	-
TNF (0 hr)	82.5±6.7	<.001
TNF (24 hr)	63.3±1.9	<.001
TNF (48 hr)	56.6±1.2	<.001
TNF (72 hr)	45.3±10.7	NS
TNF (96 hr)	41.3±6.4	NS

Colony size was determined by counting the number of cells per colony. Small colonies were defined as having less than 50 cells per colony. Results represent the mean ± SEM for 3 independent experiments done in quadruplicate. The numbers in parentheses refer to the time after the start of culture 10^{-10}M TNF was added to the cultures. NS - Not significant

TNF significantly decreased erythroid colony size at 48 hours although it did not significantly affect erythroid colony numbers.

To determine if TNF was acting directly on erythroid progenitors or required accessory cells for its effects, we tested TNF on highly enriched populations of erythroid progenitors isolated from day 5 BFU-E. As seen in Table 5, TNF significantly suppressed the growth of erythroid progenitors in 5 of 6 experiments. The size distribution of erythroid colonies formed in the presence or absence of TNF was also affected with 86.3±7.3% of erythroid colonies formed in the presence of TNF having less than 50 cells per colony compared to 38.7±3.5% of colonies formed in control cultures (p<.05).

TABLE 5. TNF Directly Affects Erythroid Progenitors Isolated From Immature BFU-E Colonies

Exp.	Control	TNF
1	279±1	117±11*
2	13±1	0*
3	90±1	64±13
4	34±4	7±2*
5	88±6	50±3*
6	10±2	5±1*

Day 5 BFU-E were isolated from methylcellulose cultures of human marrow with a finely drawn pipette. A single cell suspension was prepared and one-half of the cells were cultured with media and the other half cultured with 10^{-8}M TNF. Erythroid colonies were scored 7 days later and were small single colonies containing 8-80 cells per colony. *p<.05

We then examined the effects of TNF on the hematopoietic cell lines K562, HL60 and HEL cells to determine if TNF affected them in an analogous manner as CFU-E and BFU-E. As seen in Table 6, TNF inhibited the growth of all three cell lines although the inhibition was never 100%. HL60 and HEL cells appeared more sensitive to TNF then K562. Preincubation of HL60 cells with 10^{-8}M TNF for 15 minutes was sufficient to inhibit the growth of HL60 cells (9.35x10^5 cells vs 5.6 x 10^5 cells after 15 minutes preincubation with media versus 10^{-8}M TNF and counted 48 hours later).

TABLE 6. Effects of TNF on Hematopoietic Cell Lines

Cell Lines	% Inhibition
K562 (n=4)	30±2%
HL60 (n=5)	60±9%
HEL (n=6)	50±5%

Hematopoietic cell lines K562, HL60 and HEL cells were cultured at 4x10^4 to 1x10^5 cells per ml in RPMI 1640 and 10% fetal calf serum. TNF (10^{-8}M) was added at the start of cultures and cell number and viability determined as described in Methods. The number of experiments used to generate this data are shown in parenthesis. Results represent the mean ± the standard error of the mean for the percent inhibition of cell growth at 48 hours.

We then tested whether K562, HL60, and HEL cells expressed TNF receptors and determined the affinity of these receptors for TNF. These cell lines expressed low numbers (6,000 to 10,000 receptors) of TNF receptors per cell. These receptors were of very high affinity with a Kd of 10^{-9}M. We then determined the distribution of TNF receptors on HEL cells. The majority of HEL cells express TNF receptors with a range of 89 to 95% of the cells expressing TNF receptors in three separate experiments.

As noted above, TNF was only active if added within the first 48 hours of culture. These results suggested that as erythroid progenitor cells differentiated they lost their ability to respond to tumor necrosis factor. Therefore, we tested if differentiation of hematopoietic cell lines resulted in loss of TNF sensitivity and loss of TNF receptors. As seen in Table 7 hemin treated K562 cells were resistant to the growth inhibitory effects of TNF. Similar results were seen in three experiments. We then examined whether hemin treated K562 cells expressed TNF receptors using the fluorescence activated cell sorter. The majority of untreated K562 cells, approximately 80%, expressed TNF receptors. In contrast, few if any hemin treated K562 cells expressed TNF receptors.

TABLE 7. Effects of TNF on Control and Hemin Treated K562 Cells

Treatment	Control Cells	Hemin Treated
Media	1.2×10^7	9.4×10^6
TNF	8.7×10^6	1.3×10^7

4×10^6 control or hemin treated (0.05mM) K562 cells were cultured in the presence or absence of 10^{-8}M TNF for 48 hours, and viable cell counts determined. Similar results were seen in 3 separate experiments.

Therefore, these data show that TNF is a potent inhibitor of erythroid progenitor cell growth, inhibiting CFU-E and BFU-E in concentrations as low as 10^{-12}M. TNF appears to rapidly and irreversibly inhibit erythroid progenitor cells. TNF appears to be acting on a defined subpopulation of erythroid progenitor cells most probably early erythroid progenitors.
In addition, TNF inhibited the growth of the hematopoietic cell lines K562, HL60 and HEL cells. These cells express low numbers of high affinity receptors for TNF. Therefore small numbers of TNF molecules appear sufficient to inhibit hematopoietic cell growth.

Time course studies with bone marrow cells showed that TNF was only able to inhibit the growth of CFU-E and BFU-E if added within the first 48 hours of culture. These data suggest that cellular differentiation resulted in loss of TNF responsivity. Similarly, differentiation of K562 cells with hemin resulted in loss of TNF sensitivity and loss of expression of TNF receptors on the cell surface of the majority of cells.

TNF is most probably acting only on very early erythroid progenitor cells and that once these cells differentiate, they are no longer sensitive to TNF.

REFERENCES

Aderka, D., Fisher, S., Levo, Y., Holtman, H., Hahn, T., Wallace, D.(1985): Cachetin tumour necrosis factor production by cancer patients. Lancet 2, 1190.

Bertolini, D.R., Nedwin, G.E., Bringman, T.S., Smith, D.D., Mundy, G.R.(1986): Stimulation of bone resorption and inhibition of bone formation in vitro by human tumour necrosis factor. Nature 319, 516-518.

Broxmeyer, H.E., Williams, D.E., Lu, L., Cooper, S., Anderson, S.L., Beyer, G.S., Hoffman, R., Rubin, B.Y.(1986): The suppressive influences of human tumor necrosis factors on bone marrow hematopoietic progenitor cells from normal donors and patients with leukemia: Synergism of tumor necrosis factor and interferon-γ. J Immunol 12, 4487-4495.

Cohn, Z.A.(1978): The activation of mononuclear phagocytes. Fact, Fancy and Future. J Immunol 3, 813-816.

Iscove, N.N., Sieber, F., Winterhalter, K.H.(1974): Erythroid colony formation in cultures of mouse and human marrow: Analyses of the requirement for erythropoietin by gel filtration and affinity chromatography on agarose concanavalin A. J Cell Physiol 83, 309-320.

Kurland, J.J., Meyers, P.A., Moore, M.A.S.(1980): Synthesis and release of erythroid colony and burst potentiating activities by purified populations of murine peritoneal macrophages. J Exp Med 151, 839-852.

Old, L.J.(1985): Tumor necrosis factor. Science 230, 630-632.

Pennica, D., Nedwin, G.E., Hayflick, J.S., Seeburg, P.H., Derynck, R., Palladino, M.A., Kohr, W.J., Aggarwal, B.B., Goeddel, D.V.(1984): Human tumour necrosis factor: precursor structure, expression and homology to lymphotoxin. Nature 312, 724-729.

Roodman, G.D., Horadam, V.W., Wright, T.L.(1983): Inhibition of erythroid colony formation by autologous bone marrow adherent cells from patients with the anemia of chronic disease. Blood 62, 406-412.

Scatchard, G.(1949): The attraction of proteins for small molecules and ions. Ann NY Acad Sci 51, 660-672.

Silberstein, D.S., David, J.R.(1986): Tumor necrosis factor enhances eosinophil toxicity to Schistosoma Mansoni larvae. Proc Natl Acad Sci USA 83, 1055-1059.

Tepperman, A.D., Curtin, J.E., McCulloch, E.A.(1974): Erythropoietic colonies in cultures of human marrow. Blood 44, 659-669.

Wang, A.M., Creasey, A.A., Ladner, M.B., Lin, L.S., Strickler, J., Van Arsdell, J.N., Yamamoto, R., Mark, D.F.(1985): Molecular cloning of the complementary DNA for human tumor necrosis factor. Science 228, 149-154.

Young, N.S., Mortimer, P.P., Moore, J.G., Humphries, R.K.(1984): Characterization of a virus that causes transient aplastic crisis. J Clin Invest 73, 224-229.

Zanjani, E.D., McGlave, P.B., Davies, S.F., Banisadre, M., Kaplan, M.E., Sarosi, G.A.(1982): In vitro suppression of erythropoiesis by bone marrow adherent cells from some patients with fungal infection. Br J of Haematology 50, 479-490.

Zuckerman, K.S.(1981): Human erythroid-burst forming units: growth in vitro is dependent on monocytes, but not T lymphocytes. J Clin Invest 67, 702-709.

Résumé

L'auteur rapporte les effets du Tumor necrosis factor alpha, (TNF) produit par les macrophages activés, sur les progéniteurs érythroîdes (CFU-E, BFU-E) et sur les lignées cellulaires K562, HL60 et les cellules HEL. Le TNF inhibe nettement le développement des CFU-E et des BFU-E à de très basses concentrations : 10^{-11} à 10^{-12} M (0.2 U/ml) sans supprimer totalement la formation de colonies érythroîdes. La préincubation de la moelle osseuse avec le TNF pendant 15 minutes est suffisant pour empêcher la formation de colonies érythroîdes. L'adjonction du TNF après la mise en culture a le même effet si elle a lieu dans les 48 premières heures de la culture. Le TNF inhibe également le développement de progéniteurs érythroîdes recueillis à partir de colonies de BFU-E au 5e jour. Les colonies érythroîdes qui se développent en présence de TNF sont toujours nettement plus petites que dans les cultures sans TNF.
Le TNF (10^{-8} à 10^{-10} M) inhibe aussi le développement des lignées cellulaires, K562 et HL60 ainsi que celui des cellules HEL dont 40 à 60% sont sensibles à son action. La préincubation des cellules HL60 avec le TNF pendant 15 minutes entraine une nette diminution de leur développement. Les cellules K562, HL60 et HEL expriment des recepteurs à haute affinité pour le TNF en petit nombre (6000-10.000 récepteurs par cellule). L'analyse par cyto-fluorométrie de la fixation du TNF aux cellules HEL montre que la majorité de ces cellules ont des récepteurs pour le TNF. Par ailleurs la différentiation des cellules K562 par l'hémine entraine une perte de sensibilité au TNF et une diminution de l'expression des récepteurs au TNF sur leur surface.

Ces données suggèrent que le TNF est un inhibiteur puissant, avec une activité rapide et irréversible, des cellules hématopoîétiques. Il agit directement sur une sous population de cellules érythroîdes précoces, en particulier les CFU-E, les BFU-E et peut-être les pro-érythroblastes.

Expression of the TNF genes in response to hematopoietic growth factors

François Dautry and Dominique Weil

Laboratoire d'Oncologie Moléculaire, Institut Gustave Roussy, 94805 Villejuif Cedex, France

ABSTRACT

We have been investigating the genes involved in the proliferative response to Interleukin 2 of murine T lymphocytic cell lines. Using a differential screening approach we have identified TNF beta as one of the genes induced by IL2 in the absence of protein synthesis. We present the nucleotide sequence of a murine TNF beta cDNA and the regulation of both TNF genes by IL2 in T lymphocytic cell lines. This regulation of TNF by a growth factor is not restricted to T lymphocytes as TNF alpha is induced by IL3 in an IL3 dependent cell line.

KEYWORDS

Tumor Necrosis Factor, Lymphotoxin, cDNA, Interleukin 2, Interleukin 3

INTRODUCTION

In order to gain a better understanding of growth regulation by hematopoïetic growth factors , we have been studying gene regulation by interleukin 2 (IL2) in murine lymphocytic cell lines. Our project relies on two complementary approaches : first, the study of RNA expression of genes known to be involved in growth control, such as proto-oncogenes, second, the differential screening of a cDNA library to identify new genes or genes unsuspected to be involved in growth control. Since cell division obviously requires the expression of many genes, involved, for instance, in metabolism or DNA replication, we restricted ourselves to the primary response to IL2 (i.e. the events that do not require <u>de novo</u> protein synthesis). The differential screening approach has been used with success on fibroblasts stimulated by Platelet Derived Growth Factor (PDGF) and has led to the realisation that beside the proto-oncogenes the beta interferon and oligoAsynthetase genes are induced (Zullo 1985). This was the first indication of a direct coupling between growth factors and genes that are thought to be involved in growth inhibition.

When using this approach with hematopoïetic cells two types of experimental difficulties have to be delt with : first, in most cases, hematopoïetic growth factors are required for cell survival in vitro, second, in the case of IL2, quiescent lymphocytes do not express the high affinity IL2 receptor. In our studies, we used a subclone of the IL2 dependent T-lymphocytic cell line CTLL-2 (Baker 1979) that can be deprived of IL2 for 16 hours without a significant loss in viability. We showed previously (Dautry, F., Yu, J., Weil, D. and Dautry, A. submitted to J. Biol. Chem.) that IL2 deprivation of CTLL-2 (subclone G4) cells leads to their acccumulation in G0/G1. Restimulation with recombinant IL2 induces the accumulation of myc and pim mRNA with a maximal response between 1 and 10 hours, while a transient fos signal is observed between 20 minutes and 2 hours. Entry in S phase occurs after about 10 hours, when the abundance of myc and pim RNA is already declining.

In this report we describe the molecular characterisation of a cDNA clone that we have isolated by differential screening and subsequently identified as murine TNF beta. We then studied by Northern blot the regulation of both TNF alpha and TNF beta by IL2 in several IL2 dependent cytotoxic cell lines and by IL3 in the IL3 dependent cell line FDCP-2 (Dexter 1980).

RESULTS

With a substracted cDNA probe (Matrisian 1985) we isolated several cDNA clones corresponding to genes that are induced by IL2. Among these, clone E exhibited the most clear cut regulation, the message being undetectable in cells deprived of IL2. Figure 1 shows the DNA sequence of the coding and 3' untranslated regions of clone E. The longest open reading frame code for a protein of 202 aminoacids with an overall homology of 74% with the human TNF beta precursor (Gray 1984). In the absence of data on the murine TNF beta protein we relied on the high level of homology with the human protein to define leu 34 as being the first residue of the mature protein. The homology between the murine and human proteins can then be divided as 59% for the signal sequence and 76% for the mature protein. In the signal sequence the site of cleavage is part of a conserved strech of residues.

When clone E (TNF beta) was used as a probe on Northern blots of RNA from CTLL-2 cells first deprived of IL2 for 16 hours and then stimulated with recombinant IL2, a clear RNA accumulation was observed following stimulation (Figure 2). The level of message was highest after about 8 hours, and then declines when cells were in S phase (12 to 20 hours). The RNA level in exponentially growing cells was about half of the maximal level following stimulation. Figure 3 shows that, as could be expected from the experimental approach , TNF beta is induced by IL2 even in the presence of cycloheximide. It is

```
        GTTCTCCACATG ACA CTG CTC GGC CGT CTC CAC CTC TTG AGG GTG CTT     48
                     Met Thr Leu Leu Gly Arg Leu His Leu Leu Arg Val Leu
GGC ACC CCT CCT GTC TTC CTC CTG GGG CTG CTG CTG GCC CTG CCT CTA          96
Gly Thr Pro Pro Val Phe Leu Leu Gly Leu Leu Leu Ala Leu Pro Leu
GGG GCC CAG GGA CTC TCT GGT GTC CGC TTC TCC GCT GCC AGG ACA GCC         144
Gly Ala Gln Gly Leu Ser Gly Val Arg Phe Ser Ala Ala Arg Thr Ala
CAT CCA CTC CCT CAG AAG CAC TTG ACC CAT GGC ATC CTG AAA CCT GCT         192
His Pro Leu Pro Gln Lys His Leu Thr His Gly Ile Leu Lys Pro Ala
GCT CAC CTT GTT GGG TAC CCC AGC AAG CAG AAC TCA CTG CTC TGG AGA         240
Ala His Leu Val Gly Tyr Pro Ser Lys Gln Asn Ser Leu Leu Trp Arg
GCA AGC ACG GAT CGT GCC TTT CTC CGA CAT GGC TTC TCT TTG AGC AAC         288
Ala Ser Thr Asp Arg Ala Phe Leu Arg His Gly Phe Ser Leu Ser Asn
AAC TCC CTC CTG ATC CCC ACC AGT GGC CTC TAC TTT GTC TAC TCC CAG         336
Asn Ser Leu Leu Ile Pro Thr Ser Gly Leu Tyr Phe Val Tyr Ser Gln
GTG GTT TTC TCT GGA GAA AGC TGC TCC CCC AGG GCC ATT CCC ACT CCC         384
Val Val Phe Ser Gly Glu Ser Cys Ser Pro Arg Ala Ile Pro Thr Pro
ATC TAC CTG GCA CAC GAG GTC CAG CTC TTT TCC TCC CAA TAC CCC TTC         432
Ile Tyr Leu Ala His Glu Val Gln Leu Phe Ser Ser Gln Tyr Pro Phe
CAT GTG CCT CTC CTC AGT GCG CAG AAG TCT GTG TAT CCG GGA CTT CAA         480
His Val Pro Leu Leu Ser Ala Gln Lys Ser Val Tyr Pro Gly Leu Gln
GGA CCG TGG GTG CGC TCA ATG TAC CAG GGG GCT GTG TTC CTG CTC AGT         528
Gly Pro Trp Val Arg Ser Met Tyr Gln Gly Ala Val Phe Leu Leu Ser
AAG GGA GAC CAG CTG TCC ACC CAC ACC GAC GGC ATC TCC CAT CTA CAC         576
Lys Gly Asp Gln Leu Ser Thr His Thr Asp Gly Ile Ser His Leu His
TTC AGC CCC AGC AGT GTA TTC TTT GGA GCC TTT GCA CTG TAGATTCTAAAG        627
Phe Ser Pro Ser Ser Val Phe Phe Gly Ala Phe Ala Leu
AAACCCAAGAATTGGATTCCAGGCCTCCATCCTGACCGTTGTTTCAAGGGTCACATCCCCACAG       691
TCTCCAGCCTTCCCCACTAAAATAACCTGGAGCTCTCACGGGAGTCTGAGACACTTCAGGGGAC       755
TACATCTTCCCCAGGGCCACTCCAGATGCTCAGGGGACGACTCAAGCCTACCTAGAAGTTCTGC       819
ACAGAGCAGGGTTTTTGTGGGTCTAGGTCGGACAGAGACCTGGACATGAAGGAGGGACAGACAT       883
GGGAGAGGTGGCTGGGAACAGGGGAAGGTTGACTATTTATGGAGAGAAAAGTTAAGTTATTTAT       947
TTATAGAGAATAGAAAGAGGGGAAAAATAGAAAGCCGTCAGATGACAACTAGGTCCCAGACACA      1011
AAGGTGTCTCACCTCAGACAGGACCCATCTAAGAGAGAGATGGCGAGAGAATTAGATGTGGGTG      1075
ACCAAGGGGTTCTAGAAGAAAGCACGAAGCTCTAAAAGCCAGCCACTGCTTGGCTAGACATCCA      1139
CAGGGACCCCCTGCACCATCTGTGAAACCCAATAAACCTCTTTTCTCTGAAAAAAAAA            1197
```

Figure 1 : Nucleotide sequence of a murine TNF beta cDNA. The sequence was determined on both strands by the dideoxynucleotide sequencing technique. The first 24 nucleotides are from the partial genomic sequence of Nedospasov (1986).

therefore likely that TNF induction by IL2 does not require de novo protein synthesis.

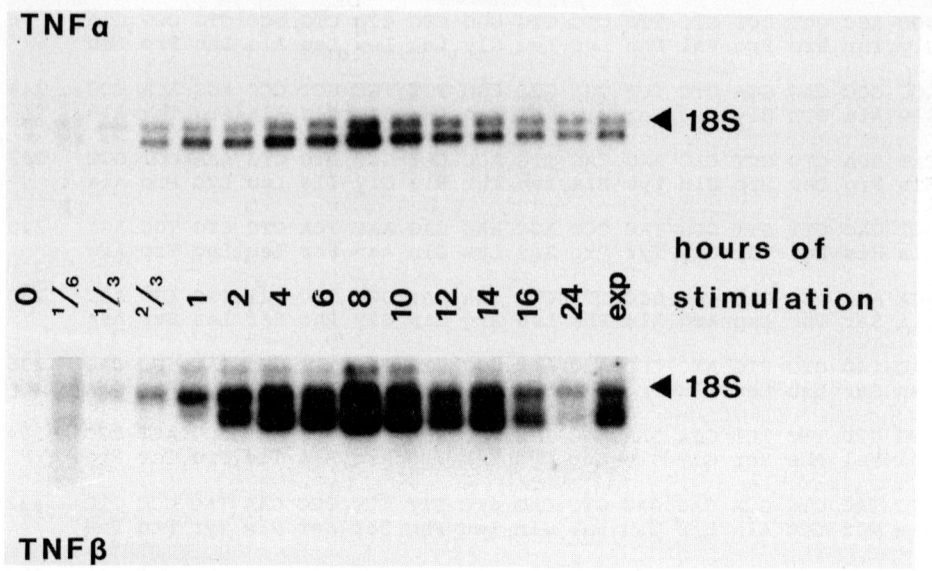

Figure 2. Northern blot analysis of TNF expression in CTLL-2 cells. Cells were first deprived of IL2 for 16 hours and then stimulated for the indicated time with recombinant IL2. 4 ug per lane of total RNA were loaded on agarose-formaldehyde gel, transfered to a nylon membrane and hybridised with a P^{32} labeled RNA probe.

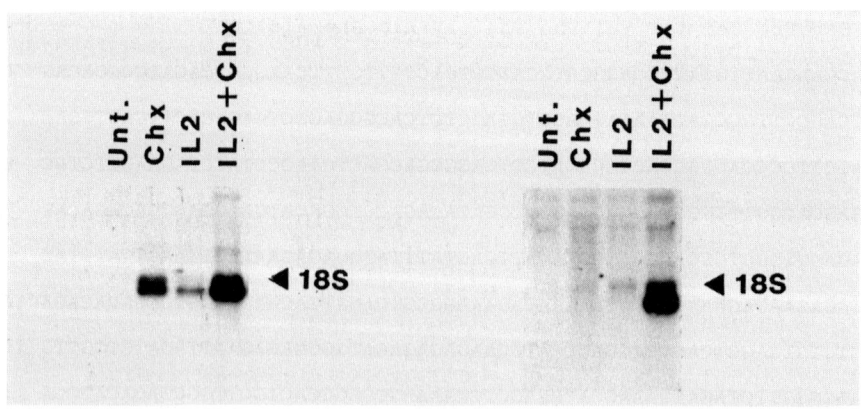

Figure 3. Northern blot analysis of TNF expression (TNF alpha on the left and TNF beta on the right) in CTLL-2 cells in the presence of cycloheximide. After 16 hours of IL2 deprivation the cells were stimulated with recombinant IL2 (5 u/ml) for 1 hour, cycloheximide (10 ug/ml) for 3 hours or a combination of both for 3 hours.

Since TNF alpha and TNF beta have, so far, undistinguishable biological activities and apparently interact with the same cellular receptor, we investigated a possible regulation of TNF alpha by IL2. As shown on figure 2, TNF alpha has a kinetics of induction by IL2 in CTLL-2 very similar to that of TNF beta. Nevertheless TNF alpha was induced more rapidly than TNF beta. Again this induction by IL2 could be detected in the presence of cycloheximide (Figure 3). In both cases, a band comigrating with the 18S ribosomal RNA could be detected. This signal could be selectively reduced under stringent hybridisation conditions, indicating only a partial homology with the probes.

Because TNF alpha has been implicated in the cytotoxicity of monocytes (Feinman 1987), and although there is no indication that it plays a role in lymphocytic cytotoxicity, there was the possibility that the induction of TNF expression by IL2 that we observed correlated with a modulation of the cytotoxic phenotype. As CTLL-2 is no longer functional, we used a different cytotoxic cell line B6.1 (Sekaly 1982) which has retained its cytoxic activity. Northern blot analysis showed that both TNF alpha and beta are induced by IL2 in B6.1, this induction being present in the presence of cycloheximide (data not shown). As the cytotoxic activity of B6.1 is independent of IL2 (Sekaly 1982), it is unlikely that the regulation of TNF by IL2 is part of the cytotoxic phenotype.

To analyse whether such a coupling between a growth factor and TNF was restricted to lymphocytes or could be detected in other systems we turned to the IL3 dependent cell line FDCP-2. Northern blot analysis showed that a WEHI-3 conditioned medium containing IL3 could induce TNF alpha but not TNF beta expression. This induction of TNF alpha was detectable in the presence of cycloheximide. An even more potent inducer of TNF alpha in FDCP-2 was the tumor promotor 4-phorbol-12-myristate-13-acetate (PMA) which simultaneously induced profound changes in cellular morphology and in particular adhesion to the culture dish (data not shown).

DISCUSSION

We describe here the molecular cloning of a murine TNF beta cDNA. The cDNA and the predicted protein are highly homologous to the previously characterised human TNF beta protein and cDNA (Gray 1984). Among the conserved features is the size of the signal sequence (34 and 33 aminoacids, respectively, for the human and murine TNF beta). This distinguishes TNF beta from TNF alpha (Pennica 1984) where the signal sequence is significantly longer (76 and 79 aminoacids, respectively, for the human and murine TNF alpha). This difference in the presequence could correspond to an important difference in the secretion of the proteins, if as suggested by several authors (for instance Feinman 1987) TNF alpha can exist as a transmembrane protein. In the 3' non coding region where there is only a limited homology with the human gene, the sequence TTATTTATTTAT which has been implicated in the control of RNA stability (Shaw 1986) is present in both genes.

TNF beta is known to be produced by lectin activated lymphocytes, while the production of TNF alpha production by lymphocytes is less documented although the cell line HUT-102 has been shown to produce it (Takeda 1986). In view of our finding that T lymphocytes express the TNF alpha mRNA, one possibility is that TNF alpha is released less efficiently from the membrane of lymphocytes than that of macrophages. More generally, the existence of two widely divergent proteins with similar biological activities strongly suggests that they differ in other essential biological parameters such as their ability to reach their target cells.

TNF alpha and beta have initialy been characterised by their cytotoxic activity against some tumor cell lines but it seems unlikely that this is their physiological function. Many other biological activities have been described, including an antiviral activity (Wong 1986), a differentiation inducing activity on monocytes precursors (Takeda 1986) and an induction of GM-CSF production by fibroblasts and endothelial cells (Munker 1986). Although most cell types seem to express the TNF receptor, no data is available on the response of lymphocytes to TNF and we cannot conclude at present whether the TNF expressed by T lymphocytic cell lines takes part in an autocrine loop. Among the neighboring cell types in the lymph node that could be the possible targets of TNF in a paracrine mechanism are the macrophages and the endothelial cells, both of which are known to be sensitive to TNF.

In the case of FDCP-2, the induction of TNF alpha by IL3 is likely to be part of an autocrine loop where the cellular response to the growth factor would include the production of a differentiation inducing factor. The final choice between proliferation and differentiation would then be controled by factors which could influence either the production or the sensitivity to TNF. We are presently investigating this system with the use of exogenous TNF and neutralising antibodies.

Beside its cytotoxic activity against some tumor cell lines TNF is also an inhibitor of some hematopoïetic precursors. In the absence of data on the body distribution of lymphocyte-produced TNF it is difficult to assess a role in the bone marrow, however chronic stimulation of lymphocytes by IL2 such as in the clinical trials of IL2 in cancer patients could lead to a significant production of TNF which could have adverse effects on hematopoïesis.

AKNOWLEDGEMENTS

We have benefited from many discussions with our colleagues in the laboratoire d'oncologie moleculaire. We are grateful to A. Dautry for discussions and help with cell culture. The TNF alpha probe was kindly provided by B. Beutler. This work was supported by CNRS and ARC.

REFERENCES

Baker, P.E., Gillis, S. and Smith, K. A. (1979) : Monoclonal cytolytic T-cell lines. J. Exp. Med. **149**, 273-278.

Dexter, T. M., Garland, J. M., Scott, D., Scolnick, E., and Metcalf, D. (1980) Growth of factor dependent Hemopoietic precursor cell lines. J. Exp. Med. **152**, 1036-1047.

Feinman, R., Henriksen-Destefano, D., Tsujomoto, M., and Vilcek, J. (1987) : Tumor necrosis factor is an important mediator of tumor cell killing by human monocytes. J. Immunol. **138**,6 35-640.

Gray, P.W. et al. : Cloning and expression of a cDNA for human lymphotoxin, a lymphokine with tumor necrosis activity. Nature **312**, 721-724.

Matrisian, L. M., Glaichenhaus, N., Gesnel, M. C., and Breathnach, R. (1985) : Epidermal growth factor and oncogenes induce transcription of the same cellular mRNA in rat fibroblasts. Embo J. **4**, 1435-1440.

Munker, R., Gasson, J., Ogawa, M., and Koeffler, H. P. (1986) : Recombinant human TNF induces production of granulocyte-monocyte colony-stimulating factor. Nature **323**, 79-82.

Nedospasov, S.A., Hirt, B., Shakhov, A.N., Dobrynin, V.N., Kawashima, E., Accolla, R.S. and Jongeneel, C.V. (1986) The genes for tumor necrosis factor and lymphotoxin are tandemly arranged on chromosome 17 of the mouse. Nucl. Acids Research **14**, 7713-7725.

Pennica, D. et al. (1984) Human tumor necrosis factor : precursor structure, expression and homology to lymphotoxine. Nature **312**, 724-729.

Sekaly, R. P., MacDonald, H.R., Zaech, P., and Nabholz (1982) : Cell cycle analysis of cloned cytolytic T cells by T cell growth factor : analysis by flow microfluorometry. J. Immunol. **129**, 1407-1415.

Shaw,G. and Kamen, R. (1986) : A conserved AU sequence from the 3' untranslated region of GM-CSF mRNA mediates selective mRNA degradation. Cell **46**, 659-667.

Takeda, K. et al. : Identity of differentiation inducing factor and tumor necrosis factor. Nature **323**, 338-340.

Wong; G. H. W. and Goeddel, D. (1986) : Tumor necrosis factors alpha and beta inhibit virus replication and synergize with interferons. Nature **323**, 819-822.

Zullo, J.N., Cochran, B. H., Huang, A. S.,and Stiles, C. D. (1985) : Platelet-derived growth factor and double stranded ribonucleic acids stimulate expression of the same genes in 3T3 cells. Cell **43**, 793-800.

Résumé

Nous avons entrepris une étude du contrôle de la croissance par l'interleukine 2 (IL2) dans des lignées lymphocytaires murines. Nous nous sommes particulièrement interessés aux gènes dont l'expression, en fait l'accumulation d'ARN messagers, augmente en réponse à l'IL2 et ce même en présence d'un inhibiteur de synthèse protéique. Par un criblage differenciel d'une banque de cADN nous avons identifié parmi les gènes régulés par l'IL2 un clone correspondant au TNF beta murin (Tumor Necrosis Factor beta, lymphotoxine). Une étude par "Northern blot" a confirmé que dans deux lignées lymphocytaires le TNF beta et le TNF alpha étaient régulés par l'IL2. De même, dans une lignée dépendante de l'interleukine 3 (IL3) l'expression du TNF alpha est régulée par l'IL3. Ce couplage entre un facteur de croissance et un facteur essentiellement connu pour son activité inhibitrice de la croissance ou d'induction de la différenciation, est probablement un élément central du contrôle de l'équilibre prolifération/différenciation.

Discussion

Chairpersons/*Modérateurs*: L.A. Rozenszajn (Israël)
G. Milon (France)

M. AGLIETTA : I have a question for Dr Mannoni. You showed an inhibitory effect of interleukin-2 on hemopoietic progenitor cells in vitro. Can you expand these data? For instance, do you know whether the effect is direct or indirect, or whether myeloid cells have IL-2 receptors?

P. MANNONI : It was too short to make comments also on the effects of IL-2. We were quite surprised to see that IL-2 could give some inhibition of leukemic growth, but we found this kind of inhibition only when we used 1000 units or more. And we thought that this kind of inhibition could be not specific and we tried to purify IL-2 by HPLC, we lost activity. My feeling is clearly that IL-2 could have an effect on some progenitors, but it is very difficult to demonstrate and it could be through the binding to low affinity receptor because to see this effect, we have to use large amounts of IL-2.

M. AGLIETTA : Just a short question to Dr Roodman. You showed that hemin is reducing the number of TNF receptors. Lu and Broxmeyer showed that hemin can enhance the BFU-E growth. My obvious question is whether the addition of hemin influences the inhibitory action of TNF on BFU-E growth?

D. ROODMAN : We have not tested hemin in our BFU-E system, so I cannot comment on that.

L. SOLBERG : I have two questions for Dr Roodman. Have you looked for TGF beta receptors on BFU-E? Have you looked for the effects of TNF alpha on BFU-E in serum free culture?

D. ROODMAN : No.

B. PRAGNELL : I have a question for Dr Roodman. What is the effect of TNF alpha on the rest of the hematopoietic compartments like CFU-GM for example?

D. ROODMAN : On CFU-GM, Dr Broxmeyer showed that TNF alpha is a suppressor. We have also looked at TNF alpha, in a long term marrow culture system which forms osteoclasts. This system is a major interest in our laboratory. TNF stimulates the formation of osteoclasts by stimulating IL-1 production. TNF stimulates IL-1 production, CSF production and CSF-GM in macrophages although that may be a controversy at present. I really do not know any direct effect of TNF alpha on lymphocytes.

Unidentified speaker : I have two questions to Dr Young. First, do you think viral infection is important for only a minority of patients or is it just lack of diagnosis sensitivity? And the second question, could you speculate on the pathogenesis of ATG effectivity in treatment?

N. YOUNG : I will try to give a short answer. My prejudice is that the virus infections are common inciting agents in bone marrow aplasia and that they are not detected because we do not have the appropriate assays. While that is my prejudice, I have to say that we have good evidence that viruses are involved in aplastic anemia in only a minority of cases. There are no serologic data linking the B19 parvovirus to classic aplastic anemia. Even in the case of Epstein-Barr virus infection, only one patient out of 30 studied retrospectively was positive by molecular assays for this virus in bone marrow. At least to the degree of sensitivity that we now possess, Epstein-Barr virus is unlikely to be commonly associated with aplastic anemia. I should add that we have looked hard for a retrovirus in aplastic anemia by electron microscopy and reverse transcriptase assays without success. In terms of the immunosuppressive effects of ATG, I mentioned that there is also a significant clinical response rate to cyclosporine. This is important mainly because it tells us more about the pathogenesis of aplastic anemia than does the response to ATG. ATG is assumed to work because it is immunosuppressive, but it in fact has other effects as well : it is quite a non-specific immunological agent, and our own data and those of others show that it acts to stimulate lymphocytes in vitro to produce growth factors, including GM-CSF. Cyclosporine, on the other hand, is a much more specific inhibitor of T cell proliferation. Whether ATG acts as an immunosuppressant or immunostimulant may become clear from planned clinical trials with hematopoietic growth factors ; if patients respond to these at the same rate as they do to ATG, it would suggest that ATG might act by a similar positive mechanism.

H. BROXMEYER : This is in response to Dr Aglietta's comment and also something for Dr Roodman. Our experiments assessing the effects of TNF alpha and TNF beta on erythroid colony formation were done in the presence of hemin. What you are doing is differentiating the cells with hemin and then looking for an effect, so the experiments are different. Question for Dr Mannoni : there must be at least somewhere between 5 and 10 groups now that have shown that purified recombinant human gamma interferon can inhibit colony formation of pretty much most of the progenitors, so I wonder if there is something different in the assay system that you are using because it is probably not the interferon. Although I guess that is possible. Do you have any comments on that?

P. MANNONI : Of course, I was aware of these reports and the only thing we have clearly done in these studies is to use bone marrow completely depleted in T cells and in monocytes and macrophages as much as possible and also, we have been using low dose of interferon and we monitored the dose of interferon we used, by using doses of interferons which clearly for example give an increase of HLA class one or HLA-DR antigens on some selected cell lines. I think the main difference between these reports and our work is about the dose and the fact that we tried to purify the stem cells as much as possible.

H. BROXMEYER : I do not think that it is the stem cell purification that is the difference because quite a few people have removed T cells and monocytes and the interferons still suppress colony formation. It is probably the dosage or the way the units are calculated before it is used that would explain the varying results between the different groups.

P. MANNONI : I agree.

L. DOUAY : I have a question for Dr Mannoni. You showed the role of G-CSF on long term blast culture ; you may be aware of a report of Dexter on human long term cultures showing the spontaneous disappearance of blast cells and my question is : did you check the production of G-CSF in such cultures because sometimes it works and sometimes it does not work. Did you check that?

P. MANNONI : Yes, we checked this by doing a long term bone marrow culture as described by Dexter, that means with a huge volume of bone marrow and the only thing we have been able to see is the fact that by using a high concentration of supernatant, it was possible to get some CSA activity, but we were not able to clearly demonstrate that this was more G-CSF or CSF-1 or GM-CSF, it is a kind of activity giving an increase in colony formation by normal bone marrow. I would like to have monoclonal antibodies against these factors to detect this activity.

M. LENFANT : It is a question to Dr Roodman. I would like to know if you have any documents on activity of TNF on hematopoietic stem cells CFU-S in mice.

D. ROODMAN : No, we don't do murine cultures.

P. RESNITZKY : Dr Young, in the light of the reported endogenous interferon production by normal and leukemic CSF-treated cells, can you elaborate on the possibility that in aplastic anemia an abnormal autocrine production of interferon may play a role in its pathogenesis?

N. YOUNG : I think that it may be very difficult to distinguish an increase in normal endogenous interferon production from pathologic interferon production. Resnitzky et al. showed beta interferon to be a product of normal bone marrow stroma. If other cells are also producing interferon, for example pathologically, these endogenous and pathological factors would synergize in their inhibitory activity on bone marrow cell proliferation. A second crucial aspect in that work, which I could not give justice to in my brief talk, is that local production may have a much greater effect than exogenously administered interferon, as shown for surface HLA expression and oligosynthetase activity. For example, when we give gamma interferon in clinical trials to patients, the number of colony forming cells do fall, but hematopoietic suppression requires thousand to millions of units of administered interferon. Interferon produced locally in the marrow might be concentrated in the membrane or in the extracellular matrix and, because of its local concentration, act much more effectively to suppressing cell proliferation. I should add that there is a report from Professor Gluckman's group (in abstract form) that beta interferon is produced by lymphocytes from aplastic patient cells. This might be a strange interferon (beta interferon from lymphocytes?) but some of our own data have suggested that the interferon molecule that we have measured might be an alpha-gamma hybrid form.

M. PLOSZCZYNSKI : Dr Young said that regarding interferon action, there is clear competence between positive growth factors and way of action of interferon. Now, on the other hand, we have some reports from Dr Roodman regarding TNF who says that there is a clear correlation between down regulation of TNF receptors and its biological action. It means that for example in course of differentiation, the number of receptors for TNF goes down and the cell becomes not responsive to TNF action. My question is if you have some idea about the biological effects of these negative growth factors. Are they due to the interaction on the receptor level or the transduction level or affect production of growth factors?

D. ROODMAN : I think it is unclear for the following reasons. If you look closely at our data, just taking the hematopoietic cell lines or CFU-E or BFU-E, although you get inhibition, you don't get a 100% inhibition. 95% of HeLa cells express TNF receptors. Although there is a heterogenous distribution of receptors per cell, only 50% of them are inhibited with a concentration of 10^{-6} M TNF. Even if you use 10^{-8} M TNF, you still get 50-60% inhibition. You won't get a 100%. There are always cells that are not going to be affected by TNF, although they express TNF receptors. So then the question is : is there data to support the fact that there are resistant cells that express TNF receptors with the same Kd as cells that are sensitive to TNF ? Data has been published using HeLa cells which shows that resistant cells have TNF receptors. Others have shown that resistant cells don't have TNF receptors, and now we have both sides of the argument in the data. If a cell has a TNF receptor and becomes resistant to TNF, it is possible that the cell has lost its TNF receptor. Alternatively, the distribution of TNF receptors is heterogenous. Maybe you will need a certain number of receptors per cell in order to be sensitive to TNF. I think it is still unclear what is the exact mechanism of resistance to TNF.

M.A. ALLIERI : The same question for Dr Mannoni and Dr Roodman. Did you study gamma interferon or TNF effect on human clonogenic leukemic cells in culture and not only on cell lines?

P. MANNONI : From patients you mean ? We tested only few patients and we tested mostly patients who probably expressed the receptor for CSF-1 at the surface of their leukemic cells. And we found a strong synergy then, between inhibitory effects of CSF-1 plus gamma interferon.

D. ROODMAN : Your question was : do we test TNF in leukemic patient marrows? No, we have not tested leukemic cells for several reasons. One, Dr Broxmeyer and several others have shown that TNF inhibits leukemic cell growth. Two, we were interested in the anemia of chronic disease which is related to our major interest, macrophage modulation of erythropoiesis. The hypothesis that we have been working on in the laboratory is that the anemia of chronic disease is the result of macrophage activation. If you think about what diseases are associated with anemia of chronic disease - chronic inflammatory diseases, antigenic stimulation, chronic infections and tumors - they all activate macrophages. Therefore, we have been interested in the effects of macrophage products on normal hematopoietic progenitors rather than on leukemic cells.

Brief reports

Communications brèves

Inhibition of the growth of normal megakaryocytic progenitor cells by recombinant human alpha and gamma-Interferons

Jürgen Greher, Arnold Ganser, Benhardt Völkers, Carmelo Carlo Stella* and Dieter Hoelzer

Department of Hematology, University of Frankfurt, FRG
* Department of Internal Medicine and Medical Therapy, University of Pavia, Italy

Interferons (IFNs) have been shown to suppress the in vitro proliferation and/or differentiation of normal human pluripotent hemopoietic progenitor cells (CFU-GEMM) (1,2), erythroid bursts (BFU-E) and granulocyte-macrophage progenitor cells (CFU-GM) (2,3,4,5). But no information is yet available concerning the effect of IFNs on human megakaryocytic progenitors (CFU-Mk) in vitro. Futhermore, the role of hemopoietic accessory cells, i.e. monocytes-macrophages and T-lymphocytes, in the mediation of the suppressive effects of IFNs is still controversial (2,5).

Therefore the aim of the present study was (a) to investigate the in vitro effect of highly purified recombinant human IFN-alpha (rIFN-alpha) and recombinant human IFN-gamma (rIFN-gamma) on normal human megakaryopoiesis and (b) to evaluate the influence of hemopoietic accessory cells on the inhibitory effect of rIFNs on in vitro hemopoiesis.

MATERIALS AND METHODS

Bone marrow cells were obtained from hematologically normal donors. Mononuclear, light density bone marrow cells (LDBMC) were separated by centrifugation on a Ficoll-Hypaque gradient (density 1.077 g/ml). In three separate experiments, the effect of rIFNs was tested on unseparated LDBMC, non adherent LDBMC (LDBMC-AC$^-$), T-cell depleted LDBMC (LDBMC-T$^-$) and non adherent, T-cell depleted LDBMC (LDBMC-AC$^-$-T$^-$) derieved from the same marrow samples.

Removal of the adherent monocytes-macrophages (AC) was done by incubation in plastic tissue culture flasks. Routinely 3% CD w14 positive cells were found after this treatment, T-cell depletion, either from LDBMC or LDBMC-AC$^-$, was achieved by rosetting with 2-amino-ethylisothiouronium bromide treated sheep red blood cells. Less than 1% CD 3 positive cells were found after the depletion.

Colony assay for hemopoietic progenitors:
1×10^5 LDBMC or separated cell fractions were plated in the presence or absence of different concentrations of rIFN-alpha and rIFN-gamma in 1ml aliquots of IMDM containing 30% human fresh frozen plasma from a single normal donor, 5% PHA leukocyte conditioned medium, 50 µM 2-mercaptoethanol, 1 U recombinant human erythropoietin /Genetics Institute, USA) and 0.9% (w/v) methylcellulose. After incubation for 14 days at 37°C in a humidified atmosphere with 5% CO_2 four culture dishes for each data point per experiment were scored with an inverted microscope.
Specific activity of rIFN-alpha (Genentech, USA) was 3.3×10^8 U/mg of protein and of rIFN-gamma (Biogen, USA) 2.4×10^7 U/mg of protein. Monoclonal antibodies EB-1 against rIFN-alpha and GZ-4 against rIFN-gamma were used.

RESULTS

When LDBMC were cultured in the presence of rIFNs throughout the entire culture period, both rIFN preparations ($10-10^4$ U/ml) showed asignificant, dose-dependent suppression of colony growth (Fig. 1); for rIFN-alpha the degree of inhibition of CFU-Mk, CFU-GEMM, BFU-E and CFU-GM was in the same range and 50% inhibition was reached with 22 U/ml (CFU-Mk), 59 U/ml (CFU-GEMM), 67 U/ml (BFU-E) and 133 U/ml

(CFU-GM). The suppressive effect of rIFN-gamma on the colony growth from CFU-Mk and CFU-GEMM was similar to that of rIFN-alpha, however CFU-GM and BFU-E appeared to be less sensitive. 50% inhibition occured at 59 U/ml (CFU-Mk), 101 U/ml (CFU-GEMM), 817 U/ml (CFU-GM) and 8819 U/ml (BFU-E).

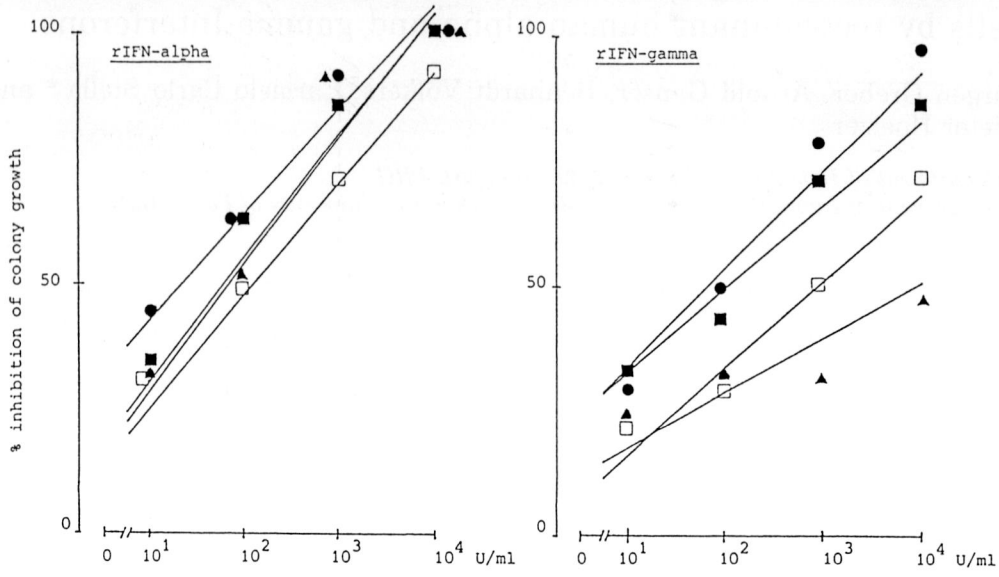

Figure 1
Effect of rIFN-alpha and rIFN-gamma on hemopoietic progenitor cells in LDBMC six experiments (CFU-GEMM five exp.)

■ CFU-GEMM r=.998 (p 0.001)   ■ CFU-GEMM r=.993 (p 0.001)
● CFU-Mk   r=.993 (p 0.001)   ● CFU-Mk   r=.991 (p 0.001)
▲ BFU-E   r=.984 (p 0.01)   ▲ BFU-E   r=.957 (p 0.05)
□ CFU-GM   r=.998 (p 0.001)   □ CFU-GM   r=.988 (p 0.01)

The addition of monoclonal antibodies against either rIFN-alpha or rIFN-gamma resulted in a neutralization of the suppressive effect of the respective rIFN on colony growth from CFU-Mk as shown in Table 1.
The inhibitory effect of rIFN-alpha, when added to LDBMC-AC$^-$, LDBMC-T$^-$ and LDBMC-AC$^-$-T$^-$, could still be demonstrated (Fig. 2), while the inhibitory effect of rIFN-gamma was significantly reduced after depletion of adherent and/or T-cells.

Table 1

Selective inactivation of suppressive effects of rIFN-alpha and rIFN-gamma on megakaryocytic progenitor cells by pre-incubation of rIFNs with their respective monoclonal antibodies*

	CFU-Mk
Control medium	9.8 ± 0.5
rIFN-alpha (10^3 U/ml)	0.7 ± 0.2, p 0.01
rIFN-gamma (10^3 U/ml)	3.2 ± 0.5, p 0.01
Anti-rIFN-alpha (EB-1)	8.8 ± 1.2
Anti-rIFN-gamma (GZ-4)	8.7 ± 0.5
EB-1 + rIFN-alpha	12 ± 0.6, ns
GZ-4 + rIFN-gamma	7.5 ± 0.6, ns
EB-1 + rIFN-gamma	4.5 ± 0.9, p 0.02
GZ-4 + rIFN-alpha	1.0 ± 0.4, p 0.00

*Colony growth per 1 x 10^5/ml LDBMC in the absence and presence of 10^3 U/ml of rIFN (preincubated for 90 min with and without 10^3 neutralizing units of monoclonal antibody against rIFN-alpha or rIFN-gamma).

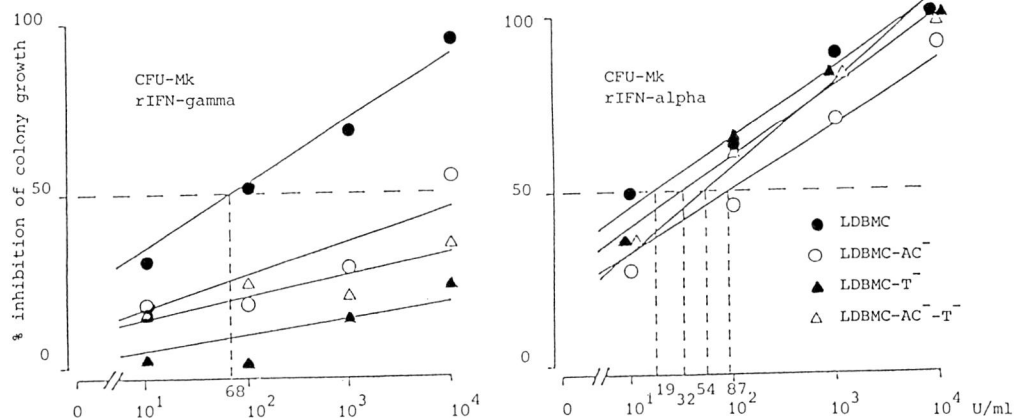

Figure 2
Influence of removing accessory cells from LDBMC on the inhibition of colony growth from CFU-Mk by rIFN.
Three separate sets of experiments were performed concomitantly on unseparated and accessory cell depleted marrow cell fractions derieved from the same marrow samples.

DISCUSSION

Both rIFN-alpha and rIFN-gamma are potent inhibitors of human megakaryopoiesis suppressing the in vitro growth of megakaryocytic progenitors (CFU-Mk). The selective inhibition of the antiproliferative effect of both rIFN preparations by the respective monoclonal antibodies confirms that the inhibitory effect was indeed due to the rIFN used.

While the inhibition of CFU-Mk by rIFN-alpha is in the same range as that of the progenitor cells CFU-GEMM, BFU-E, and CFU-GM, the results for rIFN-gamma are somewhat different. CFU-GM and especially BFU-E seem to be much more resistent to rIFN-gamma differing thereby from previous reports (2,5). One of the reasons for this difference could be the presence of recombinant human erythropoietin instead of sheep plasma derieved erythropoietin.

The experiments in which the inhibition of rIFN-gamma is abrogated by depletion of adherent cells and/or T-cells strongly indicate that the mechanims underlying the suppressive effect of the two types of rIFNs on CFU-Mk are different. Recombinant IFN-alpha apparently acts directly on the progenitor cells although regeneration of the depleted adherent and T-cells cannot be ruled out, whereas the effect of rIFN-gamma is largely mediated by accessory cells.

REFERENCES

1. Neumann HA, Fauser AA. Exp Hematol 10: 587, 1982
2. Broxmeyer HE, Lu L, Platzer E, Feit C, Juliano L, Rubin BY: J Immunol 131: 1300, 1983
3. Nissen C, Speck B, Emodi G, Iscove NN: Lancet 1: 203, 1977
4. Ortega JA, Ma A, Shore NA, Dukes PP, Merigan TC: Exp Hematol 7: 145, 1979
5. Mamus SW, Beck-Schroeder S, Zanjani ED: J Clin Invest 75: 1496, 1985

Recombinant alpha and gamma-Interferons synergistically inhibit the *in vitro* growth of hemopoietic progenitors (CFU-GEMM, CFU-MK, BFU-E, CFU-GM) from patients with myelofibrosis with myeloid metaplasia

Carmelo Carlo Stella, Mario Cazzola, Arnold Ganser, Bernhardt Völkers, Laura Dezza, Federica Meloni, Paolo Pedrazzoli and Edoardo Ascari

Department of Internal Medicine, University of Pavia, I-27100 Pavia, Italy, and Department of Hematology, University of Frankfurt, D-6000 Frankfurt am Main 70, FRG

KEY WORDS

Recombinant Interferons, Myeloproliferative disorders, CFU-GEMM

INTRODUCTION

Myelofibrosis with myeloid metaplasia (MMM) is a chronic myeloproliferative disorder due to clonal expansion of a pluripotent hemopoietic progenitor cell (Adamson et al., 1978). No definitive treatment has as yet been devised for this condition which shows a marked variability in clinical course, with about 25% of patients dying within 18 months of diagnosis (Barosi et al., 1981).

Interferons (IFNs) are a group of naturally occurring inducible proteins that exert several biologic actions (Borden et al., 1981), including inhibitory effects on the growth of hemopoietic progenitor cells from normal marrow (Broxmeyer et al., 1983; Ganser et al., submitted, 1986) as well as from patients with chronic myeloid leukemia (Oladilupo-Williams et al., 1981). Recently, it has been shown that rIFN-alpha is able to induce hematologic remission and complete suppression of Philadelphia positive cells (Talpaz et al., 1986), associated with significant slowing of the leukocyte doubling time and prolongation of remission duration (Bergsagel et al., 1986), in patients with chronic myelogenous leukemia. The dose-limiting side effects and the possible appearance of resistant clones are the two major limitations to in vivo treatment with rIFN-alpha.

The aim of this study was to analyze the effects of rIFN-alpha and rIFN-gamma used as single agents or in combination on the in vitro growth of pluripotent (CFU-GEMM) and lineage-restricted (CFU-Mk, BFU-E, CFU-GM) hemopoietic progenitor cells from patients with MMM.

MATERIALS AND METHODS

Peripheral blood cells were obtained from 18 patients and 10 normal controls, who had given informed consent. Light density mononuclear cells (LDMNC, <1.077 g/ml) were cultured at $1-5 \times 10^5$ cells/ml in Iscove's modified Dulbecco's medium supplemented with 30% human fresh frozen plasma from a single donor; 5% PHA-leukocyte conditioned medium, 5×10^{-5} M 2-mercaptoethanol, 1.0 U erythropoietin (Step III, Connaught Laboratories) and 0.9% (w/v) methylcellulose. After incubation for 14 days at 37°C in a humidified atmosphere supplemented with 5% CO_2, the cultures were examined with an inverted microscope (Carlo Stella et al., 1986).

The effect of rIFN-alpha (specific activity: 3.3×10^8 U/mg of protein, Karl Thomae Co., Biberach, F.R.G.) and rIFN-gamma (specific activity: 2.4×10^7 U/mg of protein, Biogen Research Corp., Cambridge, MA, U.S.A.) on hemopoietic progenitor cell growth was evaluated as follows. (a) LDMNC were cultured throughout the entire incubation period with varying doses of rIFN-alpha and rIFN-gamma ($10-10^4$ U/ml). (b) In some patients low doses of rIFN-alpha (10 or 10^2 U/ml) were added to cultures containing the same amount of rIFN-gamma in order to test the additive or synergistic effect of the two proteins.

RESULTS

The number (mean ± SEM) of progenitor cells grown per unit volume from the blood of the 18 MMM patients was significantly increased for CFU-GEMM (594 ± 253 vs 32 ± 3, $P \leq 0.05$), CFU-Mk (1033 ± 410 vs 53 ± 6, $P \leq 0.005$), BFU-E (4799 ± 2020 vs 267 ± 43, $P \leq 0.05$) and CFU-GM (5438 ± 2505 vs 251 ± 53, $P \leq 0.05$) as compared to the growth of normal controls.

When peripheral blood mononuclear cells were exposed to rIFN-alpha and rIFN-gamma, a statistically significant, dose-dependent suppression of CFU-GEMM, CFU-Mk, BFU-E, and CFU-GM colony growth was seen, which could be selectively blocked by respective monoclonal antibodies. Regression analysis showed that inihibition was linearly related to the logarithm of rIFN concentration over the range tested ($10-10^4$ U/ml). The rIFN-alpha and rIFN-gamma concentrations that caused 50% inhibition of colony growth were 37 and 163 U/ml for CFU-GEMM, 16 and 69 U/ml for CFU-Mk, 53 and 146 U/ml for BFU-E, and 36 and 187 U/ml for CFU-GM, respectively. In seven normal subjects a dose-dependent suppression of colony growth was also seen, and dose-response curves were not significantly different from those observed in patients with MMM.

Table 1. Percentages of growth inhibition (mean ± SE) obtained in five patients with MMM using rIFNs alone or in combination[*]

rIFNS (U/ml)	CFU-GEMM	CFU-Mk	BFU-E	CFU-GM
	% inhibition in colony growth (mean ± 1 SEM)			
alpha (10)	36±12	44±10	28±6	30±6
gamma (10)	6±3	26±7	9±3	16±6
alpha+gamma (10+10)	84±7[+]	87±6[+]	79±5[+]	72±13[+]
alpha (10^2)	54±10	74±9	57±7	59±13
gamma (10^2)	31±9	55±9	43±6	35±16
alpha+gamma (10^2+10^2)	91±9[**]	97±1[**]	92±6[**]	85±8[**]
alpha (10^3)	82±2	91±4	88±4	87±4
gamma (10^3)	72±7	80±9	76±13	65±16

* Each value represents the percentage inhibition (mean ± 1SEM) of five separate experiments.
+ Significantly different from the value obtained with 10^2 U/ml rIFN-alpha ($P \leq 0.005$) or rIFN-gamma ($P \leq 0.005$).
** Not significantly different ($P \geq 0.05$) from the value obtained with 10^3 U/ml rIFN-alpha or rIFN-gamma.

As shown in "Table 1" rIFN-alpha and rIFN-gamma used in combination at the dose of 10 U/ml exerted a marked synergism in suppressing hemopoietic progenitor cell growth. Results obtained in three normal subjects (data not shown) with combinations of rIFNs demonstrated that the percentages of growth inhibition were comparable to those obtained from patients with MMM.

DISCUSSION

Consistent with previous reports, the in vitro growth of hemopoietic progenitor cells CFU-GEMM, CFU-Mk, BFU-E and CFU-GM was homogeneously increased about 20-fold in our MMM patients as compared to normal controls.

Our data clearly indicate that the growth of circulating hemopoietic progenitors from patients with MMM was suppressed in a dose-dependent manner by rIFNs.

rIFN synergism in the inhibition of normal bone marrow-derived progenitor cell growth has been reported (Broxmeyer et al., 1985; Raefsky et al., 1985). Our data strongly suggest that such a synergistic mechanism is also effective in suppressing hemopoietic progenitors from MMM. Since it is well-known that rIFNs, especially when used at high doses, have many side effects in vivo (Quesada et al., 1986), the synergistic suppressive activity on hemopoietic progenitor cell growth has potential relevance for the use of rIFNs in vivo.

REFERENCES

Adamson, J.W., Fialkow, P.J. (1978): The pathogenesis of myeloproliferative syndromes (Annotation). Br. J. Haematol. 38, 299-303.
Barosi, G., Cazzola, M., Frassoni, F., Orlandi, E., Stefanelli, M. (1981): Erythropoiesis in myelofibrosis with myeloid metaplasia: recognition of different classes of patients by erythrokinetics. Br. J. Haematol 48, 263-272.
Bergsagel, D.E., Haas R.H., Messner, H.A. (1986): Interferon Alfa-2b in the treatment of chronic granulocytic leukemia. Sem. Oncol. XIII (suppl. 2), 29-34.
Borden, E.C., Ball, L.A. (1981): Interferon: biochemical cell growth inhibitory and immunological effects. In Progress in Hematology (vol XII), ed E.R. Brown, pp 299-339. New York: Grune and Stratton.
Broxmeyer, H.E., Lu, L., Platzer, E., Feit, C., Juliano, L., Rubin, B.Y. (1983): Comparative analysis of the influences of human gamma, alpha, and beta interferons on human multipotential (CFU-GEMM), erythroid (BFU-E) and granulocyte-macrophage (CFU-GM) progenitor cells. J Immunol 131, 1300-1305.
Broxmeyer, H.E., Cooper, S., Rubin, B.Y., Taylor, M.W (1985): The synergistic influence of human interferon-gamma and interferon-alpha on suppression of hematopoietic progenitor cells is additive with the enhanced sensitivity of these cells to inhibition by interferons at low oxygen tension in vitro. J. Immunol. 135, 2502-2506.

Carlo Stella, C., Ganser, A., Hoelzer, D. (1986): Defective "in vitro" growth of the hemopoietic progenitor cells CFU-GEM, CFU-Mk, BFU-E, and CFU-GM in the acquired immunodeficiency syndrome (AIDS). J. Clin. Invest., in press.

Oladilupo-Williams, C.K., Svet-Moldavskaya, I., Vilcek, J., Ohnuma, T., Holland, J.F. (1981): Inhibitory effect of human leukocyte and fibroblast interferons on normal and chronic myelogenous leukemic granulocytic progenitor cells. Oncology 38, 356-360.

Raefsky, E.L., Platanias, L.C., Zoumbos, N.C., Young, N.S. (1985): Studies of interferon as a regulator of hemopoietic cell proliferation. J. Immunol. 135, 2507-2512.

Quesada, J.R., Talpaz, M., Rios, A., Kurzrock, R., Gutterman, J.U. (1986): Clinical toxicity of interferons in cancer patients: a review. J. Clin. Oncol. 4, 234-243.

Talpaz, M., Kantarjian, H.M., McCredie, K., Trujillo, J.M., Keating, M.J., Gutterman, J.U. (1986): Hematological remission and cytogenetic improvement induced by recombinant human interferon alpha in chronic myelogenous leukemia. N. Engl. J. Med. 314, 1065-106

ACKNOWLEDGMENTS

Work supported by CNR (Consiglio Nazionale delle Ricerche), grant no. 86.00352.44 and from AIRC (Associazione Italiana per la Ricerca sul Cancro).

In vitro sensitivity of hematopoietic precursor cells to recombinant Interferon alpha (rIFN-alpha), recombinant Interferon gamma (rIFN-gamma) and recombinant tumor necrosis factor alpha (rTNF-alpha) in normal controls and in patients with CML : relationship to the in vivo response

Dietmar Geissler, Günther Gastl, Walter Aulitzky, Herbert Tilg, Günther Konwalinka and Christoph Huber

Department of Internal Medicine, Innsbruck University Hospital, 6020 Innsbruck, Austria

KEY WORDS

CML precursor cells, rIFN-alpha-2, rIFN-gamma, rTNF-alpha

INTRODUCTION

IFN-alpha has recently been evaluated for the treatment of CML (1, 2, 7, 8, 9). Using natural and recombinant DNA-derived IFN-alpha preparations, significant improvement of hematological parameters was obtained in the majority of patients with Ph^{1+} CML in benign phase. Even in patients with CML in blastic crisis, a transient reduction of the blast cell counts and suppression of a secondary Ph^{1+} clone were observed occasionally (9). Most notably, however, treatment with IFN-alpha was reported to increase the proportion of Ph^{1-} bone marrow cells (8).

One aim of this study was to compare the inhibitory effect of rIFN-alpha-2, rIFN-gamma and rTNF-alpha on BFU-E, CFU-E, CFU-GM and CFU-MEG colony formation in CML patients with that of normal controls. The other aim was to establish the utility of such in vitro assays in predicting the in vivo responsiveness of CML patients to IFN-alpha-2.

MATERIAL AND METHODS

Normal Controls, Patients and Treatment

Bone marrow specimens were obtained from five healthy donors and 13 patients with Ph^{1+} CML who had given informed consent. Subsequently, seven of the CML patients were treated with rIFN-alpha-2 for a median time of nine months (range: three to nine months). The interferon preparation was purchased from Boehringer-Ingelheim International (FRG) and had a specific activity of 3.2×10^8 U/mg protein. RIFN-alpha-2 was administered subcutaneously at a dose range of 1 to 8 mio. U daily.

Response Criteria

The criteria for response were defined according to Talpaz et al (7): hematologic remission was defined as normalization of peripheral blood counts and morphology; partial remission was defined as a decline in peripheral blood counts by at least

50% to less than 20,000/ml; failure was defined as a lack of cytoreductive effect on the WBC or platelet counts.

Microagar Culture System
Microagar cultures were prepared, stained and scored as previously described (3, 4, 5).

RESULTS

Effect of rIFN-alpha-2, rIFN-gamma and rTNF-alpha on Colony Formation in Normal Controls

In order to determine whether rIFN-alpha-2, rIFN-gamma and rTNF-alpha exert a different suppressive effect on erythrocytic, granulomonocytic and megakaryocytic colonies, we studied the BFU-Es, CFU-Es, CFU-GMs, CFU-MEGs and CFU-TLs from the bone marrow of five healthy donors. The results are presented in Fig. 1. A dose-dependent suppression of BFU-E, CFU-E, CFU-GM and CFU-MEG was observed in all experiments using rIFN-alpha, rIFN-gamma or rTNF. More than 85% suppression of colony growth was achieved with concentrations of 1,000 U/ml and 10,000 U/ml of rIFN-alpha-2.

It is notable that the median concentration of rIFN-alpha-2 as well as rIFN-gamma required to achieve 50% growth inhibition (IC50) was the highest for CFU-E (800 U/ml) and CFU-GM (620 U/ml), while BFU-E and CFU-MEG colony formation proved to be five- to tenfold more sensitive to rIFN-alpha-2 or rIFN-gamma. Compared to the antiproliferative effect obtained with rIFN-alpha or rIFN-gamma, rTNF-alpha produced a much lower growth inhibition of BFU-E (IC50: BFU-E 3117 U, CFU-MEG 590 U, CFU-GM 708 U).

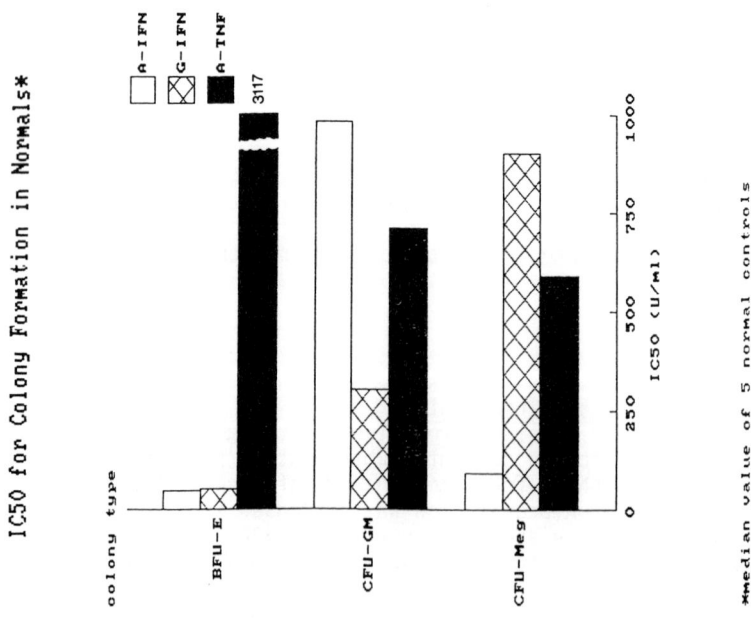

Effect of rIFN-alpha-2 on Colony Formation in Ph^{1+} CML Patients

Microagar cultures were set up with bone marrow mononuclear cells from 13 Ph^{1+} CML patients. When compared with normal controls, different patterns were seen between the various precursor cells in CML patients: while sensitivity of BFU-E and CFU-MEG in CML patients did not differ from that of healthy controls, in some of the CML cases CFU-GM revealed an increased sensitivity to inhibition with rIFN-alpha-2. IC50 CFU-GM concentrations below the 97.5% confidence limit of the normal controls were observed in four out of 13 cases (Fig. 1). Other CML patients clearly exhibited CFU-GM formation with increased resistance to inhibition by rIFN-alpha-2. As shown in Fig. 1, four of 13 CML cases required rIFN-alpha doses for inhibition of GM-CFU which were clearly outside the 97.5% confidence limit of the normal controls. It thus appears that individual CML patients differ markedly from normal controls in the sensitivity of their granulocytic-monocytic precursor cell growth to rIFN-alpha. Some of these CML patients exhibited increased, and others almost no, sensitivity to this growth regulating protein.

Antiproliferative Activity of rIFN-alpha-2: Relationship of In Vitro and In Vivo Results

In further clinical experiments, we tested whether such differences would also relate to in vivo sensitivity of CML patients to treatment with rIFN-alpha-2. Seven of the 13 Ph^{1+} CML patients were treated in a clinical phase 2 trial with IFN-

Table 1. Response of Ph^{1+} CML in relation to pretreatment IC50 (CFU-GM)

Patient Initials	IC50(CFU-GM) (U/ml)	Time of Therapy (Months)	rIFN-alpha-2 Cumulative Dose (mio. U)	$WBCC_3$ $(x10^3/\mu l)$	Response*	Cytogenetic Improvement
Z.O.	6	0	-	21.6	-	-
		3	90	15.4	PR	-
		6	179	11.5	PR	-
		9	369	8.3	CR	+**
M.G.	50	0	-	23.6	-	-
		3	86	14.5	F	-
		6	404	13.6	F	-
		9	646	10.5	PR	-
H.H.	230	0	-	40.7	-	-
		3	175	19.8	PR	-
		6	457	11.3	PR	-
		9	825	13.0	PR	-
G.D.	960	0	-	7.6	-	-
		3	112	8.6	PR	-
O.A.	10,000	0	-	48.6	-	-
		3	116	68.0	F(BC)	-
M.L.	10,000	0	-	6.7	-	-
		3	180	47.0	F	-
		6	600	58.8	F	-
		9	1132	37.0	F	-
S.H.	10,000	0	0	89.0	-	-
		3	168	112.0	F(BC)	-

*) according to the criteria of Talpaz et al (see Ref. 9)
  PR partial response, CR complete response, F failure, BC blastic crisis
**) six of eight metaphases were Ph^{1+}

alpha-2 for a median time of nine months (range: three to nine months). Clinical responses achieved in these patients in relation to the IC50 CFU-GM values are given in Table 1. As shown, two of the seven patients revealed IC50 CFU-GM values much lower than normal controls. One complete and one partial clinical response was achieved and cytogenetic improvement with significant reduction of the frequency of the Ph^{1+} clone was achieved. Two patients exhibited IC50 CFU-GM values which were in the range of normal controls. Partial clinical responses were seen in both these cases, but no evidence for reduction of the Ph^{1+} clone was obtained. Three of the seven patients had IC50 CFU-GM values which were ten times higher than those of healthy individuals. All these cases failed on treatment with rIFN-alpha-2 and two of them underwent blastic transformation of their disease within three months of treatment.

DISCUSSION

The major target cell for the neoplastic transformation in Ph^{1+} CML is a pluripotent stem cell giving rise to erythrocytic, granulocytic-monocytic, megakaryocytic and lymphocytic progenitor cells (6). The major proliferative activity during benign phase of Ph^{1+} CML is seen in the granulopoietic system and is manifested by an increased pool of granulopoietic progenitor cells in the bone marrow and peripheral blood (6). We have asked whether growth of these precursor cells as assessed by evaluation of CFU-GM, BFU-E and CFU-MEG can be inhibited by IFN-alpha, IFN-gamma and rTNF-alpha, and whether in vitro sensitivity relates to in vivo clinical response. Our results indicated that a marked heterogeneity exists among CML patients with respect to the sensitivity of various CFUs to IFN-alpha, IFN-gamma and TNF-alpha. The CFU-GMs, in particular, exhibited extremely heterogenous IC50 values, with individual patients being more sensitive and others who were almost resistant to inhibition. When patients were subsequently treated in vivo with rIFN-alpha-2, a striking correlation between these in vitro data and their subsequent response was observed in a small pilot group of seven patients. All patients with IC50 CFU-GMs below or within the range of the normal controls responded, while none of the three patients with IC50 values above the range of healthy individuals failed to do so.

Beside the potential clinical implications of these findings, which shed some light on the presumed mechanism of IFN-alpha-induced remission in CML, it appears that some of the Ph^{1+} clones exhibit increased sensitivity to growth inhibition and that in such cases reduction or eventually irradiation of this clone can be achieved. Others reveal a normal sensitivity to IFN-alpha and growth control, but no selective reduction of the Ph^{1+} cells can be obtained. A third group of patients lacks sensitivity to IFN-alpha and in these cases neither growth control nor selective cytogenetic improvement can be obtained.

REFERENCES

1. Bergsagel D.E., Haas R., Messner H.A. (1986): Interferon alpha-2b in the treatment of chronic granulocytic leukemia. Seminars in Oncology 3, 29-34.
2. Gastl G., Aulitzky W., Tilg H., Huber H., Hausmaninger H., Seewann H.L., Prinoth P., Huber C. (1987): Dose-related effectiveness of alpha-interferon in chronic myelogenous leukemia. Blut 54, 1-2.
3. Geissler D., Lu L., Bruno E., Yang H.H., Broxmeyer H.E., Hoffmann R. (1986): The influence of T lymphocyte subsets and humoral factors on colony formation by human bone marrow and blood megakaryocyte progenitor cells in vitro. J. Immunol. 137, 2508-2513.

4. Konwalinka G., Geissler D., Peschel C., Tomaschek B., Schmalzl F., Huber H., Odavic R., Braunsteiner H. (1982): A microagar culture system for cloning human erythropoietic progenitors in vitro. Exp. Hematol. 10, 71.
5. Konwalinka G., Peschel C., Geissler D., Tomaschek B., Boyd J., Odavic R., Braunsteiner H. (1983): Myelopoiesis of human bone marrow cells in a microagar culture system comparison of two sources of colony stimulating activity (CSA). Int. J. Cell Cloning 1, 401-411.
6. Silver R. T., Gale R.P. (1986): Chronic myeloid leukemia. Am J. Med. 80, 1137-1148.
7. Talpaz M., Kantarjian H.M., McCredie K., Trujillo J.M., Keating M.J., Gutterman J. (1986): Hematologic remission and cytogenetic improvement induced by recombinant human interferon alpha A in chronic myelogenous leukemia. New Engl. J. Med. 314, 1065-1069.
8. Talpaz M., McCredie K., Kantarjian H., Trujillo J., Keating M, Gutterman J. (1986): Chronic myelogenous leukemia: hematologic remission with alpha interferon. Brit. J. Haematology 64, 87-95.
9. Talpaz M., Trujillo J.M., Hittelmann W.N., Keating M.J., Gutterman J.U. (1985): Suppression of clonal evolution in two chronic myelogenous leukemia patients treated with leucocyte interferon. Brit. J. Haematology 60, 619-624.

Decrease of *in vitro* colony formation of the hematopoietic progenitor cells CFU-GEMM, CFU-MK, BFU-E and CFU-GM in the acquired immunodeficiency syndrome (AIDS)

Benhardt Völkers, Arnold Ganser, Carmelo Carlo Stella * and Dieter Hoelzer

Department of Hematology, University of Frankfurt, Frankfurt, FRG
* Department of Internal Medicine and Medical Therapy, University of Pavia, Italy

In addition to immunological derangement, cytopenias of different hematopoietic lineages have been found in the peripheral blood of a majority of patients suffering from AIDS in human immunodeficiency virus (HIV) infection (1-4). Abnormalities of the bone marrow of patients with AIDS have been observed as well including changes similar to the findings in myelodysplastic syndromes (3-7). Even in patients without hematological disorders treatment with cytostatic drugs, folic acid antagonists and the reverse transcriptase inhibitor 3'-azido-3'-deoxythymidine (AZT) is frequently complicated by unexpectedly profound thrombocytopenia, neutropenia and anemia resulting in the withdrawel of the drug.
In this study 15 patients with AIDS or AIDS-related complex (ARC) were investigated for the in vitro growth of hematopoietic progenitor cells. The aim of the study was (a) to evaluate the incidence of the hematopoietic progenitor cells in the bone marrow of patients with AIDS, (b) to evaluate the influence of accessory cells on the decreased colony formation of the hematopoietic progenitors and (c) to examine the capability of peripheral blood mononuclear cells to produce colony stimulating activity.

Materials and Methods

Bone marrow cells were obtained from hematologically normal donors and fifteen patients who met the Centers of Disease Control (CDC) case definition of AIDS and ARC. 10 patients had AIDS and 5 ARC (Table 1). Patients had not received cytotoxic therapy before being studied. Mononuclear, light density bone marrow cells (LDBMC) were separated by centrifugation on a Ficoll-Hypaque gradient (density 1.077g/ml). Adherent cell depletion was done by incubation in plastic tissue culture flasks. Not more than 1% CDw 14 positive cells were found after this treatment, T-cell depletion was achieved by rosetting with 2-amino-ethylisothioronium bromide (AET) treated sheep red blood cells. T-cell depleted cell fractions contained less than 3% CD 3 positive cells. A methylcellulose colony assay for hematopoietic progenitors (pluripotent CFU-GEMM, erythroid BFU-E, granulocyte-macrophage CFU-GM and megakaryocyte CFU-Mk) was used (8). 1×10^5 LDBMC/ml were grown in 30% human plasma, 5% PHA-leucocyte conditioned medium, 0.9% methylcellulose, 5×10^{-5} M 2-mercaptoethanol and 1 U/ml sheep erythropoietin (Connaught, Ontario, Step III) for 14 days at 37°C and 5% CO_2 in air.
Phytohemagglutinin (PHA) mononuclear cell (MNC) conditioned medium (PHA-CM) was prepared by culturing MNC of the peripheral blood of normal donors and patients with AIDS. 1×10^6 MNC/ml were cultured with 1% PHA, 10% FCS in Iscove's medium for 7 days at 37°C, 5% CO_2 in air.

Table 1. Clinical data of the patients at the time of the study

Patient	Hb g/dl	WBC $10^9/l$	PMN $10^9/l$	Ly $10^9/l$	Mo $10^9/l$	Pl $10^9/l$	CD 4 + /mm³ §	CD 8 + /mm³ ‖	CD 4 : CD 8 ratio
1	16,6	4,6	1,88	2,16	0,14	222	681	1308	0,5
2 ⁼	9,0	2,6	1,71	0,70	0,23	125	102	480	0,2
	8,1	2,7	1,79	0,76	0,05	145	133	339	0,4
3 ⁂	10,5	3,0	2,00	0,45	0,18	283	45	207	0,2
4	11,5	2,8	2,10	0,56	0,14	225	11	190	0,06
5 ⁂	10,4	2,2	1,79	0,33	0,08	160	26	107	0,2
6	12,9	4,3	2,00	1,70	0,50	165	134	595	0,2
7	10,2	2,2	1,00	0,80	0,40	45	95	464	0,2
8	11,9	6,2	2,70	2,70	0,62	263	53	813	0,06
9	16,1	5,9	3,30	1,80	0,59	291	192	637	0,3
10 ⁂	14,3	6,9	4,41	2,13	0,34	262	334	692	1,2
11	10,1	2,5	1,72	0,60	0,17	90	35	399	0,09
12 ⁂	9,3	4,2	2,39	1,34	0,46	61	213	1082	0,2
13	11,2	2,5	1 47	0,80	0,22	105	114	499	0,6
14	15,0	6,6	4,49	1,65	0,46	229	407	700	0,6
15 ⁂	11,7	2,4	0,98	1,03	0,14	174	154	637	0,2

⁂Lymphadenopathy syndrome, §Normal range: 1100-1300, ‖Normal range: 350-450, ⁼This patient was studied on two separate occasions.

Hb, hemoglobin;   WBC, white blood cells;   PMN, granulocytes;   Ly, lymphocytes;
Mo, monocytes;   Pl, platelets;   CD 4 +, CD 4 positive T-lymphocytes;   CD 8 +, CD 8 positive T-lymphocytes

Results

Initial experiments showed a reduction in the colony growth of pluripotent and committed progenitor cells. The colony growth of the pluripotent CFU-GEMM was decreased to 1.2 ± 0.3 ($\bar{x} \pm$ SEM, per 10^5 LDBMC, normal 10.9 ± 1.6) the megakaryocytic precursors CFU-Mk to 1.7 ± 0.6 (11.3 ± 1.9), the erythroid BFU-E to 17 ± 10 (85 ± 12), and the granulocyte-macrophage CFU-GM to 35 ± 10 (73 ± 5) (Figure 1).

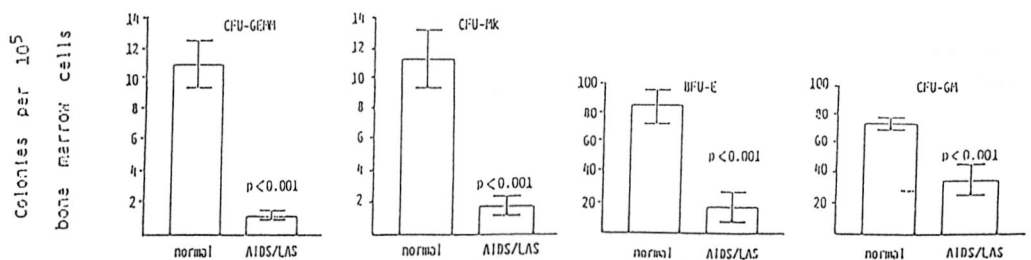

Figure 1

Incidence of the hemopoietic progenitor cells CFU-GEM, CFU-Mk, BFU-E and CFU-GM of patients with AIDS/ARC (n = 15) and normal male controls (n = 24). Cells were cultured in quadruplicates at 1×10^5 cells/ml.

To estimate the effect of bone marrow T-lymphocytes on hematopoietic progenitor cell growth, LDBMC were cultured following T-cell depletion. In addition the autologous bone marrow or peripheral blood T-lymphocytes were added back to 1×10^5 T-cell depleted bone marrow target cells. In cultures following T-cell depletion of LDBMC there was an increase in colony formation in the majority of cases studied. Co-culture of autologous T-cells and T-cell depleted LDBMC resulted in a significant decrease of growth of all colony types (Figure 2). (in detail Ref. 9.)

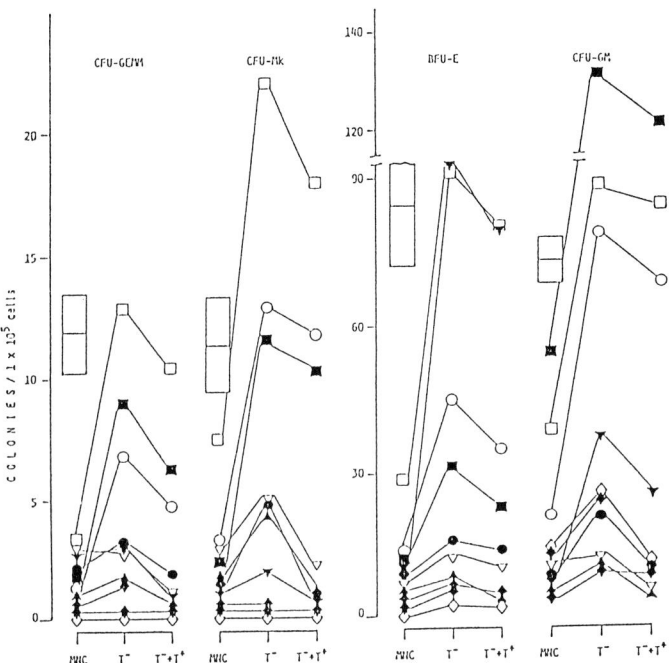

Figure 2

Hemopoietic progenitor cell CFU-GEM, CFU-Mk, BFU-E, and CFU-GM growth obtained from mononuclear bone marrow cells (MNC), T-depleted bone marrow cells (T⁻) and T-depleted plus autologous T-cells (T⁻ + T⁺) in the patients studied. Bars represent the growth (mean ± SEM) obtained in the control group per 10^5 mononuclear bone marrow cells plated. One patient (◆) was studied on two separate occasions. Blood-derived T cells were used in: □, ■, ○, ●, ◇, ◆, ▼ ; bone marrow T-cells in: ▽, ▲.

The inhibitory effect of autologous T-cells was significantly correlated to the CD 4 : CD 8 ratio in the T-cells added back to the LDBMC (Figure 3). There was no correlation between the percentage inhibition and the absolute number of CD 4 positive and CD 8 positive T-cells.

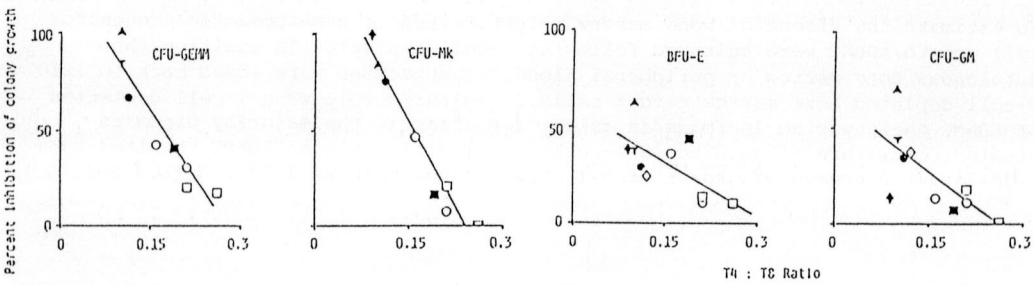

Figure 3
Correlation between the percentage inhibition of hemopoietic progenitor cell growth and the T4:T8 ratio of autologous T-cells added back to the T-cell depleted marrow cells. CFU-GEM (n = 8): r = -0.937, p < 0.001; CFU-Mk (n = 8): r = -0.972, p < 0.001; BFU-E (n = 10): r = -0.690, p < 0.05; CFU-GM (n = 10): r = -0.702, p < 0.05.

Each symbol represents one patient; the symbols correspond to those used in Figure 2. In two patients (□, ○), the effect of both peripheral blood T-cells and bone marrow T-cells was analyzed.

The colony growth of T-cell depleted and adherent cell depleted LDBMC stimulated by PHA-CM were tested in the colony assay described above. The preliminary data of three different experiments show no difference in the stimulating potential of PHA-CM of AIDS patients MNC compared to PHA-CM from MNC of normal persons (Tab. 2)

Table 2. Colony growth of LDBMC stimulated by PHA-CM from AIDS patients (percentage of normal standards)

	I	II	III
BFU-E	90,6%	178,1%	92,8%
CFU-GM	78,0%	111,0%	105,5%

Discussion

A significant decrease in the incidence of all types of progenitor cells was found in the bone marrow of the patients studied, which by itself could explain the peripheral blood cytopenias found in patients with AIDS. However the increased number of colonies observed after T-cell depletion of the bone marrow cells and the decreased number of colonies after adding back autologous T-cells, correlated to the CD4/CD8 ratio of the T-cells are indicating that reduced colony growth is caused by a defect in regulatory mechanisms. The observed defect in regulatory control mainly involves T-cells and is probably caused by an imbalance of the T-cell subpopulations. Involvement of the monocytes cannot be ruled out but could not be observed in our assay system. Depletion of monocytes did not result in increased colony formation (data not shown) nor were the mononuclear cells incapable of producing colony stimulating factors. The failure to completely restore colony formation in the majority of cases by T-cell depletion refers to the possibility that progenitor cells are lost due to direct infection by HIV.

References

1. Spivak, J.L. et al. JAMA 250:3084, 1983;   8. Messner, H.A. et al. J.Cell.Physiol.
2. Spivak, J.L. et al. AM.J.Med. 77:224, 1984   (Suppl.) 1:45, 1982
3. Schneider, D.R. et al. Am.J.Clin.Pathol. 84:144, 1985
4. Treacy, M. et al. Br.J.Haematol. 65:289, 1987   9. Carlo Stella, C. et al. J.Clin.Invest.
5. Osborne, B.M. et al. Hum.Pathol. 15:1048, 1984   (in press)
6. Geller, S.A. et al. Arch.Pathol.Lab.Med. 109:138, 1985
7. Castella, A. et al. Am.J.Clin.Pathol. 84:425, 1985

Peripheral blood adherent cells from AIDS patients inhibit normal T colony growth through decreased expression of Interleukin-2 receptors and production of Interleukin-2

Yanto Lunardi-Iskandar, Vassilis Georgoulias, Daniel Vittecoq, Marie-Thérèse Nugeyre, Anne-Marie Bertoli, Adlen Ammar, Corinne Clémenceau, Françoise Barre-Sinoussi, Jean-Claude Chermann, Léon Schwarzenberg and Claude Jasmin

INSERM U268, Hôpital Paul Brousse, BP200, 94804 Villejuif Cedex, France et Département de Virologie, Institut Pasteur, Paris, France

Key words: T-cell colonies, AIDS, Interleukin 2, Interleukin 2 receptors.

Introduction

Human immunodeficiency virus 1 (HIV-1) is etiologically associated with AIDS (1). HIV-1 is capable to infect $CD4^+$ lymphocytes (2-3) but also lymphoblastoid B (4), dendritic (5) cells and macrophages (6). We and others have previously reported that T-cell colony growth in AIDS and LAS (lymphadenopathy syndrome) patients is impaired (7-10). However, it is yet unknown whether the observed low plating efficiency of T-CFC from these patients is due to a decrease of the T-CFC pool or/and the presence of some inhibition mechanisms.

Since macrophages display an important accessory role for normal T-cell colony formation (11) and could be infected by the HIV-1 we studied their effect on T-cell colony formation from T-cell colony formation from T-CFC of AIDS patients.

Patients and methods

Patients: All AIDS patients fulfilled the CDC criteria (12) and were HIV infection was verified by both enzymo-immunoassay and Western blot. Normal cells were obtained from healthy HIV sero-negative blood donors. Peripheral blood mononuclear cells (PBMC) and adherent (AT) cells were obtained by standard methods (7,10).

Methods:
T-cell colonies: 5×10^5 PBMC/ml were seeded in methylcellulose (0.8% v/v) in the presence of PHA and Interleukin 2 (IL2) as already reported (7).
Conditioned media: They were prepared by incubating 10^6 cells/ml for 24 hr as described (13).
Infection of a HIV-1 permissive cell line: infection experiments were performed as previously described (1) using the CEM.A.301 human leukemic T cell line. Infection was monitored by the presence of reverse transcriptase activity (RT) (1).

Expression of IL2-R and production of IL2: Cell surface expression of IL2-R on mitogen-activated T cells was studied by indirect immunofluorescence using the anti-Tac monoclonal antibody. Supernatant of 48 hr-cultured mitogen-activated normal T cells were tested for IL2 activity using the CTLL-2 murine cell line as described (14).

RESULTS

In 5 out of 12 AIDS patients adherent cell-depletion of PBMC enhanced plating efficiency, whereas the inverse was observed in normal controls. Moreover, both irradiated patients A+ cells as well as media conditioned by unstimulated patients A+ cells (LCM-A+) inhibited the normal T-cell colony formation in a dose-dependent manner (Fig.1 and 2). The patients LCM-A+ were also capable to inhibit the expression of IL2-R on normal activated T cells (Fig.3). In addition, these LCM-A+ could inhibit (66-82% inhibition) of IL2 production by normal mitogen-stimulated T-cells whereas, normal LCM-A+ allowed only a 24% inhibition. Patients' LCM-A+ did not contain detectable reverse transcriptase activity or could not infect the HIV 1-permissive T-cell line CEM throughout the 24 days culture period.

DISCUSSION

Our findings indicate that in some AIDS patients adherent cells may supress the T-cell colony formation, explaining thus, the observed low plating efficiency. This may be supported by the fact that adherent cell-depletion allow a higher plating efficiency as well as the observation that patients A+ cells could inhibit normal T-cell colony formation. This inhibition is, probably, due to an impaired expression of IL2-R on activated normal T cells and to a decreased production of IL2 by these cells. It is well known the role of IL2/IL2-R system for in vitro proliferation of mature T cells (15). Since IL2 requires the presence of high affinity IL2-R (16) in order to mediate its activity, it is reasonable to hypothesize that the impaired expression of IL2-R is more crucial for the inhibitory effect of A+ cells. This inhibitory activity of patients A+ is mediated by some humoral factors which produced spontaneously, since it was also detected in media conditioned by unstimulated A+ cells. These inhibitory activities do not seem to be HIV since we could not detect either RT activity or infect the HIV-permissive T cell line CEM. Moreover, no virus particles could be detected in the ultracentrifugated pellet by electron microscopy studies (not shown) and by western blot using an antiserum against the HIV proteins.

The isolation and biochemical characterization of inhibitory factors produced by immunocompetent cells during the HIV infection could add to our knowledge concerning the physiopathology of the immunedeficiency which present these patients. In addition, it could be proved a useful target for therapeutic manipulations.

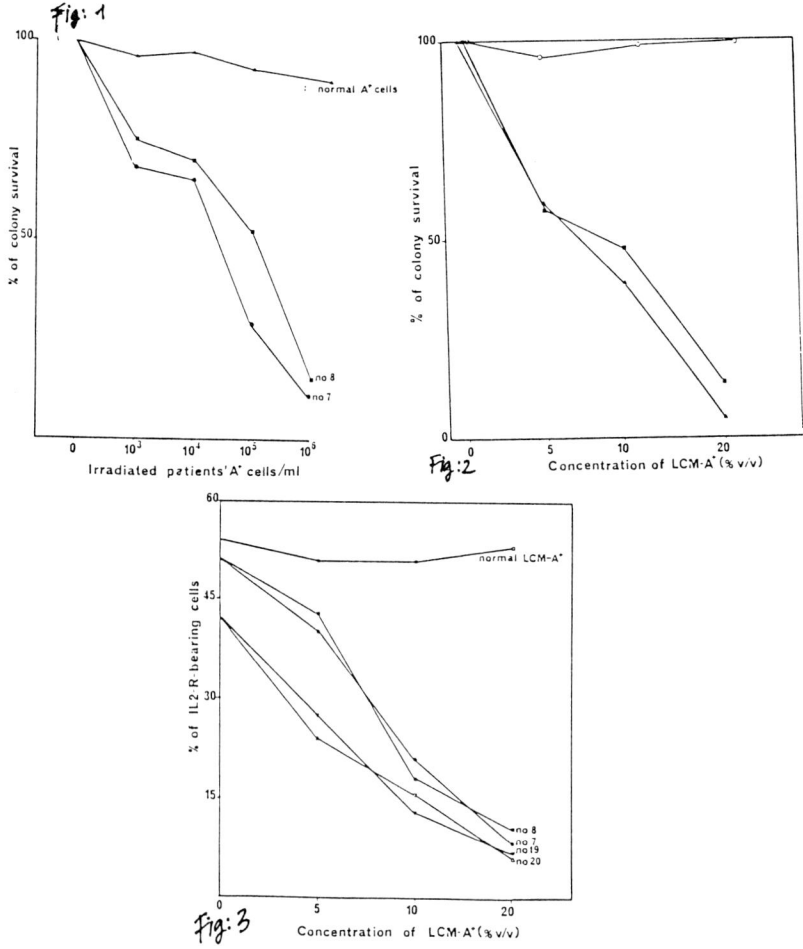

Figure 1: Effect of adherent cells from AIDS patients on normal T-cell colony growth. Normal PBMC (500,000/ml) were seeded in methylcellulose in the presence of increasing concentration of patients' irradiated A+ cells as described in Patients and Methods.

Figure 2: Effect of media conditioned by unstimulated adherent cells from AIDS patients on normal T-cell colony formation. Normal PBMC (500,000/ml) were seeded in methylcellulose in the presence of increasing concentrations of conditioned media prepared as described in Patients and Methods. (▲——▲) LCM from patient 7's adherent cells; (■——■) LCM from patient 8's adherent cells; (O——O) LCM from normal adherent cells.

Figure 3: Effect of media conditioned by AIDS patients adherent cells on the IL2-R expression on normal activated T cells: Normal PBMC (1,000,000/ml) were stimulated with PHA 1% and PMA 10 ng/ml in the presence of increasing concentration of LCM-A+ or LCM-A+n. Cells were incubated as described in Patients and Methods and IL2-R expression on washed cells was determined by indirect immunofluorescence.

REFERENCES
1. Barre-Sinoussi F. Chermann JC et al: Science 204:868.1983.
2. Klatzmann D. Champagne E et al: Nature 312:767.1984.
3. Dalgleish A.G. Beverly P.C.L et al: Nature 312:763.1984.
4. Montagnier L. Guest J. Chamaret S: Science 25:63.1984.
5. Armstrong J.A. Horne R: Lancet ii:370.1984.
6. Salahuddin S.Z. Rose R.M et al: Blood 68:281.1986.
7. Lunardi-Iskandar Y. Georgoulias V et al:Clin.Exp.Immunol 60:285.1985
8. Winkelstein A. Klein R.S et al:J.Immunol.134:151.1985.
9. Raoni M.R. Winkelstein A et al: Blood 64:105.1985.
10. Lunardi-Iskandar Y. Georgoulias V et al:Blood 67:1063.1986
11. Goupe de Laforest P. Lasmayons-Riou N et al:Exp.Hematol 8:361.1980
12. Centers for disease task force MMWR 31:365.1982.
13. Georgoulias V. Jasmin C : Leuk.Res.9:357.1985.
14. Bertoglio J.A. Boisson N et al:Lymphokine Res 1:12.1984
15. Cantrell D.A. Smith K.A: Science 224:1321.1984
16. Robb RJ. Munck A. Smith K.A:J.Exp.Med.154:1455.1981.

In vitro inhibition of hematopoiesis by HNK_1, CD8, DR positive T cells from allogeneic bone marrow transplantation and normal subjects

Giovanna Vinci, Jean-Paul Vernant, Mishal Zohar, Annie Henri, Henri Rochant, Janine Breton-Gorius and William Vainchenker

INSERM U91, Service d'Hématologie Clinique, Hôpital Henri Mondor, 94010 Créteil and Laboratoire de Cytométrie du CNRS, Hôpital Paul Brousse, 94800 Villejuif, France

Key Words: Hematopoiesis - In vitro inhibition - Bone marrow transplantation - T lymphocytes -

INTRODUCTION

Allogeneic bone marrow transplantation (BMT) represents a unique tool for the study of "in vivo" hematopoietic reconstitution by pluripotent stem cells in man. The development of "in vitro" colony assays for hematopoietic progenitors provides opportunities for the analysis of modifications in the compartments of human committed stem cells after BMT. However, numerous studies have demonstrated that T lymphocytes function as critical components of in vitro hematopoiesis both in normal and in pathological conditions (Torok-Storb et al., 1981; Mangan et al., 1982; Torok-Storb et al., 1982; Zoumbos et al., 1985). In this study, we have investigated the role of T lymphocytes on hematopoiesis following allogeneic BMT and compared it to their role in the regulation of normal hematopoiesis.

MATERIAL AND METHODS

Subjects

Peripheral venous blood was obtained from 11 patients after allogeneic BMT and from two patients after autologous BMT. Bone marrow samples were obtained from BMT donors.

Samples

Studies were performed on Ficoll separated light-density blood and marrow cells (LDC) from patients and normal donors respectively.

Isolation of sheep red cell rosetting cells (E^+C)

E^+C were obtained by rosetting LDC with AET-treated sheep red cells and non rosetting cells (E^-C) kept for culture.

Cytoxicity assay

The assay was performed on blood LDC from patients using anti-CD3, anti-CD8 and HNK_1 monoclonal antibodies (MoAbs). Baby rabbit serum was used as a source of complement.

Immunofluorescent labeling and cell sorting

Indirect immunofluorescent labeling was performed on E^+C from normal marrows either with the HNK_1 MoAb alone or in double fluorescent labeling with an anti-HLA-DR MoAb. Binding was revealed either by a goat Fab'_2 fragment against mouse Ig conjugated to fluorescein or anti µ chain or γ chain antibodies conjugated respectively to fluorescein and rhodamine. E^+C were separated, under sterile conditions, into a positive and negative fraction by cell sorting using a FACS 440 with one laser beam. The sorting rate was 1.000 cells/sec.

Assays of CFU-GM and BFU-E

CFU-GM and BFU-E growth was investigated by the plasma clot technique. The stimulating factors were either 10% supernatant from the Mo cell line or 5% Mo medium plus 1.5 IU/ml of porcine erythropoietin for CFU-GM and BFU-E growth respectively. Cultures were performed in triplicate at different cellular concentrations and scored between days 11 and 13. Blocking experiments were performed by adding different anti-HLA-class II MoAbs in culture.

RESULTS AND DISCUSSION

In patients who had allogeneic BMT, the following results have been found:

a) A reduction of colony growth. The numbers of CFU-GM (1-40/ml) or BFU-E (1-20/ml) were lower than in normal subjects (200-500 CFU-GM/ml and 50-600 BFU-E/ml), without any correlation with the clinical status of the patients. The hypothesis of a possible inhibition of hematopoiesis by T lymphocytes was subsequently tested.
b) An inhibition of hematopoiesis by E^+C. In fact, marked differences in the number of CFU-GM were observed in four cases before and after E^+C depletion, whereas in two others, minor differences were present. E^+C added back to E^-C, were able to elicit a marked inhibition of CFU-GM growth in patients in whom E^+C depletion increased colony growth.

We subsequently characterized the cells that suppressed CFU-GM and BFU-E growth. The phenotype of these inhibitory cells was investigated by complement-mediated lysis with anti-CD3, anti-CD8 and HNK_1 MoAbs. In all six cases, the cells with the HNK_1 surface marker were involved in the suppression of CFU-GM growth, whereas CD8 and CD3 complement-mediated lysis had only a moderate effect. Five cases of autologous BMT, one syngeneic BMT and two normal subjects were also studied, but cytotoxicity assays did not affect colony growth.

In addition these inhibitory T cells expressed the HLA-DR antigen. In four additional cases, complement-mediated cytotoxicity was performed with anti-CD8, anti-CD4, HNK_1 and anti-DR MoAbs on the E^+C. HNK_1 or HLA-DR complement-mediated cytotoxicity produced a suppression of the inhibitory effect mediated by E^+C. HNK_1 positive sorted LDC both inhibited CFU-GM and BFU-E growth **(Table 1)**.

	CFU-GM/10^6 ℓ	BFU-E/10^6 ℓ
LDC + HNK_1 + complement (HNK_1^- LDC)	75 ± 5	55 ± 5
HNK_1^- LDC + HNK_1^+ LDC	36 ± 3	34 ± 2
(ratio 4:1 between HNK_1^- LDC and HNK_1^+ cells)		

c) HLA-DR restriction of the inhibition mediated by T cells. This inhibition by T cells was genetically restricted to HLA-class II antigens. In three cases, E^+C from each patient were added back in coculture to E^-C from the patient himself, bone marrow donor, members of his family and normal unrelated subjects. CFU-GM and BFU-E growth inhibition was only found in the subjects who had a common HLA-DR haplotype with the patient. These results demonstrate that after allogeneic BMT a subset of $CD3^+$, $CD8^+$, HNK_1^+, $HLA-DR^+$ T lymphocytes is able to inhibit hematopoiesis with a genetic restriction.

Subsequently, we have investigated whether these cells play the same role in normal hematopoiesis since previous results have demonstrated that HNK_1^+ T cells are present in normal marrow in low concentration (less than 1%) (Abo et al., 1983).

- Effect of normal marrow HNK_1^+ E^+C on colony growth. Marrow E^+C were stained with HNK_1 MoAb by indirect immunofluorescent labeling. HNK_1^+ E^+C were obtained by cell sorting. This fraction represented about 3% of the E^+C. CFU-GM and BFU-E colony growth was inhibited when the sorted HNK_1^+ E^+C were added back to the E^-C at different ratios from 1/6 to 1/100. The degree of inhibition was higher for CFU-GM than for BFU-E growth and was observed up to a ratio 1/100 in one case. In another experiment, colony growth was entirely restored at a ratio 1/200.

- Effect of HNK_1^+, DR^+ E^+C on CFU-GM and BFU-E growth (**Table 2**). E^+C from two normal marrows were stained by double indirect immunofluorescent labeling using both HNK_1 and anti-DR MoAbs. HNK_1^+, DR^+ E^+C were sorted as the positive fraction, whereas the other E^+C were included in the negative fraction. The cells from the positive fraction were added back in co-culture to the E^-C at a ratio 1/12, whereas cells from the negative fraction were added back in a 1/1 ratio (**Table 2**). In these two cases, only the positive fraction was able to significantly inhibit colony growth. The HLA-DR antigen was the only HLA-class II antigen implicated in this inhibition of colony growth. In fact, blocking experiments performed by adding in culture different anti-HLA-class II MoAbs to the $CD3^+$ HNK_1^+ or $CD8^+$ HNK_1^+ E^+C, demonstrated that only an anti-HLA-DR MoAb was able to prevent the growth inhibition.

Table 2

	CASE 1		CASE 2
E^-C	456 ± 36		238 ± 12
E^-C + HNK_1^+ DR^+ E^+C (ratio 12:1 between E^-C and HNK_1^+ DR^+ E^+C)	324 ± 6	A	170 ± 5
E^-C + HNK_1^- DR^- E^+C (ratio 1:1)	ND		236 ± 6
E^-C	142 ± 16		144 ± 2
E^-C + HNK_1^+ DR^+ E^+C (ratio 12:1)	66 ± 8	B	96 ± 4
E^-C + HNK_1^- DR^- E^+C (ratio 1:1)	ND		140 ± 6

ND : Not determined   A: $CFU\text{-}GM/10^5$ ℓ B: $BFU\text{-}E/10^5$ ℓ

These results provide some informations about the negative regulation of hematopoiesis.

First, a subset of T lymphocytes with CD3, CD8, HNK$_1$ and DR phenotype is able to inhibit "in vitro" hematopoiesis after allogeneic BMT. In addition this effect is genetically restricted to the HLA-class II antigens, especially the DR locus.

Second, the same subset of T lymphocytes present in normal marrow at low concentration (about 1%), is able to inhibit CFU-GM and BFU-E colony growth in a similar way.

In conclusion, the $CD3^+$, $CD8^+$, HNK_1^+, DR^+, $CD2^+$ cells may represent a subset of T cells which is able to negatively regulate hematopoiesis both in normal and in allogeneic BMT.

The immunological reasons of their expansion in allogeneic BMT remain unknown. However, these T cells, despite an identical or close phenotype with some NK cells, have no natural killer activity. NK cells are also able to inhibit hematopoiesis but differ from this subset of T cells by their needed preactivation with target cells (Herrmann et al., 1987) and the absence of genetic restriction in their inhibitory activity. Further studies will be required to understand the mechanisms by which the $CD2^+$, $CD3^+$, $CD8^+$, HNK_1^+, DR^+ subset is able to negatively regulate hematopoiesis.

REFERENCES

Abo, T., Miller, C.A., Gartland, G.L., and Balch, C.M. (1983) Differentiation stages of human natural killer cells in lymphoid tissues from fetal to adult life. J. Exp. Med. 157, 273-284.
Herrmann, F., Schmidt, R.E., Ritz, J., and Griffin, J.D. (1987):In vitro regulation of human hematopoiesis by natural killer cells: analysis at a clonal level. Blood 69, 246-254.
Mangan, K.F., Chikkappa, G., Bieler, L.Z., Scharfman, W.B., and Parkinson, D.R. (1982) Regulation of human blood erythroid burst-forming unit (BFU-E) proliferation by T lymphocyte subpopulations defined by Fc receptors and monoclonal antibodies. Blood 59, 990-996.
Torok-Storb, B., Martin, P.J., and Hansen, J.A. (1981): Regulation of in vitro erythropoiesis by normal T cells. Evidence for two T-cell subsets with opposing function. Blood 58, 171-174.
Torok-Storb, B., and Jansen, J.A. (1982) Modulation of in vitro BFU-E growth by normal Ia-positive T cells is restricted by HLA-DR. Nature 298, 473-474.
Zoumbos, N.C., Gascon, P., Djeu, J.Y., Trost, S.R., and Young, N.S. (1985) Circulating activated suppressor T lymphocytes in aplastic anemia. N. Engl. J. Med., 312, 257-265.

ACKNOWLEDGEMENT

We thank A.M. Dulac for typing the manuscript.

Immunoglobulin-Binding Factors (IBF) are potent inhibitors of the growth of tumor B cells

Claire Mathiot, Sébastian Amigorena, Janine Moncuit, Jean-Luc Teillaud and Wolf Herman Fridman

Laboratoire d'Hématologie et Laboratoire d'Immunologie Cellulaire et Clinique (INSERM U255), Institut Curie, Paris, France

KEY WORDS

Cell Growth, Fcγreceptor positive T cells, Immunoglobulin-Binding Factors (IBF), Tumor B cells.

INTRODUCTION

Immunoglobulin-Binding Factors (IBF) are a family of lymphokines secreted by T cells which express receptors for the Fc part of Immunoglobulin (FcR). IBF are, at least in part, structurally related to FcR (Fridman, 1987). They are able to regulate immunoglobulin (Ig) production by normal and tumor B cells in an isotypic manner ; IgG-BF, IgE-BF, IgD-BF and IgA-BF have been described which specifically bind to IgG, IgE, IgD and IgA respectively and which up or down regulate *in vitro* Ig production (Fridman, 1987 ; Yodoï, 1983 , Ishizaka, 1985 ; Adachi, 1986). Interestingly, FcγR bearing T cells are strongly expanded in patients with lymphoproliferative diseases such as myeloma or B chronic lymphocytic leukemia (B-CLL) (Kay, 1983 ; Hoover, 1981). We investigated therefore whether IgG-BF produced by FcR positive T cells, beside controlling the Ig secretion, could act on the growth of tumor B cells such as hybridoma B cells cultured *in vitro*.

MATERIALS AND METHODS

Cells

T2D4 mouse hybridoma T cells were obtained by fusing alloactivated T cells (ATC) with cells from the BW5147 lymphoma T cell line. T2D4 cells express FcγR and secrete IgG-BF. The cell line D10C5 has been derived from T2D4 cells by cloning in limiting dilution and characterized as an FcγR negative, IgG-BF non producer cell line. UN2 hybridoma B cells secrete IgG2a, K antibodies anti Sheep Red Blood Cells (SRBC).

Reagents

Mouse IgG-BF was isolated from T2D4 cell-free supernatant by adsorption onto and elution from rabbit IgG-coated Sepharose beads. This IgG-BF preparation induced a strong inhibition of IgG secretion (60%) in an *in vitro* secondary antibody response of normal mouse spleen cells. A similar IgG-BF preparation derived from supernatants of alloantigen activated T cells (ATC) was fractionated on a Non-Equilibrium pH Gradient Electrophoresis (NEPHGE) gel and resolved in four distinct molecular entities (with different apparent isoelectric points of 4.7, 6.5, 7.7 and 8.4) which suppressed the *in vitro* antibody response of spleen cells (Blank, 1986).

Assays

A co-culture technique was adapted from the soft agar method described by Coffino et al.(1972) ; Rat embryo fibroblasts were plated in 60 mm tissue culture dishes. At day 1, the supernatant was carefully removed and the cells were overlaid with 5 ml of culture medium (10% FCS-RPMI 1640 containing 0.26% agarose). After the base layer had gelled at $4^{\circ}C$, 1 ml of the same mixture (but with 20% FCS) containing 700 T2D4 or D10C5 cells was added dropwise in each plate. At day 5, T2D4 or D10C5 clones were enumerated under an inverted microscope (x125). Plates where difference in the number of colonies between the two cell lines were higher than 15% were discarded. Each selected plate was then overlaid with 1 ml of the same agarose solution (with 20% FCS) containing 400 UN2 hybridoma B cells. The growth of UN2 clones was then examined every day between day 8 and day 13. In experiments where UN2 cells were cultured in presence of IgG-BF, the same basic procedure was used except that UN2 cells mixed with IgG-BF or Phosphate Buffered Saline (PBS) adjusted at a final dilution of 1 : 100 were plated at day 1 directly on the 5 ml base layer. All these experiments were performed in triplicate. In addition, 2 x 10^5 UN2 cells were incubated with IgG-BF eluted from NEPHGE gels in fresh liquid culture medium (10% FCS-RPMI 1640) in flat-bottomed 24 well plates. Cells were harvested at different times and counted. Viability was determined by a trypan blue exclusion test. The inhibition of Ig secretion was analyzed by an indirect PFC assay as previously described.

RESULTS

Inhibition of UN2 hybridoma B cells growth by IgG-BF producing T2D4 cells in soft agar co-cultures

In four different experiments, UN2 hybridoma B cells were first plated over a soft agar layer containing clones of either T2D4 cells or D10C5 cells. The cloning efficiency of UN2 cells was lowered in presence of T2D4 clones compared to that of D10C5 (a representative experiment is depicted Fig. 1). Furthermore, the UN2 clones that had been nevertheless growing in the presence of T2D4 cells underwent a morphological modification between day 4 and 7, leading to their necrosis.

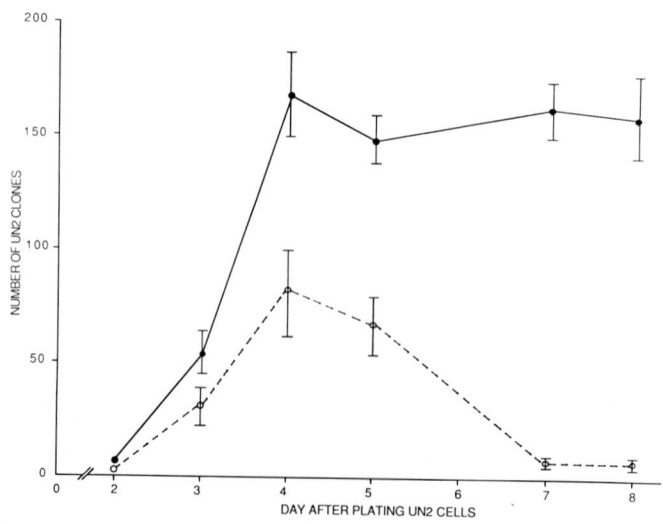

Figure 1 : Inhibition of UN2 hybridoma B cells growth in a soft agar co-culture experiment with T2D4 (O --- O) or D10C5 (● — ●) cells plated in the underlay. UN2 clones larger than 10 cells were enumerated.

Inhibition of the growth of UN2 hybridoma B cells in soft agar by IgG-BF

In other experiments, UN2 cells were directly plated with IgG-BF (Fig. 2) ; the number of UN2 clones was drastically lowered in presence of IgG-BF indicating that IgG-BF is directly responsible for the cytostatic effect observed in Figure 1. It should be stressed that only the cloning efficiency was affected in this situation. No necrosis of the UN2 clones which grew in presence of IgG-BF was observed ; however, the growth of some of these clones appeared to be slightly delayed in presence of IgG-BF.

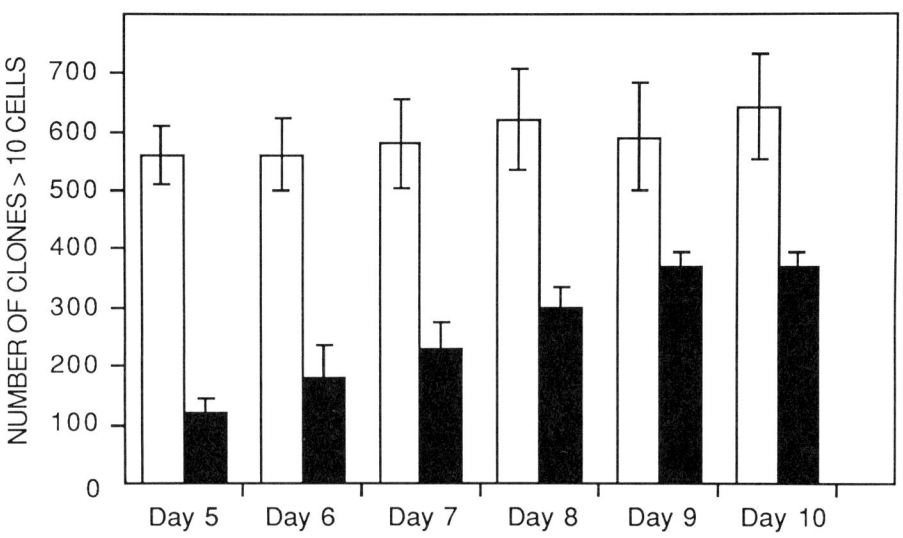

Figure 2 : Inhibition of UN2 hybridoma B cells growth induced by IgG-BF in a soft agar experiment. UN2 cells were plated in presence of 1/100 diluted PBS or IgG-BF

Inhibition of UN2 hybridoma B cell growth in liquid medium by IgG-BF separated according to their charges

We investigated whether the various IgG-BF molecules separated according to their charges and suppressing the IgG production of normal and hybridoma B cells were also able to inhibit the growth of these latter cells. Since the soft agar technique did not allow to quantitate easily the Ig production, we tested the effects of such IgG-BF on UN2 cells cultured in liquid medium. The four IgG-BF molecular entities were able to block the UN2 cells growth as well to inhibit the IgG secretion by these cells (Fig. 3A,B,C,D) ; these inhibitions were not due to a direct cytotoxic effect of IgG-BF on UN2 hybridoma B cells. PFC inhibition could be detected as early as after a 2h incubation (Fig.3A) ; after a 6h incubation, up to 55% of PFC inhibition were obtained whereas no significant difference in cell counts between IBF treated or untreated cell samples could be observed (Fig. 3B,C). Furthermore, when the percentage of inhibition of the cell proliferation reached 50%, more than 90% of PFC were inhibited (Fig. 3A,B,C).

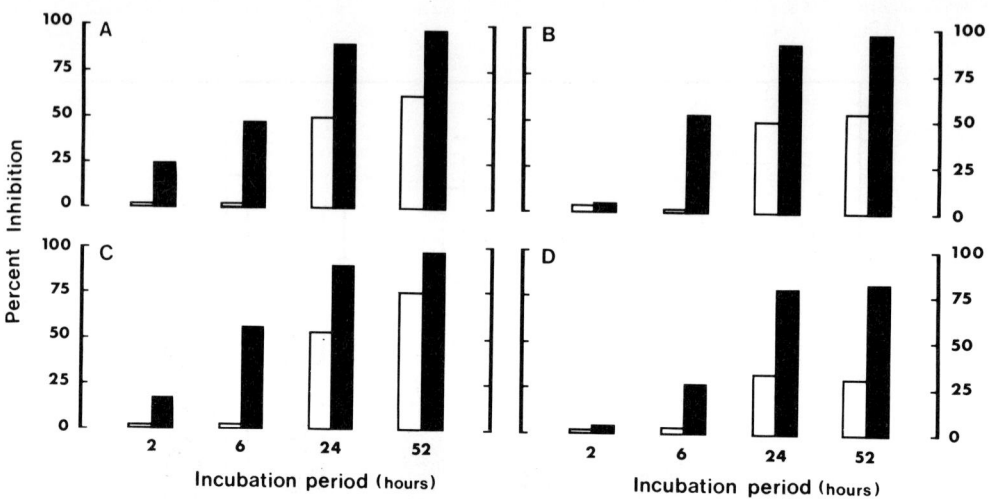

Figure 3 : Inhibitory effect of IgG-BF on cell proliferation ( ) and Ig secretion ( ) by UN2 hybridoma B cells. The IgG-BF preparations were derived from ATC and corresponded to apparent isoelectric points of 4.7 (A), 8.4 (B), 6.5 (C) and 7.7 (D). Cells were harvested, counted and equal numbers from each cell sample were tested in a PFC assay.

CONCLUSION

IgG-BF can exert a potent cytostatic effect on hybridoma B cells beside inhibiting the IgG production by these cells. This effect appears to be specific of the presence of either purified IgG-BF or IgG-BF-secreting T cells. It is unlikely that contaminant molecules could account for this growth inhibition since no dissociation between the inhibition of IgG secretion and the cytostatic effect could be observed when using the different IgG-BF molecules purified on NEPHGE gels. Furthermore, D10C5 cells which are derived from T2D4 cells and which have been selected for the absence of FcγR expression and IgG-BF production did not exert any inhibitory effect on UN2 cell growth. IBF could therefore represent potent immunomodulators of tumor B cells with potential clinical applications. Indeed, one can hypothesize, in view of our data demonstrating the immunoregulatory and growth inhibitory role of IBF, that the increase in FcR positive T cells in some lymphoproliferative diseases (myeloma and B-CLL) represents a physiological attempt to control these diseases. In the long term, it should be possible to *in vivo* manipulate with IBF the growth of myeloma cells or B cells from CLL as well as their Ig production. Whether the inhibitory role of IBF on cell growth could also be observed with cells of the hematopoietic lineage is of particular interest. Indeed, it has been reported that recombinant interferon gamma (IFNγr), interferon alpha (IFNαr) and tumor necrosis factor (TNFr) could exert antiproliferative activities on hematopoietic stem cells derived from normal or various patients, indicating that lymphokines could act on such cells. Moreover, it has been reported that FcγR positive T cells from human patients with aplastic anemia suppress the *in vitro* colony formation (CFUc) (Bacigalupo, 1980) ; this suggests that IgG-BF could play a major role in this latter situation by acting as an inhibitory factor in the regulation of hematopoiesis.

REFERENCES

Adachi, M. and Ishizaka, K. (1986) : IgG-Binding Factors from mouse T lymphocytes. Proc. Acad. Natl. Sci. USA. 83, 7003-7007.

Bacigalupo, A., Podesta, M., Mingari, M.C., Moretta, L., Vanlint, M.T. and Marmont, A.(1980) : Immuno suppression of hematopoiesis in aplastic anemia: activity of Tγ lymphocytes. J. Immunol. 125, 1449-1453.

Blank, U., Fridman, W.H., Daëron, M., Galinha, A., Moncuit, J. and Néauport-Sautès, C. (1986) : Size and charge heterogeneity of murine IgG-Binding Factors (IgG-BF). J.Immunol. 136, 2975-2982.

Coffino, Ph., Baumal, R., Laskov, R., Scharff, M.D. (1972) : Cloning of mouse myeloma cells and detection of rare variants. J. Cell Physiol. 79, 429-440.

Fridman, W.H., Teillaud, J.L., Amigorena, S., Daëron, M., Blank, U. and Néauport-Sautès, C. (1987) : The isotypic circuit : immunoglobulins, Fc receptors and immunoglobulin-binding factors.Intern.Rev.Immunol.2, 221-240.

Hoover, R.G., Gebel, H.M., Dieckgraefe, B.K., Hickman, S., Rebbe, N.F., Hirayama, N., Ovary, Z., Lynch, R.G. (1981) : Occurence and potential significances of increased numbers of T cells with Fc receptors in myeloma. Immunol. Rev. 56, 115-139.

Ishizaka, K. (1985) : Twenty years with IgE : from the identification of IgE to regulatory factors for the IgE response. J. Immunol. 135, 1-10.

Kay, N.E., Oken, M.M., Perri, R.T. (1983) : The influential T cell in B cell neoplasm. J. Clin. Oncol. 1, 810-816.

Yodoï, J., Adachi, M., Teshijawara, K., Migama-Inuba, M., Masuda, T. and Fridman, W.H. (1983) : T cell hybridomas coexpressing Fc receptors (FcR) for different isotypes. II : IgA-induced formation of suppressive IgA-binding factors by a murine T hybridoma bearing FcγR and FcαR. J.Immunol. 131, 303-310.

ACKNOWLEDGEMENTS

The authors wish to thank Ms. I. Lefranc for typing the manuscript. This work was supported by Institut National de la Santé et de la Recherche Médicale (INSERM), by Institut Curie and by a grant from the Ligue Nationale Française contre le Cancer. S. Amigorena is a Ph.D. student supported by a fellowship from the Ministère de l'Education Nationale.

Perspectives

Inhibitors of hemopoiesis : the accomplishments and the prospects

Maurice Tubiana

Institut Gustave Roussy, 94805 Villejuif Cedex, France

To be asked to discuss the prospects of inhibitors of hemopoietic cells is an honour that I fully appreciate. Many others would have certainly been more competent than myself, nevertheless I accepted because I felt that interaction between fundamental research and clinicians may be useful.

The history of inhibitory factors started with the pioneering work on "chalones". The great merit of this work was the clear vision of the necessity of a balance between stimulatory and inhibitory factors. A car cannot be driven with only a clutch, without a brake or several brakes. However to realize this basic requirement is not sufficient to understand how a brake works. The early papers on "chalones" were hampered by the lack of an appropriate methodology. The chalones were not chemically purified and to a great extent this was due to the absence of reproducible and sensitive bioassays. The modern era of the study of hemopoietic stimulators began in 1965 with the reports by Sachs and Metcalf who described simultaneously but independently the culture in soft agar of hemopoietic granulocytic progenitors.

I shall not attempt to write the history of the inhibitors of hemopoietic cell proliferation. This was reviewed in several papers as well as during this meeting. Moreover it is probably too early to do so. However the accomplishments which have been presented during this two-day meeting are impressive. The hemopoietic system is a very complex one and a large number of control mechanisms is required for a steady state production as well as for an increase in the end cell production when needed as for the regeneration of the bone marrow after an insult. Several types of short range and long range regulators should therefore be expected.

Four different growth factors (CSF) have been purified and synthesized, many others probably exist. The number of inhibitors should be at least as large ; several families have been identified, currently the most promising ones are those of small peptides. One of these has been purified and chemically characterized : the hemoregulator identified by Paukovits and Laerum. Another inhibitory small molecule acting on multipotential stem cells is being studied by E. Frindel, M. Guigon and M. Lenfant. These two small molecules are probably chemically distinct but their mode of action is still unknown as well as their relative specificities. Several other substances have inhibitory effects such as those described by the Manchester group (Lord).

Finally a large number of biologically active substances have, among other properties, an inhibitory effect on the proliferation of some hemopoietic cells. Let us quote for example lymphokines, interferons, tumor growth factor β (TGF β), tumor necrozing factor (TNF), prostaglandins, lactoferrin. During these two days we heard of new and interesting data along these lines.

For the inhibitors which have not yet been purified and identified, the first goal should, obviously, be to do so because without this first step one cannot go any further in their characterization. The second step is their production either by synthesis or by molecular cloning and recombinant techniques. The production of the inhibitors will enable one to check their biological activity because when a factor is purified by chemical or physical means, it is impossible to exclude the possible presence of trace amounts of another substance which might be the active one. However, the presence of "helper" substances may be necessary and without them the purified molecule may be inactive.

The third step is the study of the mode of action, in particular the identification of the target cells and the study of the relationship between the concentration and the effect. At low concentration the specificity is generally much higher than at high concentration. Moreover the nature of the effect may even be reversed by the concentration, for example a substance may stimulate cell division at low concentration and inhibit it at high concentration. A small modification of the molecule may reverse the effect from inhibition to stimulation (i.e. from monomer to dimer - Laerum).

Several growth regulatory peptides have been found that inhibit the replication of certain cells yet fail to inhibit and sometimes stimulate the growth of other cells (Todaro). The interplay between various types of molecules and cells is often quite complex. In the immune system, for example after the introduction into the body of an antigen, several types of cells and factors are involved in the chain of reactions which provokes the proliferation of "killer T lymphocytes" : the macrophages which secrete IL 1, the helper T lymphocytes which release IL 2, in the presence of the antigen and IL 1, the specific pre-T killer cells which are activated by the antigen and under this influence acquire receptors for IL 2 and thus can be electively triggered into proliferation by IL 2. Hence 2 different types of cells and 3 types of molecules play a part in the recruitment of lymphocyte killers into proliferation. Moreover the progression of T-lymphocytes through the cell cycle is regulated by the successive appearance of surface receptors for different growth factors, for example exposure of T-lymphocytes to antibodies against transferrin receptors stops the cells in S-phase. The regulation of T-lymphocyte proliferation illustrates the complexity and the precision of the systems which specifically control cell proliferation.

The side-effects of the inhibitors should be investigated. Experience has shown that the biologically active molecules generally have several effects, often on various types of cells. It seems that nature has only a limited imagination and when an active molecule has been produced, it is present not only in several species but also in various types of biological chains of reactions. Side-effects therefore are likely to be observed and should be systematically searched for.

Only when all these studies are completed, a fourth step can be envisaged : administration to patients. The prospects of the clinical use of the hemopoietic inhibitors are extremely promising.

The cycle specific cytotoxic drugs are widely used in medical oncology and are among the most effective. The limiting factor is bone marrow toxicity. E. Frindel and her group proposed about 10 years ago that keeping stem cells out of cycle could protect them from phase specific drugs. They have shown that in effect, blocking hemopoietic stem cells in Go by a physiological inhibitor in mice with or without experimental tumors, it was possible to increase the survival of these mice when given lethal doses of cytosine-arabinoside. This protection is nearly as large as that afforded by bone marrow grafting. In France, approximately 20 000 patients are submitted each year to aggressive courses of chemotherapy with a curative intent. It is estimated that, when technically feasible, bone marrow grafting is justified in at least 3 000 patients each year. This underlines the usefulness of research in this field. Another interesting avenue for research is the identification of those diseases in which inhibitory factors play an important role. Fascinating data along these lines have been mentioned here.

However the prospects of inhibitors can even be more important. In normal cell lines proliferation and differentiation are coupled. Interaction between the factors which induce cell proliferation and those which induce cell differentiation are complex, slight changes in the micro-environment may alter the balance between these two processes. In neoplastic cells proliferation and differentiation are uncoupled and in the clonal subline of a stem cell, a small but permanent advantage for proliferation over differentiation may result in leukemia. Inhibitors which have an effect on this clone might not only slow down the growth rate but may even contribute to reequilibrate proliferation and differentiation. This is why the study of inhibitors is so important for cancer research. Currently this type of research is the most advanced in hematology ; it is to some extent linked with the fifth step, i.e. the analysis of the mechanisms of action.

In the long run this last step will probably be the most crucial. The mechanisms of action of growth factors and the normal mitogenic pathway are relatively well understood. The growth factor binds to a specific membrane receptor that serves as a transducer of the external signal. An intracellular messenger system transmits the mitogenic signal from the receptor to the nucleus and ultimately leads to the initiation of DNA synthesis and cell division. Conversely very little is known about the mechanism of action of growth inhibitory factors although a few of them have been purified, in particular tumor inhibitory factors (TIF 1 and 2).

One may hypothesize that they could act at the level of the membrane receptor like an antagonist of the growth factor. However for hemopoietic tissues, the chemical configuration of the growth factors (CSF) and the inhibitors are so different that this hypothesis is unlikely. Inhibitors may bind to specific receptors but none of these putative receptors have yet been characterized, although their existence is very likely. Some inhibitors may act also as kinase antagonists, down regulate the number of membrane growth receptors, or block at some level the intracellular signal system triggered by the growth factors.

In view of the paucity of our knowledge in this field, it is urgent to investigate the effects caused by the inhibitors. The study of growth inhibitors may in the long run be the most effective biological approach to cancer treatment. This is why I would once again urge all those who participate in these studies to concentrate their efforts on the purification, synthesis and labeling of the inhibitors. However this effort toward the characterization of the inhibitors should not discourage the scientists from reporting their data even if the inhibitors are not completely purified. Progress in science is a complex process and in such an important field, one should never overlook an opportunity for further advances. This is why this meeting, and the workshop which is going to take place after it, are so important because they will facilitate and promote communication between those who are engaged in this field.

Author Index
Index des auteurs

Aglietta, M. 11, 59
Amigorena, S. 343
Ammar, A. 335
Arrenbrecht, S. 221
Ascari, E. 59, 321
Aulitzky, W. 325
Axelrad, A.A. 79

Bagnara, G.-P. 205, 209
Baillou, C. 151
Barre-Sinoussi, F. 335
Bergamaschi, G. 59
Bertoli, M. 335
Bicknell, D.C. 139
Biljanović-Paunović, L. 133
Bonsi, L. 209
Breton-Gorius, J. 101, 339
Broxmeyer, H.E. 139
Brunelli, M.-A. 205
Buzzi, M. 205, 209

Carlo Stella, C. 59, 317, 321, 331
Catani, L. 205, 209
Cazzola, M. 59, 321
Cerami, A. 197
Chen, W. 217
Chermann, J.C. 335
Chu, J. 217
Clémenceau, C. 335
Cooper, S. 139
Cork, M. 267
Craig, V. 73
Croizat, H. 79

Dainiak, N. 93, 151
Dautry, F. 301
Del Rizzo, D. 79
Dezza, L. 59, 321
Djukanović, L. 133
Douay, L. 151
Durkin, J.P. 201

Eriksen, J.A. 51
Eskinazi, D. 79

Fasciotto, B. 201
Fetsch, J. 55
Fletcher, J. 67
French, A. 69
Frickhofen, N. 213
Fridman, W.H. 343
Fu-Lu, L. 227

Gaggioli, L. 205, 209
Ganser, A. 317, 321, 331
Gastl, G. 325
Gavosto, F. 11
Geissler, D. 325
Georgoulias, V. 335
Gorin, N.C. 151
Grant, B.W. 111
Greher, J. 317
Guarini, A. 205, 209
Guigon, M. 31, 241, 271
Gutterman, J. 139

Heimpel, H. 213
Heit, W. 213
Henri, A. 101, 339
Hestdal, K. 51
Hoelzer, D. 317, 331
Hofmann, M.C. 221
Huber, C. 325

Irvine, A.E. 69, 73

Jakobsen, S.E. 51
Jasmin, C. 335
Johansen, J.H. 51

Kanazir, D. 201
Kieffer, N. 101
Kobari, L. 151
Konwalinka, G. 325
Kreczko, S. 93
Krsmanovic, V. 201

Laporte, J.-P. 151
Laerum, O.D. 21, 31, 51

Lauria, F.	205, 209	Sauter, C.	221
Lenfant, M.	271	Sawatzki, G.	63
Liu, L.	217	Schanche, J.S.	31, 51
Lombard, M-N.	271	Schwarzenberg, L.	335
Lord, B.I.	227	Slater, K.	67
Lu, L.	139	Solberg, L.A. Jr.	111
Lunardi-Iskandar, Y.	335	Song, Y.	217
		Sotty, D.	271
Mann, K.G.	111	Spooncer, E.	227
Marini, M.	205, 209	Steinberg, H.N.	163
Mathiot, C.	343	Stewart, S.	79
Maurer, H.R.	55	Stojanović, N.	133
Meloni, F.	321	Strauss, P.R.	93
Metcalf, D.	3		
Milenković, P.	133	Teillaud, J.-L.	343
Mitjavila, M.T.	101	Thomson, A.	73
Moncuit, J.	343	Tilg, H.	325
Morris, T.C.M.	69, 73	Tricot, G.	139
Moses, H.L.	111	Tubiana, M.	351
		Tucker, R.F.	111
Nadal, C.	271	Tveterras, T.	51
Najman, A.	151		
Nečas, E.	263	Vadhan, S.	139
Nugeyre, M.-T.	335	Vainchenker, W.	101, 339
		Van der Gaag, H.	79
Olofsson, T.B.J.	177	Valvassori, L.	209
		Vernant, J.-P.	339
Paukovits, J.B.	31	Villeval, J.-L.	101
Paukovits, W.R.	21, 31, 51	Vinci, G.	101, 339
Pavlović-Kentera, V.	133	Vittecocq, D.	335
Pedrazzoli, P.	59, 321	Völkers, B.	317, 321, 331
Pezzutti, G.	79		
Piacibello, W.	11	Wan, J.	217
Pojda, Z.	227	Wdzieczak-Bakala, J.	271
		Weil, D.	301
Raghavachar, A.	213	Whitfield, J.P.	201
Ralph, P.	139	Williams, D.E.	139
Raspadori, D.	205, 209	Wolpe, S.D.	197
Rich, I.N.	63, 213	Wu, K.	217
Riches, A.	267		
Rizzoli, C.	209	Young, N.S.	279
Rochant, H.	339		
Roodman, G.D.	289	Zinzani, P.-L.	205, 209
		Znojil, V.	263
Sassa, S.	197	Zohar, M.	339